# PROBLEM EXERCISES FOR GENERAL CHEMISTRY

THIRD EDITION

# PROBLEM EXERCISES FOR GENERAL CHEMISTRY

G. Gilbert Long
Forrest C. Hentz
*North Carolina State University–Raleigh*

JOHN WILEY & SONS
NEW YORK CHICHESTER BRISBANE TORONTO SINGAPORE

**Library of Congress Cataloging-in-Publication Data**

Long, G. Gilbert (George Gilbert), 1929–
Problem exercises for general chemistry.

Includes index.
1. Chemistry—Problems, exercises, etc.
I. Hentz, Forrest C., 1933– II. Title.
QD42.L59 1986 546′.076 86-13347
ISBN 0-471-82840-8

*Printed in the United States of America*

20 19 18 17 16 15 14 13

# *Preface*

In writing this *third edition*, we have adhered to our objectives of providing a large number and variety of exercises that are pertinent to almost every general chemistry course, in the style used for examinations in large general chemistry programs and on professional school aptitude examinations—for example MCAT and VCAT. The exercises emphasize the importance of problem-solving, equation writing, and the mastery of notation and symbolism that are the "tools of the chemical trade" and fundamental to the understanding of chemistry.

In this edition we have added 'Drill Sheets' throughout the book. These may be assigned as homework to be turned in and graded. Most of the sheets have one or two entries that the professor can furnish data so as to make each homework assignment distinctive from semester to semester. The back sides of these sheets are blank and the professor may ask to show here how a specific problem was calculated, the 'Set-Up'. In addition, we have expanded upon the material in Chapters 7 and 8, Chemical Reactions.

The following order is generally followed within a chapter: Introduction-to-Subsection, Illustrative Examples on the Subsection, Exercises-on-Subsection. This order is repeated with a chapter for any new topic. Each chapter closes with a large collection of exercises covering the topics of all of the subsections. Answers to all of the exercises are given at the end of the chapter.

An acid-base table, along with acid dissociation constants, and a table of reduction potentials have been included in appropriate chapters; a concise section on significant figures and thermodynamic tables have been included as appendices. The thermodynamic tables have been given not only for the convenience in the use of this book, but also because values do vary somewhat from one book to another. Answers to the exercises are consistent with this tabulated data.

G. GILBERT LONG
FORREST C. HENTZ

# Contents

# 1
# *Introductory Concepts*

Chemistry is the study of *matter,* the transformations that matter undergoes, and the *energy changes* that accompany these transformations. Such a study requires a wide variety of laboratory measurements. Many of these are similar to ones that you use daily: items purchased at the grocery store may be counted in dozens (measure of *number*) or weighed in pounds or ounces (measure of *mass*); fuel is frequently purchased by the gallon (a measure of *volume*); *linear* measurements are used to express distances between towns (miles) or to indicate the size of a sheet of paper, *e.g.* 8-1/2 × 11 inches. To express a measured quantity, it is necessary to indicate **both** the **numerical value** and an **appropriate unit.** If someone asks a price of 15 (unit unspecified) for an object, it certainly makes a big difference whether the price is 15 dollars or 15 cents. Also, it would be ridiculous to ask the attendant to put 10 acres of gasoline in the tank of an automobile (an inappropriate unit). Indeed, the unit is just as important as the numerical value!

## UNITS, CONVERSIONS BETWEEN UNITS, AND SIGNIFICANT FIGURES

Although we readily recognize units such as gallons and feet, they are not commonly used in most of the rest of the world. The system more widely used throughout the world is the *metric system,* a modified form of which is used in the sciences and is called the *Système International d'Unités* (abbreviated, *SI*). In the SI system seven fundamental units (see Table 1) are defined and all measurements are made by using either these units or units derived directly from them.

Numerical quantities are most conveniently expressed when one or two digits is followed by a decimal point and then an appropriate number of digits beyond the decimal point; thus, 62,500 meters is more conve-

TABLE 1. FUNDAMENTAL SI UNITS

| Quantity | SI Unit | Symbol | SI/English Conversion Factor |
|---|---|---|---|
| Mass | kilogram* | kg | $\frac{0.4536\ \text{kg}}{1\ \text{pound}}$ or $\frac{453.6\ \text{g}}{1\ \text{pound}}$ |
| Length | meter | m | $\frac{0.02540\ \text{m}}{1\ \text{inch}}$ or $\frac{2.540\ \text{cm}}{1\ \text{inch}}$ |
| Amount | mole | mol | $\frac{\text{number of items}}{6.023 \times 10^{23}\ \text{number of items/mole}}$ |
| Time | second | s | The same in both systems |
| Temperature | kelvin** | K | See p. 133 |
| Electric Current | ampere | A | The same in both systems |
| Luminous Intensity | candela | cd | The same in both systems |

*For experiments in the laboratory the kilogram usually represents an inconveniently large mass. Laboratory quantities of chemicals are measured in gram quantities; the gram is one-thousandth of a kilogram. **The centigrade (or Celsius) temperature, °C, is also accepted as an alternate to the Kelvin temperature.

TABLE 2. PREFIXES USED AS DECIMAL MULTIPLERS FOR UNITS

| Prefix | Abbreviation | Multiplication Factor |
|---|---|---|
| exa | E | $10^{18}$ |
| peta | P | $10^{15}$ |
| tera | T | $10^{12}$ |
| giga | G | $10^{9}$ |
| mega† | M | $10^{6}$ |
| kilo† | k | $10^{3}$ |
| deci | d | $10^{-1}$ |
| centi† | c | $10^{-2}$ |
| milli† | m | $10^{-3}$ |
| micro† | μ | $10^{-6}$ |
| nano | n | $10^{-9}$ |
| pico | p | $10^{-12}$ |
| femto | f | $10^{-15}$ |
| atto | a | $10^{-18}$ |

†Prefixes that are most commonly used in chemistry and should be memorized by the student.

niently expressed by its equivalent 62.5 km. The SI system uses prefixes to indicate large or small quantities. Hence, one thousandth of a gram, 0.001 g, would be more simply expressed as 1 mg. For purposes of comparison with other data, it may be desirable to express this mass as 1,000 μg. All of these refer to exactly the same mass.

## DERIVED UNITS

Upon inspection of Table 1, one notes that no units are specified for the very important quantities *area* and *volume.* This is because suitable units for these can readily be derived by multiplication of fundamental units. Area can be expressed by squaring a linear unit, the meter or the meter with an appropriate prefix. Likewise, volume can be represented by cubing a linear unit; the volume of a cube having 2-cm edges is 8 $cm^3$ [a $cm^3$ is also referred to as a *cubic centimeter* (*cc*)]. The *liter* (L) has been used as a measure of volume in the metric system for many years and is exactly 1 $dm^3$. Due to its wide usage, the liter is an accepted part of the SI. Thus, 1 milliliter (1 mL) is equal to 1 $cm^3$ (or 1 cc); carry out the arithmetic to check this yourself. With respect to the English system, the liter is just a little larger than a quart; the relationship between these two units is 0.9461 L = 1 quart.

Various other derived units have been developed to describe particular properties of matter. For instance, for comparing the "heaviness" of different substances, *density* (*d*) is defined as mass per unit volume, which for liquids and solids (in the SI and metric systems) is usually expressed in $g/cm^3$. Thus, a cube of wood, 2.0 cm on an edge, weighing 6.0131 g, would have a density of:

$$d = \frac{6.0131 \text{ g}}{(2.0 \text{ cm})^3} = 0.75 \text{ g/cm}^3$$

If the above arithmetic had been carried out on a calculator, one would have obtained additional decimal places, *i.e.* 0.751638. It is necessary to *round off* this number to two *significant figures,* 0.75 (or equivalently, using *exponential* or *scientific notation,* to $7.5 \times 10^{-1}$), so that the calculated answer shows an *accuracy* that is in agreement with the measured numbers. See pp. 349-351 in the Appendix.

A wide variety of important numerical conversions can be made on the basis of a knowledge of the SI and English systems. The successful chemistry student must learn to handle such conversions easily, since the problem-solving techniques are exactly the same as those used in solving many chemical problems. Carefully read the section, "Reporting Numerical Results", in the Appendix (pp. 446–448) and then study the following example problems, paying particular attention to the units of each factor and how they cancel, as well as the number of significant figures reported in the final answer.

## Example Problems Involving Units, Conversions Between Units, and Significant Figures

### 1-A. Conversion of Metric Volume to English Volume: The "Factor-Label" or "Unit Conversion" Method

A soft drink bottler distributes its product in 2.00-L bottles. What is the volume in quarts?

(A) 1.89 qt (B) 2.00 qt (C) 2.11 qt (D) 8.00 qt (E) 9.39 qt

***Solution*** 2.00 L → ? qt

This is a metric-to-English volume conversion. In the factor-label (or dimensional-analysis) method of problem solving, the original quantity, 2.00 L, is multiplied by one or more unity factors, cancelling units until the desired unit, quarts, is obtained. A unity factor is a ratio in which the numerator and denominator are equivalent.

In the section, "Derived Units", the connection, 0.9461 L = 1 qt, was given. By dividing both sides of this relationship by 1 qt, we have the ratio,

$$\frac{0.9461\ \text{L}}{1\ \text{qt}} = 0.9461\ \text{L/qt} = 1;$$

similarly, by dividing both sides by 0.9461 L we obtain the inverse ratio,

$$\frac{1\ \text{qt}}{0.9461\ \text{L}} = 1.057\ \text{qt/L} = 1,$$

as available unity factors. We see that the second of these will permit cancellation of the "L" from 2.00 L, and simultaneously interject "qt" into the numerator.

***Answer*** Thus, (2.00 L) (Unity Factor) = ? qt

or

$$(2.00\,\cancel{\text{L}})\left(\frac{1\ \text{qt}}{0.9461\,\cancel{\text{L}}}\right) = 2.11\ \text{qt} \qquad [\text{C}]$$

### 1-B. Mileage-Volume of Gasoline Required: The "Factor-Label" Method

A car is rated with a highway mileage of 41 miles per gallon of gasoline. Given that 1 mile = 1.6093 kilometers, how many liters of gasoline will be needed for a highway trip of 500. kilometers?

(A) 74 (B) $3.0 \times 10^3$ (C) 2.0 (D) 5.2 (E) 29

### *Solution*

Again, one first writes down the numerical value and the unit(s) of the given condition, here 500. km. Then one writes down the unit(s) required by the answer. Compare these two sets of units, and from other available data, write down the sequence of steps that will be needed. In this case these are:

$$\text{km} \rightarrow \text{mi} \rightarrow \text{gal} \rightarrow \text{qt} \rightarrow \text{L}$$

The steps used will depend upon the unity factors available since each step requires the use of a unity factor. Thus, for our problem, the set-up takes the form:

$$(500.\ \text{km})\,(\text{Factor A})\,(\text{Factor B})\,(\text{Factor C})\,(\text{Factor D}) = ?\ \text{L}$$

It is worth noting that the word "per" used in the problem indicates a division line. The required unity factors are:

$$\left(\frac{1\ \text{mi}}{1.6093\ \text{km}}\right),\ \left(\frac{41\ \text{mi}}{1\ \text{gal}}\right),\ \left(\frac{1\ \text{gal}}{4\ \text{qt}}\right),\ \text{and}\ \left(\frac{1\ \text{qt}}{0.9461\ \text{L}}\right).$$

These are now used directly or inverted so that cancellations leave only liters (L) as the final unit. This leads to the set-up:

$$(500.\ \cancel{\text{km}})\left(\frac{1\ \cancel{\text{mi}}}{1.6093\ \cancel{\text{km}}}\right)\left(\frac{1\ \cancel{\text{gal}}}{41\ \cancel{\text{mi}}}\right)\left(\frac{4\ \cancel{\text{qt}}}{1\ \cancel{\text{gal}}}\right)\left(\frac{0.9461\ \text{L}}{1\ \cancel{\text{qt}}}\right) = 29\ \text{L} \qquad [\text{E}]$$

### *Answer*

If you used a calculator to carry out the indicated arithmetic, the calculator read-out would have been 28.67782235. This answer was rounded off to 2 significant figures so that it would agree in accuracy with the datum (41 mi/gal) used in the calculation. See the section in the Appendix (pp. 349-351) entitled "Reporting Numerical Results" for a discussion of significant figures and rounding off of numbers. Here, the factor, 4 qt/1 gal, is exact to any desired accuracy, *i.e.*, a "defined number", as opposed to a "measured number".

## 1-C. Conversion of Area in English Units to Metric Units: The "Factor-Label" Method–Unit Factors Raised to a Power

A typical freshman chemistry exam consists of pages measuring 8-1/2 × 11 inches (or a rather imposing 93-1/2 square inches, *i.e.* 93-1/2 $\text{in}^2$). What is the approximate area (reported to one significant figure) of one side of such a page in square meters, *i.e.* $\text{m}^2$?

(A) 0.001 $\text{m}^2$ (B) 2 $\text{m}^2$ (C) 0.01 $\text{m}^2$ (D) 0.06 $\text{m}^2$ (E) 0.0006 $\text{m}^2$

### *Solution*

If the unity factor relating square meters and square inches were known, all that would be required would be to multiply the given area in square inches by this unity factor. However, there are no unity factors involving square measure in our table. Some related linear factors that do appear are:

$$\frac{2.540 \text{ cm}}{1 \text{ inch}}, \text{ and } \frac{10^{-2} \text{ m}}{1 \text{ cm}} \text{ or } \frac{100 \text{ cm}}{1 \text{ m}}.$$

Linear unity factors cannot be directly used, but raising any of these to a power generates a new unity factor. Since the problem deals with area, these factors will need to be squared to give unity factors involving area, *i.e.*

$$\left(\frac{2.540 \text{ cm}}{1 \text{ inch}}\right)^2 \text{ and } \left(\frac{10^{-2} \text{ m}}{1 \text{ cm}}\right)^2 \text{ or } \left(\frac{100 \text{ cm}}{1 \text{ m}}\right)^2.$$

Using these factors appropriately, the following steps are needed:

$$\text{in}^2 \rightarrow \text{cm}^2 \rightarrow \text{m}^2$$

$$(93\text{-}1/2 \cancel{\text{in}^2})\left(\frac{2.540 \cancel{\text{cm}}}{1 \cancel{\text{in}}}\right)^2\left(\frac{1 \text{ m}}{100 \cancel{\text{cm}}}\right)^2 = 0.060322 \text{ m}^2$$

**Alternate Solution.** The result would have been exactly the same if the linear English measurements had first been converted to meters and then multiplied together to get the area in $m^2$. Thus,

$$\left(8.5 \text{ in} \times \frac{2.540 \text{ cm}}{1 \text{ in}} \times \frac{1 \text{ m}}{100 \text{ cm}}\right)\left(11 \text{ in} \times \frac{2.540 \text{ cm}}{1 \text{ in}} \times \frac{10^{-2} \text{ m}}{1 \text{ cm}}\right) = 0.060322 \text{ m}^2$$

***Answer***

The first two zeroes are not significant figures, but merely hold the decimal point. Rounding to one significant figure gives 0.06 $m^2$ or equivalently $6 \times 10^{-2}$ $m^2$. (Each page of the chemistry test now looks much less imposing, doesn't it?) **[D]**

You should note that $\left(\frac{2.540 \text{ cm}}{\text{in}}\right)^2 = \frac{(2.540)^2 \text{ cm}^2}{1^2 \text{ in}^2} = \frac{6.452 \text{ cm}^2}{1 \text{ in}^2}$.

## 1-D. Mass- and Volume-Unit Conversion

The density of lead, 11.2 g/$cm^3$, when expressed in lb/$ft^3$ is:

(A) 2.60 (B) 699 (C) 11.2 (D) 0.179 (E) 0.753

***Solution***

$$\text{g/cm}^3 \rightarrow \text{lb/cm}^3 \rightarrow \text{lb/in.}^3 \rightarrow \text{lb/ft}^3$$

$$\left(\frac{11.2 \cancel{\text{g}}}{1 \cancel{\text{cm}^3}}\right)\left(\frac{1 \text{ lb}}{453.6 \cancel{\text{g}}}\right)\left(\frac{2.540 \cancel{\text{cm}}}{1 \cancel{\text{in}}}\right)^3\left(\frac{12 \cancel{\text{in}}}{1 \text{ ft}}\right)^3 = 699.1813973 \text{ lb/ft}^3$$

*Answer*

699 lb/ft$^3$ (It is rounded off to 3 significant figures, since the original density was given to 3 significant figures.) [B]

## 1-E. Volume to Area and Thickness Conversion

A latex semigloss enamel is advertised as having a coverage of 450. ft$^2$/gal when used on sealed surfaces. What is the average thickness of the coat of paint in millimeters?

(A) 10.1 (B) 188 (C) 0.0905 (D) 0.378 (E) 1.01

*Solution*

$$\text{Thickness} = \frac{\text{Vol}}{\text{Area}} = \frac{\text{gal} \rightarrow \text{qt} \rightarrow \text{L} \rightarrow \text{mL} \rightarrow \text{cm}^3 \rightarrow \text{mm}^3}{\text{ft}^2 \rightarrow \text{in}^2 \rightarrow \text{cm}^2 \rightarrow \text{mm}^2}$$

$$\frac{(1\ \text{gal})\left(4\ \frac{\text{qt}}{\text{gal}}\right)\left(0.946\frac{\text{L}}{\text{qt}}\right)\left(10^3\ \frac{\text{mL}}{\text{L}}\right)\left(1\ \frac{\text{cm}^3}{\text{mL}}\right)\left(10\ \frac{\text{mm}}{\text{cm}}\right)^3}{(450\ \text{ft}^2)\left(12\frac{\text{in}}{\text{ft}}\right)^2\left(2.540\ \frac{\text{cm}}{\text{in}}\right)^2\left(10\ \frac{\text{mm}}{\text{cm}}\right)^2}$$

$$= 0.0905125267\ \text{mm}$$

*Answer*

0.0905 mm (3 significant figures). The accuracy implied is 1 part in 905, or about 0.1%. This is comparable in order of magnitude to the accuracy indicated in the data, *i.e.*, the area (~0.2%) and in the qt → L conversion factor (~0.1%). [C]

## 1.F. Significant Figures in Addition, Subtraction, and Division

A mixture is prepared by mixing 7.65 g of A and 6.95 g of B. The total weight of the mixture can be reported to ___?___ significant figures; the ratio, wt of A/wt of B, in the mixture can be reported to ___?___ significant figures; and the difference, A - B, can be reported to ___?___ significant figures. The numbers required to fill the blanks are, respectively,

(A) 3, 3, 3 (B) 4, 3, 2 (C) 4, 3, 3 (D) 4, 4, 3 (E) None of these

*Solution*

In the summation, both weights are known to the nearest 0.01 g; therefore the sum, 14.60 g, is likewise known to the hundredths place, and one significant figure has been gained in the addition. In the ratio, the quotient, 7.65/6.95 = 1.100719424, must be rounded to 1.10, or 3 significant figures. The difference, 0.70 g, is also good to the hundredths place, but contains only two significant figures due to the loss of the units place in the subtraction.

*Answer*

4, 3, 2 [B]

## Exercises on Units, Conversions Between Units, and Significant Figures

1. One cm is the same as:

   (A) 100 mm (B) 2.54 in (C) 1 mL (D) 0.1 m (E) none of these

2. A velocity of 98 ft/sec corresponds (in units of km/sec) to:

   (A) about $3 \times 10^{-2}$ km/sec (B) about 2 km/sec
   (C) about 0.005 km/sec (D) about $5 \times 10^{2}$ km/sec
   (E) about 0.8 km/sec

3. One $mm^3$ is equivalent to how many μL?

   (A) 1 (B) $10^3$ (C) $10^{-3}$ (D) 0.1 (E) $10^2$

4. Suppose that a standard snail's pace ($P_{sn}$) is measured to be 0.040 ft/min. Measured in cm/sec, $P_{sn}$ would be:

   (A) 0.020 (B) 0.010 (C) 0.079 (D) 0.19 (E) 0.60

5. Two samples were weighed using different balances:

   (1) 3.529 g (2) 0.40 g

   How should the **total** weight of the samples be reported?

   (A) 3.929 g (B) 3 g (C) 3.9 g (D) 3.93 g (E) 4 g

6. A 0.060-g sample of Al wire was found to weigh 56.0 mg after treatment with hydrochloric acid. How should the percent weight **loss** be reported?

   (A) 6.7% (B) 15% (C) 7% (D) 6.67% (E) 10%

7. Which of the following contains 4 significant figures?

   (A) 0.0005 g (B) 0.0050 mL (C) 0.0500 m (D) 0.5000 cm
   (E) all of these

8. Add 5.17 g, 0.251 g, and 0.0023 g. The answer to the correct number of digits after the decimal point is:

   (A) 5.4233 g (B) 5.423 g (C) 5.42 g (D) 5.4 g (E) 5 g

9. A metallic sphere ($V = 4/3\ \pi r^3$) has a diameter of 0.200 in and weighs 0.0066 oz. What is the density of the metal in $g/cm^3$?

   (A) 18 (B) 2.7 (C) 0.18 (D) 3.6 (E) 5.6

10. It has been estimated that there is $4 \times 10^{-6}$ mg of gold per liter of sea

## DRILL ON UNIT CONVERSION

***Student's Name*** ____________________

Convert each of the following measured numbers to corresponding values for the units indicated in the blanks. Be sure to report your answers to the correct number of significant figures. You may be asked to turn in this sheet.

| | | | | |
|---|---|---|---|---|
| 502 mm | km | in | ft | mi |
| mm | km | in | 29 ft | mi |
| g | 2.395 kg | mg | lb | tons |
| g | kg | mg | lb | 8.9 tons |
| $8.312 \times 10^4$ mL | L | $cm^3$ | $in^3$ | qt |
| mL | L | $cm^3$ | 452 $in^3$ | qt |
| 3.45 g/mL | kg/L | $g/cm^3$ | $lb/ft^3$ | $tons/mi^3$ |
| | | | | |

water. At a price of \$19.40 per gram of gold, what would be the value of the gold in 1.00 cubic kilometer of the ocean?

(A) \$8 (B) \$50 (C) \$50,000 (D) \$80,000 (E) \$500,000

## MATTER: SYMBOLS, FORMULAS, EQUATIONS, AND THE LAWS OF CHEMICAL COMBINATION

All *matter* is made up of *atoms* of some 100 *elemental substances* (*elements*). Atoms are electrical in nature, and all atoms of a given element have the same number of positive charges on their cores (nuclei). This number of charges is called the *atomic number* of that particular element. Different elements have different average masses ("weights") of their atoms. This average weight of the atoms of an element is called its *atomic weight.* Atomic weights are not absolute masses of atoms, but are **relative** weights based upon the assignment that the most common type of carbon atom has an atomic weight of **exactly** 12 *atomic mass units* (*amu*). Rounded to two significant figures, the atomic weight of the element silicon is 28 amu; that of molybdenum is 96 amu. This means that the average silicon atom has a mass that is 2-1/3 times greater than carbon, 28 amu/12 amu = 2-1/3; similarly, a molybdenum atom is 96 amu/12 amu = 8 times as heavy as the standard carbon atom. Thus 1 g of carbon, 2-1/3 g of silicon, and 8 g of molybdenum each contain the same number of atoms. See Example 1-G.

As a shorthand notation, each element is assigned a *symbol. Compounds* are *pure substances* that are made up of two or more elements combined in explicit whole-number ratios of atoms and are assigned *formulas* by designating the relative numbers of each type of atom present, *e.g.,* $NH_4Cl$ indicates that (in the compound ammonium chloride) for every nitrogen atom there are combined 4 hydrogen atoms and 1 chlorine atom; $Ca(NO_3)_2$ is the formula for calcium nitrate and shows that for every calcium atom there are 2 nitrogen atoms and 6 oxygen atoms. Sometimes formulas are written by considering a compound as consisting of units of other compounds. The units are separated from one another in the formula by a dot (or period) with a numerical multiplier preceding the unit (the multiplier one is omitted). The most common instance of this use today is with compounds that contain water of crystallization. Thus, $CuSO_4 \cdot 5H_2O$ indicates a compound containing one $CuSO_4$ unit and five $H_2O$ units; it represents a combination of 1 copper, 1 sulfur, 9 oxygen, and 10 hydrogen atoms.

The **ratio** of combined atoms in a given compound cannot be altered—to do so would change the identity of the substance. This is a statement of the *Law of Definite Composition* (also called the *Law of Definite Proportions*). In terms of weights it means that for a specified compound (*e.g.* $MgCl_2$) the weight ratio, wt. Mg/wt. Cl, has a fixed value. The source of the sample, or whether we have a large or a minute amount of $MgCl_2$, is unimportant; this weight ratio has only the **one, fixed value.** Similarly, weight ratios of any one element to the weight of the compound, *e.g.* wt. Cl/wt. $MgCl_2$, also have fixed values for a given compound; these are referred to as *fractional parts by weight* and when multiplied by 100 are *percents by weight.* Thus, if two samples of pure compounds have different weight percentages of magnesium, they **cannot be the same compound.**

In many cases two or more compounds between the same elements ($N_2O$, NO, $N_2O_3$, $NO_2$, and $N_2O_5$) are known to exist. The weight ratios of the constituent atoms bear a simple numerical relationship to one another:

$$\frac{\left(\dfrac{\text{wt. N in Compound 1}}{\text{wt. O in Compound 1}}\right)}{\left(\dfrac{\text{wt. N in Compound 2}}{\text{wt. O in Compound 2}}\right)} = \begin{cases} \text{An integer } = 1, \text{ or a} \\ \text{decimal fraction that} \\ \text{is expressible as a} \\ \text{ratio of small whole numbers} \end{cases}$$

This is referred to as the *Law of Multiple Proportions.*

Chemical changes are shown by writing formulas of *reactants* (compounds and/or elements) followed by an arrow (indicating change) and the formulas of the *products.* In a chemical change compounds may be destroyed and new ones formed, but the total mass remains constant (*Law of Conservation of Mass*). Furthermore, individual atoms retain their identity; this means that every atom that is present in a reactant must also appear in a product and *vice versa, i.e.* the equation must be *balanced.* **Equation balancing is fundamental to understanding chemical reactions.** A systematic first approach to balancing equations by conserving atoms is illustrated in detail in Example Problems 1-H and 1-I.

## Example Problems Involving Symbols, Formulas, Equations and the Laws of Chemical Combination

### 1-G. Relative Atomic Weights

Element X reacts with oxygen to produce a pure sample of $X_6O_{11}$. In an experiment it is found that 1.0000 g of X produces 1.2082 g of $X_6O_{11}$. If we take the atomic weight of oxygen to be 16.00 amu, what is the atomic weight of X?

(A) 41.92 (B) 219.9 (C) 140.9 (D) 1.817 (E) 170.2

### *Solution*

Assume that the 1.2082-g sample contains N $X_6O_{11}$ molecules. Find the mass of each atom, *i.e.*, g/atom, in terms of **N**. Since atomic weights are *proportional* to the absolute atomic masses, the ratio $AW_X/AW_O$, and then $AW_X$, may be calculated;

$$\text{mass of X-atom} = \frac{1.0000 \text{ g}}{6\text{N atoms}}; \quad \text{mass of O-atom} = \frac{0.2082 \text{ g}}{11\text{N atoms}};$$

$$\frac{\text{mass of X-atom}}{\text{mass of O-atom}} = \frac{AW_X}{AW_O} = \frac{1.0000 \text{ g}}{6\text{N}} \times \frac{11\text{N}}{0.2082 \text{ g}} =$$

***Calculator*** $AW_X/AW_O = 8.805635607; AW_X = 140.8901697$

### *Answer*

$AW_X = 140.9$ (Here we have 4 significant figures, limited by the mass of the combined oxygen and the atomic weight of oxygen, which had been rounded off to 16.00. [C]

## 1-H. Equation Balancing

In the balanced equation for the oxidation of ammonia,

$$\_\_\_\_ NH_3 + \_\_\_\_ O_2 \rightarrow \_\_\_\_ N_2O + \_\_\_\_ H_2O,$$

the coefficient for $H_2O$ is:

(A) 1 (B) 2 (C) 3 (D) 4 (E) 5

### *Solution*

Many equations can readily be balanced by inspection, *i.e.*, spot balanced. Start with an atom that appears only in one reactant and one product, *e.g.*, N, and balance the "equation" with respect to N by inserting tentative coefficients in front of the two substances containing this atom. Thus,

$$\underline{\;2\;} NH_3 + \_\_\_\_ O_2 \rightarrow \underline{\;1\;} N_2O + \_\_\_\_ H_2O$$

Then adjust coefficients for the next atom, H, so that the "equation" is balanced with respect to H,

$$\underline{\;2\;} NH_3 + \_\_\_\_ O_2 \rightarrow \underline{\;1\;} N_2O + \underline{\;3\;} H_2O$$

Tentative coefficients have now been determined for all of the products, and all that remains to be done is to add up the number of oxygen atoms in the products to determine the coefficient for $O_2$,

$$\underline{\;2\;} NH_3 + \underline{\;2\;} O_2 \rightarrow \underline{\;1\;} N_2O + \underline{\;3\;} H_2O \qquad [C]$$

### *Answer*

$2\ NH_3 + 2\ O_2 \rightarrow N_2O + 3\ H_2O$. The coefficient 1 is understood; hence, it is not necessary to write it in the final equation.

## 1-I. Equation Balancing

The complete combustion of ethane, $C_2H_6$, yields carbon dioxide and water as products as indicated in the equation,

$$\_\_\_ C_2H_6 + \_\_\_ O_2 \rightarrow \_\_\_ CO_2 + \_\_\_ H_2O$$

The ratio of the number of molecules of $O_2$ required to molecules of ethane burned is:

(A) 1 (B) 1-1/2 (C) 2 (D) 3 (E) 3-1/2

### *Solution*

It is first necessary to balance the equation; then, the ratio, $O_2/C_2H_6$, can readily be determined. Start by balancing the carbon atoms by finding appropriate tentative coefficients for ethane and $CO_2$,

$$\underline{1}\ C_2H_6 + \_\_\_ O_2 \rightarrow \underline{2}\ CO_2 + \_\_\_ H_2O$$

Next, balance the "equation" with respect to hydrogen atoms by adding the coefficient 3 in front of the water,

$$\underline{1}\ C_2H_6 + \_\_\_ O_2 \rightarrow \underline{2}\ CO_2 + \underline{3}\ H_2O$$

On the basis of the coefficients found for the products, count the number of oxygen atoms required, *i.e.*, 7 oxygen atoms. Oxygen, however, exists as $O_2$ molecules when a reactant; hence, 7/2 $O_2$ molecules would be required,

$$\underline{1}\ C_2H_6 + \underline{7/2}\ O_2 \rightarrow \underline{2}\ CO_2 + \underline{3}\ H_2O$$

The expression must be cleared of fractions by multiplying everything through by 2 as given in the answer. It is clear at this point that the ratio is 7/2 to 1 or 3-1/2. [E]

### *Answer*

$2\ C_2H_6 + 7\ O_2 \rightarrow 4\ CO_2 + 6\ H_2O$. The ratio, $O_2/C_2H_6$, is equal to 7/2, which reduces to 3-1/2.

## 1-J. Laws of Chemical Combination

Two samples containing only vanadium and oxygen are analyzed. Sample #1 consists of 0.6281 g O/g V; sample #2 consists of 0.7851 g O/g V. If sample #2 is the compound $V_2O_5$, what is the formula of sample #1?

(A) $VO_2$ (B) $V_2O_3$ (C) VO (D) $VO_3$ (E) $V_2O_5$

### *Solution*

The weight ratios are distinctly different. The Law of Definite Composition tells us that sample #1 and sample #2 cannot be the same compound, and therefore cannot have the same formula, $V_2O_5$. Using the Law of Multiple Proportions to look for a small whole number ratio, we find

$$\frac{(\text{wt. O/wt. V})_{\#1}}{(\text{wt. O/wt. V})_{\#2}} = \frac{0.6281}{0.7851} = 0.8000, \text{ expressible as } \frac{4}{5}.$$

Therefore, since there is 4/5 as much mass of oxygen per fixed mass of vanadium in #1 as in #2, there has to be 4/5 as many **atoms** of O per **atom** of V in #1 as in #2. Since the ratio in #2 is given as O/V = 5/2, the atom ratio in #1 must be 5/2 × 4/5 = 2/1. Compound #1 must then be $VO_2$. Had the ratio **not** been expressible as a ratio of small whole numbers, #1 would likely **not have been a compound**, but a mixture. [A]

## Exercises on Symbols, Formulas, Equations, and the Laws of Chemical Combination

11. Which of the following elements has the greatest atomic weight?

    (A) manganese (B) potassium (C) phosphorus (D) tin (E) bromine

12. The formula of calcium nitrite, $Ca(NO_2)_2$, indicates that for every one dozen calcium atoms there will be ______ dozen oxygen atoms. *Fill in the blank.*

    (A) one (B) two (C) four (D) five (E) six

13. ___ $Ag_2S$ + ___ Al → ___ Ag + ___ $Al_2S_3$

    For the reaction above, how many silver atoms will be formed when one aluminum atom reacts?

    (A) one (B) two (C) three (D) four (E) seven

14. Given two pure substances (I and II) containing elements A and B. Substance I has a mass of 2.25 g and contains 0.900 g of B; substance II is analyzed as 40.0% B and 60.0% A by weight. These data

    (A) illustrate the law of multiple proportions
    (B) prove that either I or II is an element
    (C) illustrate the Law of Definite Composition
    (D) prove that either I or II is a mixture
    (E) show that the formulas of I and II are AB and $AB_3$

15. Acetylene, $C_2H_2$, along with the by-product $Ca(OH)_2$, is produced by adding $CaC_2$ to water. The coefficient preceding the formula for water ($H_2O$) in the balanced equation (simplest integral coefficients) is:

    (A) 1 (B) 2 (C) 3 (D) 4 (E) 5

16. A 40.0-mg sample of $X_2O_5$ contains 22.5 mg of O. What is the atomic weight of X?

    (A) 4.98 (B) 17.5 (C) 9.75 (D) 31.1 (E) 35.0

17. The following reaction (unbalanced) is important in the recovery of gold from ores:

    ___ Au + ___ KCN + ___ $O_2$ + ___ $H_2O$ → ___ $KAu(CN)_2$ + ___ KOH

    The balanced equation shows that for every molecule of $O_2$ used, ______

## DRILL ON EQUATION BALANCING

***Student's Name*** ______________________

Make equations out of each of the following sets of reactants and products by supplying the correct coefficient for each substance. You may be asked to turn in this sheet.

| | | |
|---|---|---|
| ___ $SiO_2$ + ___ HF | $\rightarrow$ | ___ $SiF_4$ + ___ $H_2O$ |
| ___ $XeF_6$ + ___ $SiO_2$ | $\rightarrow$ | ___ $SiF_4$ + ___ $XeOF_4$ |
| ___ $O_2$ + ___ $Fe_2(CO)_9$ | $\rightarrow$ | ___ $CO_2$ + ___ $Fe_2O_3$ |
| ___ $B_2H_6$ + ___ $LiAlH_4$ | $\rightarrow$ | ___ $LiBH_4$ + ___ $Al(BH_4)_3$ |
| ___ $SiO_2$ + ___ $Ca_3(PO_4)_2$ | $\rightarrow$ | ___ $P_2O_5$ + ___ $CaSiO_3$ |
| ___ $SiO_2$ + ___ $Ca_3(PO_4)_2$ + ___ C | $\rightarrow$ | ___ CO + ___ $CaSiO_3$ + ___ $P_4$ |
| | $\rightarrow$ | |

atoms of gold reacted. The number required to fill the blank is:

(A) 1 (B) 2 (C) 2.5 (D) 3 (E) 4

---

## MIXTURES

Mixtures consist of two or more pure substances, and unlike compounds, their percent compositions are variable. A very common type of mixture used in chemistry is the *homogeneous* mixture, the *solution*. The composition of a mixture is frequently given in terms of **percent (parts per hundred)** of the various components; in the sciences this is always understood to be percent by weight unless specifically noted. Thus, 5.0% NaCl means

$$\frac{5.0\text{ g NaCl}}{100\text{ g of mixture}} \text{ or } \frac{5.0\text{ lb NaCl}}{100\text{ lb of mixture}} \text{ or } \frac{5.0\text{ ton NaCl}}{100\text{ ton of mixture}} \text{ or}$$

similar in other mass units. If the indicated division is carried out, one obtains 0.050 g NaCl/g mixture, which is frequently referred to as the **fractional parts** of NaCl, or $\frac{\textbf{parts of a component}}{\textbf{1 part of a mixture}}$. In recent years trace components (much less than 1%) in a mixture frequently have been found to be very important. Neither fractional parts nor percentage are very convenient units for reporting amounts of trace constituents since a number of zeroes preceding the first significant figure would need be shown, *e.g.* 0.0000070% Hg. This can be remedied by using units based upon much larger amounts of mixture, **parts per million (ppm)** or **parts per billion (ppb)**. Thus, 0.0000070% Hg corresponds to 0.000000070 g Hg/g sample; in a much larger sample such as a million grams, there would be a million times as much mercury,

$$\frac{0.000000070\text{ g Hg}}{1\text{ g sample}} \times \frac{1{,}000{,}000}{1{,}000{,}000} = \frac{0.070\text{ g Hg}}{1{,}000{,}000\text{ g sample}} = 0.070\text{ ppm Hg.}$$

This could just as well be expressed as 70. ppb Hg, meaning

$$\frac{70\text{ g Hg}}{1{,}000{,}000{,}000\text{ g sample}}.$$

All of the above units are very convenient unity factors for converting "mass of a mixture to mass of a component" or the reverse, "mass of a component to mass of a mixture". If the mixture is a solution, the **mass** of the solution can be converted to the corresponding **volume** of solution by use of the the **solution density**, grams of *solution*/mL of *solution*.

## Example Problems Involving Mixtures

### 1-K. Mass of a Component Needed to Prepare a Given Quantity of a Mixture

A detergent contains 10. % phosphorus in the form of the compound $Na_6P_6O_{18}$. The compound itself contains 30. % phosphorus. How many tons of $Na_6P_6O_{18}$ should be ordered to prepare 38-ton batch of detergent?

(A) 3.8 (B) 0.88 (C) 13 (D) 0.13 (E) None of these

#### *Solution*

The percentages given in the problem give us four unity factors which can be expressed in any suitable mass units we choose. Since the problem deals with tons, it would be well to express these factors in tons. The available factors then would be:

$$\frac{10 \text{ tons P}}{100 \text{ tons deterg.}}, \quad \frac{100 \text{ tons deterg.}}{10 \text{ tons P}}, \quad \frac{30 \text{ tons P}}{100 \text{ tons } Na_6P_6O_{18}}, \text{ and } \frac{100 \text{ tons } Na_6P_6O_{18}}{30 \text{ tons P}}$$

There is no single factor that permits us to convert tons of detergent to tons of $Na_6P_6O_{18}$. The desired calculation can be made however in two steps; first convert the tons of detergent required to tons phosphorus required, and then convert the tons of P to tons of $Na_6P_6O_{18}$.

$$\text{tons of deterg.} \rightarrow \text{tons of P} \rightarrow \text{tons of } Na_6P_6O_{18}$$

Using the appropriate unity factor for the transformation in each step, the set-up becomes:

$$(38 \text{ tons deterg.})\left(\frac{10 \text{ tons P}}{100 \text{ tons deterg.}}\right)\left(\frac{100 \text{ tons } Na_6P_6O_{18}}{30 \text{ tons P}}\right) =$$

#### *Answer*

13 tons of $Na_6P_6O_{18}$ would need to be ordered. The answer is good to two significant figures since all of the data in the problem are good to two significant figures. [C]

### 1-L. Use of Percent by Weight

An ore contains 0.20% by weight of the mineral *calaverite*, $AuTe_2$, a gold compound containing 43.56% Au. How many tons of the ore would have to be processed to yield 1.0 lb of pure gold?

(A) 0.11 (B) 0.57 (C) 1.1 (D) 2.3 (E) 5.7

#### *Solution*

Percent by weight data may be read as parts of the component per 100 parts of the whole in **any** convenient mass units. Thus, 100 lb of the mineral $AuTe_2$ contains 43.56 lb of Au; about 2-1/3 lb of $AuTe_2$ will be needed to yield 1 lb of Au. Furthermore, since the mass of the ore is 500 times greater than the mass of *calaverite* it contains, a much greater mass of ore will be needed, as compared to the mass of $AuTe_2$ required. Therefore, the mass of ore required will be about 500 × 2-1/3, or ~1200 lb, or ~0.6 ton.

$$1.0\ \cancel{\text{lb Au}}\left(\frac{100\ \cancel{\text{lb AuTe}_2}}{43.56\ \cancel{\text{lb Au}}}\right)\left(\frac{100\ \cancel{\text{lb ore}}}{0.20\ \cancel{\text{lb AuTe}_2}}\right)\left(\frac{1\ \text{ton ore}}{2000\ \cancel{\text{lb ore}}}\right) =$$

***Answer***

0.57 ton. (Our answer has 2 significant figures since the quantity of gold specified, and the % composition of the ore were given only to 2 significant figures.) [B]

### 1-M. Mass of a Component from Density and Percent by Weight Data

A sulfuric acid solution has a density of 1.84 g/cm$^3$ and contains 98% sulfuric acid ($H_2SO_4$) by weight. How many milliliters of this solution should be taken in order to supply 1000 g of $H_2SO_4$?

(A) $1.8 \times 10^3$ (B) $1.9 \times 10^3$ (C) $7.5 \times 10^2$ (D) $5.5 \times 10^2$
(E) None of these

***Solution*** g $H_2SO_4$ → g solution → mL solution

$$\left(1000\ \cancel{\text{g H}_2\text{SO}_4}\right)\left(\frac{100\ \cancel{\text{g soln}}}{98\ \cancel{\text{g H}_2\text{SO}_4}}\right)\left(\frac{1\ \text{mL soln}}{\cancel{1.84\ \text{g soln}}}\right) =$$

***Answer*** $5.5 \times 10^2$ (rounded off to 2 significant figures) [D]

## Exercises on Mixtures

18. The compound $Na_3PO_4$ contains 42% sodium. How many grams of a mixture containing 75% $Na_3PO_4$ and 25% $K_3PO_4$ would be needed to supply 10. g of sodium?

    (A) 24 g (B) 18 g (C) 95 g (D) 53 g (E) 32 g

19. A solution prepared by dissolving 26.0 g of a substance in 101 mL of pure water has a density of 1.14 g/cc. What is the volume of this solution?

    (A) 89 mL (B) 101 mL (C) 111 mL (D) 127 mL (E) 145 mL

20. How many grams of KOH are contained in 50.00 mL of a KOH solution that has a density of 1.46 g/cm$^3$ and contains 45% KOH by weight?

    (A) 33 (B) 15 (C) $1.6 \times 10^2$ (D) 1.6 (E) 56

21. A solution is prepared by dissolving 25 g of substance X in 100 mL of pure water. The resulting solution has a density of 1.136 g/cc. How many milliliters of the solution should be taken in order to provide 2 g of X?

    (A) 9 mL (B) 4.0 mL (C) 7 mL (D) 8 mL (E) 0.8 mL

## DRILL ON CALCULATIONS INVOLVING A MIXTURE

***Student's Name*** ______________________

Given the following data for several different sodium hydroxide (NaOH) solutions (mixtures of NaOH and water), calculate answers to fi!! in the blanks in the table. You may be asked to turn in this sheet.

| g NaOH | g $H_2O$ | % NaOH (by weight) in solution | density of the solution g/mL | mL of solution | mass of solution grams |
|---|---|---|---|---|---|
| | | 15.00 | 1.1662 | 25.00 | |
| 17.19 | 25.81 | | | 30.00 | |
| | | 4.50 | 1.0501 | | 50.00 |
| | 36.26 | | | 33.33 | 42.17 |
| 8.88 | | 9.50 | | 85.00 | |

22. A mining company supplies a concentrated ore that is 11% *chalcocite* ($Cu_2S$) by weight. $Cu_2S$ itself contains 79.86% copper by weight. How many tons of ore should be purchased in order to produce 600 tons of an alloy containing 90. % Cu?

    (A) $6.1 \times 10^3$ (B) $7.6 \times 10^3$ (C) $3.9 \times 10^3$ (D) 74 (E) 47

23. A certain *n*-type semiconductor consists of very pure germanium containing 0.50 ppm arsenic. What is the % As in the semiconductor?

    (A) 0.50% (B) 0.0050% (C) 0.000050%
    (D) 0.00000050% (E) 0.000000050%

---

## ENERGY AND TEMPERATURE CHANGES

Simultaneously with changes in matter, *energy* is released or absorbed. The basic SI unit of energy is the *joule*, J ($10^7$ *ergs* = 1 J), but commonly the *calorie* (4.184 J) is still used. One form of energy is *heat*. Associated with heat energy is an intensity factor called *temperature*. On the still commonly used *Fahrenheit scale* the freezing point of water is assigned as 32°F, and the normal boiling point of water is 212°F, which means there are 180 Fahrenheit degrees between these two points. The *Celsius scale* (*centigrade*) is more commonly used in science; on this scale the freezing point of water is defined as 0°C, and the normal boiling point as 100°C. This means that the centigrade degree is the larger of the two; in fact, the number of Fahrenheit degrees required to cover a given temperature range will be 1.8 times the number of centigrade degrees required. Also, it is convenient to be able to convert from a temperature on one scale to the corresponding temperature on the other. From the above definitions,

$$\text{Celsius temperature} = \left(\frac{100}{180}\right)(\text{Fahrenheit temperature} - 32)$$

The *heat capacity* is the amount of heat needed (usually in calories) to change the temperature of a sample by one degree centigrade. If the amount of substance specified is one gram, then this is called the *specific heat*. For a substance of mass $m$ grams with a specific heat of $C$ J/g–°C undergoing a temperature change $\Delta T$ °C (where $\Delta T = T_{final} - T_{initial}$), the number of joules absorbed ($q > 0$) or liberated ($q < 0$) is given by

$$q = mC\Delta T$$

It is very convenient to remember that the specific heat of liquid water is very near 4.184 J/g–°C, or 1 cal/g–°C, at any temperature.

## Example Problems Involving Energy and Temperature Changes

### 1-N. Calculation of the Energy Change from an Observed Temperature Change

A coffee cup containing 100. $cm^3$ of liquid water at 100°F is heated to final temperature of 180°F. The number of joules absorbed by the water is:

(A) $6.0 \times 10^4$ J (B) $1.9 \times 10^4$ J (C) $6.0 \times 10^{-10}$ J (D) $1.1 \times 10^3$ J (E) $1.9 \times 10^{-10}$ J

***Solution*** If we change from °F to °C,

$$q = mC\Delta T$$

$$= (1.00\ \text{cm}^3)\left(\frac{1.00\ \text{g}}{1\ \text{cm}^3}\right)\left(\frac{4.184\ \text{J}}{\text{g-°C}}\right)\left(\frac{5°\text{C}}{9°\text{F}}\right)(180°\text{F} - 100°\text{F})$$

***Answer***

$1.9 \times 10^4$ J (2 significant figures). Note that although the volume and the temperatures are given to 3 significant figures, a significant figure is lost in the subtraction, (180°F - 100°F) = 80°F.

### 1-O. Final Temperature When Two Samples at Different Temperatures are Mixed

100.0 g of liquid water at 95.0°C is mixed with 135.0 g of liquid water at 13.6°C. Assuming that no heat is lost to the surroundings, what would be the resultant temperature of the mixture?

(A) 48.2°C (B) 42.4°C (C) 38.7°C (D) 35.5°C (E) 54.3°C

***Solution***

The assumption is that $q_{hot\ water} + q_{cold\ water} = 0$, *i.e.* 100% of the heat lost by the 95.0-degree water will be gained by the 13.6-degree water. Rearranging and substituting that $q_i = m_i C_i \Delta T_i$ where $\Delta T = T_{final} - T_{initial}$, gives:

$$q_{hot} = -q_{cold}$$

$$m_{hot} C_{hot} \Delta T_{hot} = -m_{cold} C_{cold} \Delta T_{cold}$$

$$(100.0\ \text{g})\left(\frac{4.184\ \text{J}}{\text{g-°C}}\right)(T_f - 95.0°\text{C}) = -(135.0\ \text{g})\left(\frac{4.184\ \text{J}}{\text{g-°C}}\right)(T_f - 13.6°\text{C})$$

Now rearrange the above equation and solve for the unknown, $T_f$.

***Answer***

$T_f$ = 48.2°C (three significant figures). If one of the containers had contained isopropyl alcohol instead of water, the specific heat for the alcohol would have been needed.

[A]

## Exercises on Energy and Temperature Changes

24. An **increase** in temperature of 15°C corresponds to an **increase** in temperature of ______ °F.

    (A) 15°F (B) 8.3°F (C) 27°F (D) 59°F (E) 32°F

25. A 200. g metal bar requires 1.200 kcal to change its temperature from 0°C to 100°C. What is the specific heat of the metal in J/g-°C?

    (A) 69.9 (B) 0.251 (C) 10.0 (D) 0.0837 (E) None of these

26. Experimentally it is found that 163 kJ of thermal energy (*heat*) is evolved when 10.0 g of $Al_2O_3$ is formed from the elements. To what final temperature would 1.00 gallon of water at 20°C be heated by absorbing this quantity of heat?

    (A) 30°C (B) 10.3°C (C) 29.2°C (D) 29°C (E) 20.02°C

27. A 10.0-kg piece of iron at 50.0°C is placed in 1.00 liter of water at 10.0°C. The iron and water will come to the same temperature. What is that temperature, assuming that no heat is lost to the outside? (Specific heat of iron = 0.481 J/g-°C)

    (A) 13.4°C (B) 31.4°C (C) 43.1°C (D) 25.0°C (E) 21.5°C

---

## GENERAL EXERCISES ON INTRODUCTORY CONCEPTS

28. One $cm^3$ is the same as:

    (A) 100 mm (B) 1 mL (C) 0.4 $in^3$ (D) 1 L (E) None of these

29. If the atomic weight of carbon had been assigned exactly 100 amu, instead of 12 amu, the atomic weight of oxygen would have been:

    (A) 104 (B) 133 (C) 16 (D) 128 (E) None of these

30. Given the substances: nitrogen, manganese, nickel, silver, silicon, and magnesium. Arranged left to right in order of **increasing** atomic weight, the chemical symbols are:

    (A) N, Mg, Si, Mn, Ni, Ag
    (B) N, Mg, S, Mn, Ni, Si
    (C) Ni, Mn, Si, Mg, N, Ag
    (D) N, Mn, Sc, Mg, Ni, Si
    (E) N, Mg, Sc, Mn, Ni, Ag

31. A sample of water taken from a certain stream contains 30. ppm of dissolved $O_2$. How many grams of dissolved oxygen are present in 1.0 L of the water (remember that the density of water is 1.0 g/mL)?

    (A) 0.030 g (B) $3.0 \times 10^{-6}$ g (C) $9.4 \times 10^{-4}$ g
    (D) 0.096 g (E) 0.94 g

32. The liter, a unit of volume in the metric system, most closely approximates the ______ in the English system.

    (A) gallon (gal) (B) cubic foot ($ft^3$) (C) pint (pt)
    (D) volumetric flagon ($f$) (E) quart (qt)

33. Which of the following masses is reported to four significant figures?

    (A) 2.0110 mg (B) $17.4 \times 10^{-4}$ g (C) 0.0020 g (D) 3.030 kg
    (E) $3 \times 10^4$ g

34. A mixture of X and Y weighing 14.00 g is found to contain 13.50 g of X. The percentage of Y (by weight) in this mixture should be reported to __?__ significant figure(s). Fill in the blank.

    (A) one (B) two (C) three (D) four (E) eight

35. Which of the following pairs could be used to illustrate the Law of Multiple Proportions?

    (A) $Au_2O$ and $Au_2O_3$ (B) $O_2$ and $O_3$ (C) $Au_2O_3$ and $Fe_2O_3$
    (D) $H_2O$ and $OH_2$ (E) $Au_2O_3$ and $AuCl_3$

36. Which of the following are chemical **elements**: O, $O_2$, $O_3$, P, $P_4$, S, $S_8$?

    (A) all are chemical elements (B) only O, P, and S
    (C) only $O_2$, P, and S (D) only $O_2$, $O_3$, $P_4$, and $S_8$
    (E) only $O_2$, $P_4$, and $S_8$

37. The formula $(NH_4)_2HPO_4$ indicates that for every dozen combined oxygen atoms in this compound, there are ______ dozen combined hydrogen atoms. The number required to fill the blank is:

    (A) 2 (B) 1-1/4 (C) 9 (D) 8 (E) 2-1/4

38. Under appropriate conditions, ammonia ($NH_3$) reacts with oxygen as represented by the following **unbalanced** equation:

    $$\_\_\_ NH_3 + \_\_\_ O_2 \rightarrow \_\_\_ NO + \_\_\_ H_2O$$

    The ratio of the coefficients of $NH_3/O_2$ in the balanced equation would be:

    (A) 2/1 (B) 3/2 (C) 1/1 (D) 3/5 (E) 4/5

39. Vitamin $B_{12}$ has a molecular formula that is $C_{63}H_{88}CoN_{14}O_{14}P$ and a molecular weight of 1355.42. Which of the following is an element that is present in Vitamin $B_{12}$?

    (A) potassium (B) calcium (C) nickel (D) phosphorus (E) copper

40. At one time gold sold for $400./Troy ounce. If the density of gold is 19.3 g/$cm^3$, calculate the cost of 4.54 mL of gold at that price. (1 Troy pound = 12 Troy ounces = 373 g.)

    (A) $0.353 (B) $1,130 (C) $33.38 (D) $980. (E) $3.03

41. If you could count individual atoms at the rate of one atom per second, about how many years would be required to count $6.02 \times 10^{23}$ atoms?

    (A) $10^2$ yr (B) $10^3$ yr (C) $10^9$ yr (D) $10^{12}$ yr (E) $10^{16}$ yr

42. The planet Earth has a volume of $1.1 \times 10^{21}$ $m^3$ and an average density of 5.5 $g/cm^3$. What is the mass of the earth in kilograms?

    (A) 5.0 mg (B) $5.5 \times 10^{20}$ kg (C) $4.2 \times 10^{30}$ kg (D) $2.0 \times 10^{23}$ kg (E) $6.1 \times 10^{24}$ kg

43. A compound of formula $AB_3$ contains 40% A by weight. The atomic weight of A must be:

    (A) one-half that of B. (B) equal to that of B.
    (C) three times that of B. (D) one-third that of B. (E) twice that of B.

44. A mixture of $NH_4VO_3$ and $NH_4Cl$ contains 66.7% $NH_4VO_3$ by weight. What weight of this mixture should be taken in order to provide 1.50 g of $NH_4VO_3$?

    (A) 1.00 g (B) 1.50 g (C) 2.00 g (D) 2.25 g (E) 3.00 g

45. The density of the element gold, Au, is to be calculated from the following data:

    | | |
    |---|---|
    | Mass of a golden coin | = 13.512 g |
    | Volume of the coin and water | = 25.1 mL |
    | Volume of the water alone | = 24.4 mL |

    The density of the metal in the coin equals 13.512 g/(25.1 mL - 24.4 mL); the answer to the correct number of significant figures should be reported as:

    (A) 19.303 g/mL (B) 19.30 g/mL (C) 19.3 g/mL (D) 19 g/mL (E) $2 \times 10^1$ g/mL

46. Consider the compound $X_2Y_5$ that is 60.0% X and 40.0% Y by weight. From these data, it can be concluded that the ratio of the atomic weight of X to the atomic weight of Y, $(AW_X/AW_Y)$, is:

    (A) 1.50 (B) 0.667 (C) 1.33 (D) 3.75 (E) 2.67

47. Atom Z is found to be 12.0 times as heavy as a carbon atom. You wish to prepare a compound that contains 2 atoms of carbon for every 1 atom of Z. If 1.00 g of carbon is used, how many grams of Z are required?

    (A) 12.0 g (B) 48.0 g (C) 6.00 g (D) 3.00 g (E) 24.0 g

48. 200. g of copper shot at 25.0°C is added to 200. mL of water at 80.0°C. The final temperature of the mixture is 75.3°C. Assuming that the specific heat of water is 1.0 cal/g-°C and that no heat is lost to or gained from the surroundings, what is the specific heat of copper metal in cal/g-°C?

(A) 0.093 (B) 0.10 (C) 1.0 (D) 10 (E) None of these

49. Three different samples containing only elements A and B are analyzed:

Sample #1 weighs 24.00 g and contains 3.00 g of B
Sample #2 is 12.5% B by weight
Sample #3 is composed of 11.375 lb of A and 1.625 lb of B

It is likely that:

(A) all three samples are identical compounds.
(B) only samples #1 and #2 are identical compounds; #3 is a different compound.
(C) only samples #2 and #3 are identical compounds; #1 is a different compound.
(D) each of the three samples is a different compound.
(E) at least one of the samples is a mixture.

50. Iron pyrites ($FeS_2$) reacts with $O_2$ to produce $Fe_2O_3$ and $SO_2$. One million $FeS_2$ formula units would consume ______ million $O_2$ molecules upon reaction. The number needed to fill the blank is:

(A) 1-1/2 (B) 2 (C) 2-3/4 (D) 1-2/3 (E) 2-2/3

51. For the reaction (unbalanced),

$$\_\_\_ Fe_2O_3 + \_\_\_ NH_3 \rightarrow \_\_\_ Fe_3O_4 + \_\_\_ N_2 + \_\_\_ H_2O$$

the coefficients of $NH_3$ and $Fe_3O_4$ in the balanced equation would be in the ratio ______, respectively.

(A) 3:2 (B) 1:2 (C) 9:2 (D) 1:3 (E) None of these

52. The sweet taste of saccharin is detectable at 10 ppm. A certain product is to have 25 ppm saccharin added to it. How many pounds of saccharin should be added to a vat containing 2.6 tons of unsweetened product?

(A) $4.8 \times 10^{-9}$ lb (B) $2.5 \times 10^{-4}$ lb (C) 0.13 lb
(D) 0.0052 lb (E) $6.5 \times 10^{-5}$ lb

53. It was necessary to make the following calculation from the three experimental measurements shown in the following set-up:

$$\frac{36.24}{12.00} - 2.4723 = ?$$

The answer, reported to the correct number of significant figures, would be:

(A) 0.5 (B) 0.55 (C) 0.548 (D) 0.5477 (E) 0.54770

54. Mercury has a density of 13.6 g/mL. What volume, in cubic inches, would 34.0 lb of mercury occupy?

(A) 46.2 $in^3$ (B) 69.3 $in^3$ (C) 227 $in^3$ (D) 448 $in^3$
(E) $1.16 \times 10^3$ $in^3$

55. A 2.0000-g sample of element X reacted with oxygen to form 2.5392 g of the compound $XO_2$. Taking 16.00 amu for the atomic weight of oxygen, determine the identity of element X.

(A) Sn (B) Si (C) Co (D) Ti (E) C

---

## ANSWERS TO CHAPTER 1 PROBLEMS

| | | | | | |
|---|---|---|---|---|---|
| 1. E | 10. D | 19. C | 28. B | 37. E | 46. D |
| 2. A | 11. D | 20. A | 29. B | 38. E | 47. C |
| 3. A | 12. C | 21. A | 30. A | 39. D | 48. A |
| 4. A | 13. C | 22. A | 31. A | 40. B | 49. A |
| 5. D | 14. C | 23. C | 32. E | 41. E | 50. C |
| 6. C | 15. B | 24. C | 33. D | 42. E | 51. D |
| 7. D | 16. D | 25. B | 34. B | 43. E | 52. C |
| 8. C | 17. E | 26. A | 35. A | 44. D | 53. C |
| 9. B | 18. E | 27. B | 36. A | 45. E | 54. B |
| | | | | | 55. A |

# 2
# *The Mole Concept and Its Use*

Atoms are so small (diameters of approximately $10^{-8}$ cm) and have so little mass (averaging about $10^{-22}$ g) that individual atoms can neither be counted out nor weighed. Instead of dealing with individual atoms, 'counting groups' consisting of some very large number of individuals are used.* The 'counting unit' most commonly used in the sciences is the *gram mole* (usually simply called the *mole*) and consists of 6.02252 $\times$ $10^{23}$ individuals. This number, also referred to as *Avogadro's Number* ($N_0$), is the number of atoms in exactly 12 g of carbon-12, the most common type of carbon atom. The mole represents an extremely large number of individuals and, although it is usually used to count atoms, molecules, and various chemical entities, it could be used to count anything that exists in extremely large numbers. The mole concept can be used with other mass units. The milligram mole (or *millimole*) is the number of atoms in exactly 12 mg of carbon-12, 6.02 $\times$ $10^{20}$ atoms; the kilogram mole is the number of atoms in 12 kg of carbon-12, 6.02 $\times$ $10^{26}$ atoms; the pound mole is the number of atoms in 12 pounds of carbon-12, 273 $\times$ $10^{26}$ atoms; the ton mole is the number of atoms in exactly 12 tons of carbon-12, 5.46 $\times$ $10^{29}$ atoms. It should be remembered that although these other mole quantities are frequently used, the term mole always refers to the gram mole.

## NUMBER-MASS-MOLE RELATIONSHIPS

A chemical formula can be understood to represent the number of individual atoms of each type combined in a single formula unit or the number of moles of atoms of each type combined in a mole of the substance. The

*A very common "counting group" used in this country is the dozen, which consists of 12 individuals.

sum of the masses of the individual atoms in amu is the *Formula Weight, FW*, of the species; if the masses of moles of the various atoms making up the unit are expressed in grams and added together, this will also yield the formula weight but in terms of grams/mole instead of amu/formula unit. These will have the same numerical values. Indeed, any ratio of moles is exactly equal to the ratio of the individual chemical particles.

You must be able to make the following interconversions quickly and accurately,

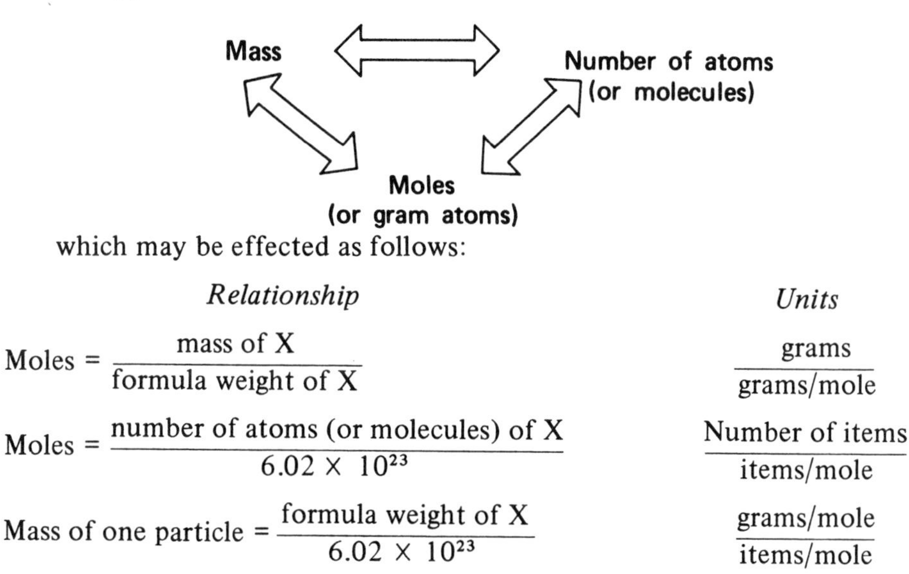


which may be effected as follows:

| *Relationship* | *Units* |
|---|---|
| $\text{Moles} = \dfrac{\text{mass of X}}{\text{formula weight of X}}$ | $\dfrac{\text{grams}}{\text{grams/mole}}$ |
| $\text{Moles} = \dfrac{\text{number of atoms (or molecules) of X}}{6.02 \times 10^{23}}$ | $\dfrac{\text{Number of items}}{\text{items/mole}}$ |
| $\text{Mass of one particle} = \dfrac{\text{formula weight of X}}{6.02 \times 10^{23}}$ | $\dfrac{\text{grams/mole}}{\text{items/mole}}$ |

Note that if you know any two of the variables in one of the above relationships, you can rearrange the equation and solve for the third quantity. Thus, if the formula weight of a compound is 80.0 g/mole and you are given 0.300 mole of the compound, the mass of the compound is obtained by rearranging the first relationship to:

$$\text{Mass of X} = (\text{Moles of X})\,(\text{FW of X})$$

$$\text{Mass of X} = (0.300 \text{ moles of X}) \left(\frac{80.0 \text{ g of X}}{1 \text{ mole of X}}\right)$$

$$\text{Mass of X} = 24.0 \text{ g}$$

### Example Problems Involving Number-Mass-Mole Relationships

#### 2-A. The Mole as a Number

Which of the following samples contains the greatest number of atoms?

(A) 1.0 g of gold, Au (B) 1.0 g of water, $H_2O$ (C) 1.0 g of helium, He
(D) 1.0 g of octane, $C_8H_{18}$
(E) All of the above contain the same number of atoms.

### *Solution*

The convenience of the mole is that it is a number, and that the more moles you have, the more **specified** particles you have; the simple multiplication of the number of particles by the number of units per particle will give you the total number of units. Remembering that moles = g/(g/mole) and that (moles $\times N_0$) = number of particles (where $N_0$ represents Avogadro's number), we have:

(A) $\frac{1.0}{197} \times N_0$ Au atoms $\rightarrow \frac{1.0}{197} N_0$ total **atoms** $\rightarrow 0.0051 N_0$ atoms

(B) $\frac{1.0}{18} \times N_0$ $H_2O$ molecules $\rightarrow \frac{3.0}{18} N_0$ total **atoms** $\rightarrow 0.17 N_0$ atoms

(C) $\frac{1.0}{4.0} \times N_0$ He atoms $\rightarrow \frac{1.0}{4.0} N_0$ total **atoms** $\rightarrow 0.25 N_0$ atoms

(D) $\frac{1.0}{114} \times N_0$ $C_8H_{18}$ molecules $\rightarrow \frac{26}{114} N_0$ total **atoms** $\rightarrow 0.23 N_0$ atoms

### *Answer*

Thus, since $N_0$ is a fixed number, the greatest number of atoms is [C]
Multiplying with $N_0 = 6.0 \times 10^{23}$ gives $1.5 \times 10^{23}$ total atoms.

## 2-B. Number-Mass Relationship

A bottle contains $x$ atoms of carbon weighing 6.00 g. Another bottle, containing an equal number of nickel atoms, is desired. What mass of nickel should be taken?

(A) 52.7 g (B) 1.23 g (C) 23.7 g (D) 12.3 g (E) 29.4 g

### *Solution*

Grams C → moles C → moles Ni → grams Ni

$$\left(6.00\ \cancel{g\ C}\right)\left(\frac{1\ \cancel{mole\ C}}{12.0\ \cancel{g\ C}}\right)\left(\frac{1\ \cancel{mole\ Ni}}{1\ \cancel{mole\ C}}\right)\left(\frac{58.7\ g\ Ni}{1\ \cancel{mole\ Ni}}\right) =$$

### *Answer*

The calculator result is 29.35 g, which must be rounded off to 29.4 g of Ni. Note that the atomic weights used in this calculation were chosen to have the same number of significant figures (3) as warranted by the data. [E]

## 2-C. Percent by Atom to Mass in a Sample

A single razor blade contains a total of eighty-four hundred billion billion ($8.4 \times 10^{21}$) atoms, 57% of which are iron atoms, 14% chromium atoms, and 29% carbon atoms. What mass of carbon does the blade contain?

(A) 0.049 g (B) 0.14 g (C) 2.1 g (D) 0.17 g (E) 0.024 g

### *Solution*

Total atoms → C atoms → moles of C → grams of C

$$\left(8.4 \times 10^{21} \cancel{\text{atoms}}\right)\left(\frac{29 \cancel{\text{C atoms}}}{100 \cancel{\text{atoms}}}\right)\left(\frac{1 \cancel{\text{mole of C}}}{6.02 \times 10^{23} \cancel{\text{C atoms}}}\right)\left(\frac{12.0 \text{ g C}}{\cancel{\text{mole C}}}\right) =$$

### *Answer*

The answer shown on the calculator would be $4.85581395 \times 10^{-2}$ g, which would have to be rounded off to $4.9 \times 10^{-2}$ g of C so as to be in accord with the 2 significant figures given for %C. [A]

## 2-D. Atomic Mass to Moles

The absolute mass of atom X is $2.40 \times 10^{-22}$ g. What fraction of a mole (gram-atom) of X is represented by a sample of X weighing 10.5000 g?

(A) 0.105 (B) $4.38 \times 10^{22}$ (C) 0.0727 (D) 0.264 (E) 1.74

### *Solution*

Grams of X → atoms of X → moles of X

$$\left(10.5000 \cancel{\text{g X}}\right)\left(\frac{1 \cancel{\text{atom X}}}{2.40 \times 10^{-22} \cancel{\text{g X}}}\right)\left(\frac{1 \text{ mole of X}}{6.02 \times 10^{23} \cancel{\text{atoms of X}}}\right) =$$

### *Answer*

The calculator result would show $7.2674419 \times 10^{-2}$ moles, which would have to be rounded off to 3 significant figures; hence the correct answer is 0.0727 moles of X. [C]

## 2-E. Avogadro's Number and the Pound Mole

How many nitrogen atoms are present in 0.50 pounds of ammonium nitrate, $NH_4NH_3$?

(A) $1.7 \times 10^{24}$ (B) $7.5 \times 10^{21}$ (C) $3.8 \times 10^{21}$ (D) $3.4 \times 10^{24}$
(E) $3.0 \times 10^{23}$

### *Solution*

There are two alternative ways of thinking through the problem. The pounds of the given compound can be converted to grams; then by use of the formula weight the grams of compound to moles of compound; then by reading the formula convert moles of compound to moles of nitrogen. Finally, using Avogadro's number convert moles of N to the number of nitrogen atoms. Or schematically,

$$\text{lbs cpd} \rightarrow \text{g cpd} \rightarrow \text{moles cpd} \rightarrow \text{moles N} \rightarrow \text{atoms N}$$

The FW of $NH_4NO_3$ = $2AW_N + 3AW_O + 4AW_H = 80.$

$$(0.50 \text{ lb cpd}) \left[\frac{454 \text{ g cpd}}{1 \text{ lb cpd}}\right]\left[\frac{1 \text{ mol cpd}}{80. \text{ g cpd}}\right]\left[\frac{2 \text{ mole N}}{1 \text{ mole cpd}}\right]\left[\frac{6.02 \times 10^{23} \text{ N atoms}}{1 \text{ mole N}}\right] =$$

The second approach involves converting pounds of compound to pound-moles of compound by using the formula weight, 80. pounds per pound-mole. Avogadro's number is not applicable to the pound-mole, but refers only to the gram mole (mole). The next step then would be to remember that a pound-mole is 454 times the size of a mole and convert the lb-mole to mole. The remainder of the solution is then the same as before.

$$\text{lbs cpd} \rightarrow \text{lb-moles cpd} \rightarrow \text{moles cpd} \rightarrow \text{moles N} \rightarrow \text{atoms N}$$

$$(0.50 \text{ lb cpd}) \left[\frac{1 \text{ lb-mole cpd}}{80. \text{ lb cpd}}\right]\left[\frac{454 \text{ mole cpd}}{1 \text{ lb-mole cpd}}\right]\left[\frac{2 \text{ mole N}}{1 \text{ mole cpd}}\right]\left[\frac{6.02 \times 10^{23} \text{ N atoms}}{1 \text{ mole N}}\right] =$$

Notice that in both instances exactly the same numbers were multiplied and divided.

***Answer*** $3.4 \times 10^{24}$ nitrogen atoms **[D]**

## Exercises on Number-Mass-Mole Relationships

1. Avogadro's number of krypton atoms (AW = 83.8) weighs:

   (A) 83.8 amu (B) 1.00 g (C) 83.8 g (D) $5.04 \times 10^{25}$ g (E) $13.9 \times 10^{-23}$ g

2. The mass of $3.01 \times 10^{21}$ atoms of mercury is:

   (A) 1.00 g (B) 2.00 g (C) $5.00 \times 10^{-5}$ g (D) 200 g (E) none of these

3. The **total** mass of 1.00 dozen carbon atoms is:

   (A) 144 g (B) 1.00 g (C) $2.39 \times 10^{-22}$ g (D) $2.00 \times 10^{-23}$ g (E) $1.67 \times 10^{-24}$ g

4. Which of the following contains the greatest number of **atoms**?

   (A) 0.1 mole of $F_2$ (AW = 19.0) (B) 100 g Pb (AW = 207)
   (C) Five billion $O_2$ molecules (AW = 16.0)
   (D) 4 g of H (AW = 1.0)
   (E) 6 g of He (AW = 4.0)

5. The weight in grams of a single urea molecule, $(NH_2)_2CO$, is:

   (A) 60.0 g (B) $6.02 \times 10^{-23}$ g (C) $3.61 \times 10^{25}$ g
   (D) $9.97 \times 10^{-23}$ g (E) $7.31 \times 10^{-23}$ g

6. Convert 5.0 moles of $SO_3$ (FW = 80) to grams of $SO_3$.

   (A) 16 g (B) 80 g (C) 400 g (D) 625 g (E) 5.0 g

7. Which of the following has the largest mass?

   (A) 16.0 g of $O_2$ (B) 16.0 g of $O_3$ (C) $6.02 \times 10^{23}$ oxygen atoms
   (D) 32 amu of oxygen (E) 1.00 mole of $O_2$

8. How many **total** atoms would be in 1.00 lb-mole of $CO_2$?

   (A) $6.02 \times 10^{23}$ (B) $18.06 \times 10^{23}$ (C) 3 (D) $8.19 \times 10^{26}$
   (E) $1.86 \times 10^{25}$

9. One mole of $K_2Co(SO_4)_2 \cdot 6H_2O$ weighs:

   (A) 437 g (B) 341 g (C) 398 g (D) 347 g (E) 405 g

10. Atom Z is found to be 12.0 times as heavy as a carbon atom. You wish to prepare a compound that contains four atoms of carbon for every one of Z, *i.e.*, $ZC_4$. How many grams of Z would be needed to combine with 1.00 g of carbon?

    (A) 12.0 (B) 48.0 (C) 6.00 (D) 3.00 (E) 24.0

---

## THE MOLE CONCEPT—APPLICATIONS TO PURE SUBSTANCES

Since the mole is a fixed number, the ratio of moles of *constituent atoms* to moles of *compound* must be numerically equal to the number of atoms in a single molecule (or *formula unit*) of that compound. An understanding of this fact is of considerable importance (and convenience) in solving problems dealing with the constitution and formulation of pure substances, *i.e.*, with respect to the hypothetical compound $A_xB_yC_z$, interconversions as follow:

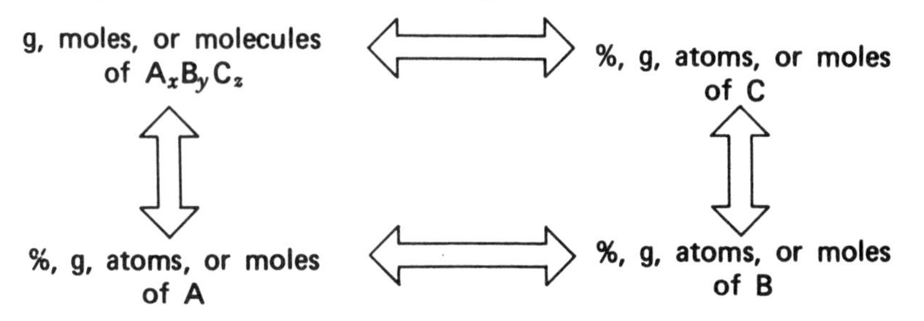


The mole is a very convenient unit; once the number of moles of compound, or the number of moles of one of the atoms in a compound, is known, the moles, grams, and numbers of all of the constituents can be readily obtained. For example, if one has 0.125 mole of a substance with the formula $Na_2S_2O_3 \cdot 5H_2O$ (equivalent in composition to either $Na_2S_2O_3\ (H_2O)_5$ or $Na_2S_2O_8H_{10}$), one has in chemical combination:

| | | | | |
|---|---|---|---|---|
| 0.250 mole Na | or | $1.51 \times 10^{23}$ Na atoms | or | 5.75 g Na |
| 0.250 mole S | or | $1.51 \times 10^{23}$ S atoms | or | 8.02 g S |
| 1.00 mole O | or | $6.02 \times 10^{23}$ O atoms | or | 16.0 g O |
| 1.25 mole H | or | $7.53 \times 10^{23}$ H atoms | or | 1.26 g H |

Or if just the water of crystallization is of interest, one could say there is

0.625 mole $H_2O$ or $3.76 \times 10^{23}$ $H_2O$ molecules or 11.3 g $H_2O$.

Problems may conveniently be divided into two categories: (1) the formula of the compound is known and, when given the weight or number of moles of a constituent, information about any other constituent or the compound may be calculated by the relationship†:

$$\begin{pmatrix}\text{g of sub-}\\\text{stance}\\\text{sought}\end{pmatrix} = \begin{pmatrix}\text{moles of}\\\text{substance}\\\text{given}\end{pmatrix}\left(\frac{\text{moles of substance sought}}{\text{moles of substance given}}\right) \times \left(\frac{\text{g of substance sought}}{\text{mole of substance sought}}\right)$$

and (2) the composition (weight relationship or %) is known, which allows

†The middle term on the right-hand side of the equation is called the *stoichiometric link*. When dealing with calculations involving a single substance, it may be read directly from the formula of the substance; hence, it is always visible. It relates moles of whatever is being sought (atoms, molecules, formula units) to moles of something given (atoms, molecules, formula units).

us to first calculate the number of moles of each constituent and then to find the simplest whole-number relationship between constituents, *i.e.*, the simple empirical formula of the compound.

The ratio of combined atoms in a given compound cannot be altered—to do so would be to change the identity of the substance. A formula showing the simplest whole number ratio of combined atoms is the simplest or empirical formula; however, in certain cases, it is possible to determine that a particular combination of atoms are bound together in a neutral unit, a *molecule*. The mass of a molecule (the *molecular weight, MW*) is either equal to, or is some whole-number multiple of the formula weight. In such cases the *molecular formula* is frequently used. For example:

| Empirical Formula | Observed Molecular Weight | Molecular Formula |
|---|---|---|
| CH | 26 | $C_2H_2$, acetylene |
| CH | 78 | $C_6H_6$, benzene |
| $CH_2O$ | 30 | $CH_2O$, formaldehyde |
| $CH_2O$ | 60 | $C_2H_4O_2$, acetic acid |
| $CH_2O$ | 180 | $C_6H_{12}O_6$, glucose |
| SN | 92 | $S_2N_2$, disulfur dinitride |
| SN | 184 | $S_4N_4$, tetrasulfur tetranitride |

## Example Problems Involving The Mole Concept—Applications to Pure Substances

### 2-F. Grams of Compound A to Atoms of Component B

How many oxygen atoms are there in a 42-g sample of ammonium dichromate, $(NH_4)_2Cr_2O_7$ (FW = 252)?

(A) $7.0 \times 10^{23}$ (B) $1.0 \times 10^{23}$ (C) $6.0 \times 10^{23}$ (D) $1.4 \times 10^{22}$
(E) $2.5 \times 10^{22}$

***Solution***

$$\text{g compound} \xrightarrow{a} \text{moles compound} \xrightarrow{b} \text{moles O} \xrightarrow{c} \text{atoms O}$$

*Step a:* Conversion of given amount of compound to moles.

*Step b:* Use the *stoichiometric link,* 7 moles of oxygen per 1 mole of compound (cpd). Note that the stoichiometric link may be read directly from the formula.

*Step c:* Convert moles of substance sought (in this case, oxygen) to the units asked for (in this case, number).

$$\left(42 \cancel{\text{g cpd}}\right)\left(\frac{1 \cancel{\text{mole cpd}}}{252 \cancel{\text{g cpd}}}\right)\left(\frac{7 \cancel{\text{mole O}}}{1 \cancel{\text{mole cpd}}}\right)\left(\frac{6.02 \times 10^{23} \text{ atoms O}}{1 \cancel{\text{mole O}}}\right) =$$

### *Answer*

The calculator would show the result $7.0233... \times 10^{23}$ atoms, which would have to be rounded off to 2 significant figures to $7.0 \times 10^{23}$ oxygen atoms.

[A]

## 2-G. Atoms of Component A to Grams of Component B

Given a sample of ammonium dichromate that contains $8.03 \times 10^{23}$ atoms of nitrogen. How many grams of hydrogen does the sample contain?

(A) 3.03 (B) 0.337 (C) 0.674 (D) 0.189 (E) 5.39

### *Solution*

Atoms N → moles N → moles H → grams H

$$\left(8.03 \times 10^{23} \cancel{\text{atoms N}}\right)\left(\frac{1 \cancel{\text{mole N}}}{6.02 \times 10^{23} \cancel{\text{atoms N}}}\right)\left(\frac{8 \text{ mole H}}{2 \cancel{\text{mole N}}}\right)\left(\frac{1.01 \text{ g H}}{1 \cancel{\text{mole H}}}\right) =$$

### *Answer*

The calculator result of 5.388903654 g of hydrogen must be rounded off to 5.39 g of hydrogen.

[E]

## 2-H. Moles, Gram, and Atoms to Empirical Formula

A given sample of a compound contains 0.667 mole of nitrogen atoms, 2.688 g of hydrogen, $2.01 \times 10^{23}$ chromium atoms, and contains half as many oxygen as hydrogen atoms. What is the simplest (empirical) formula for the compound?

(A) $(NH_4)_2Cr_2O_7$ (B) $(NH_4)_2CrO_4$ (C) $(NH_4)HCrO_4$
(D) $(NH_4)_3CrO$ (E) $NH_4Cr(OH)_4$

### *Solution*

Given data (%, no. atoms, etc.) → moles† for each atom → simplest whole-number mole ratio → simplest formula

---

†In a molecule (or formula unit) of a compound there must be at least **one** of each kind of atom, or in a mole of the compound there must be at least **one** mole of each type of atom present. Therefore, the whole-number ratios are obtained by putting the

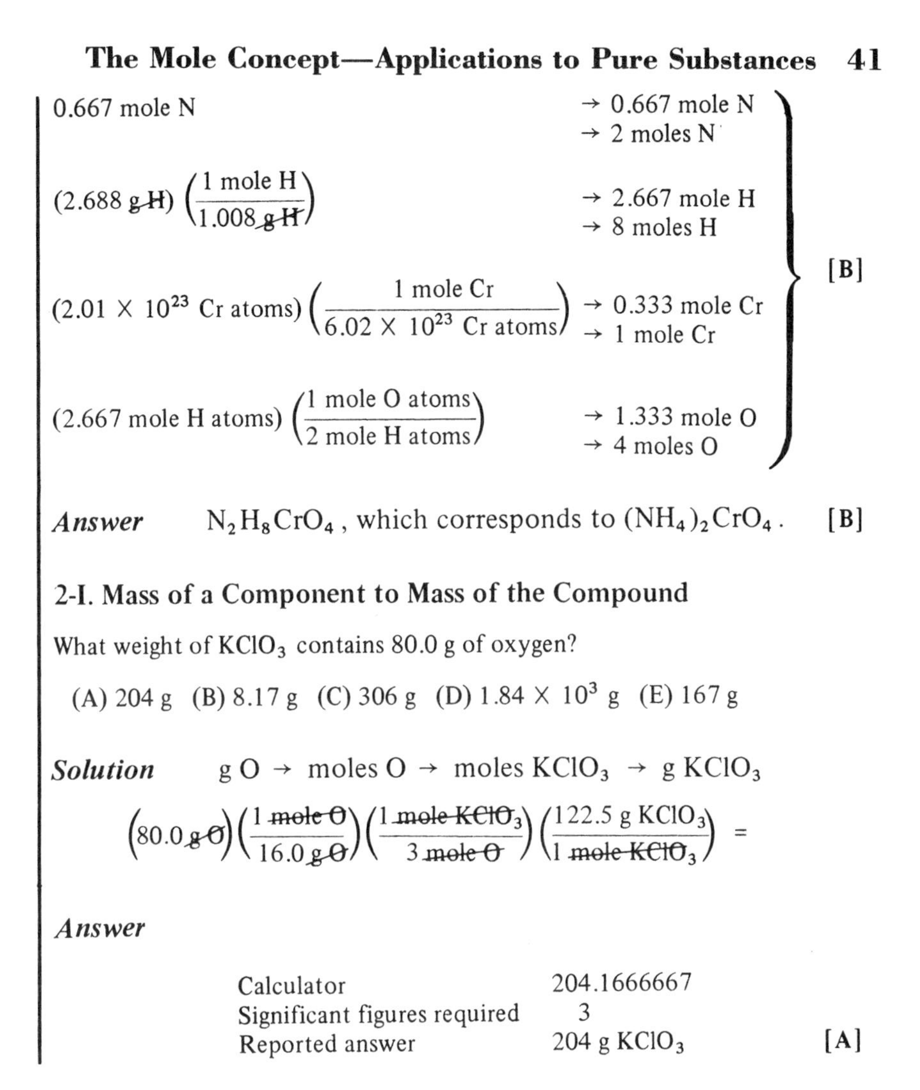

$0.667 \text{ mole N} \rightarrow 0.667 \text{ mole N} \rightarrow 2 \text{ moles N}$

$(2.688 \cancel{\text{g H}}) \left(\dfrac{1 \text{ mole H}}{1.008 \cancel{\text{g H}}}\right) \rightarrow 2.667 \text{ mole H} \rightarrow 8 \text{ moles H}$

$(2.01 \times 10^{23} \text{ Cr atoms}) \left(\dfrac{1 \text{ mole Cr}}{6.02 \times 10^{23} \text{ Cr atoms}}\right) \rightarrow 0.333 \text{ mole Cr} \rightarrow 1 \text{ mole Cr}$

$(2.667 \text{ mole H atoms}) \left(\dfrac{1 \text{ mole O atoms}}{2 \text{ mole H atoms}}\right) \rightarrow 1.333 \text{ mole O} \rightarrow 4 \text{ moles O}$

[B]

*Answer* $N_2H_8CrO_4$, which corresponds to $(NH_4)_2CrO_4$. [B]

## 2-I. Mass of a Component to Mass of the Compound

What weight of $KClO_3$ contains 80.0 g of oxygen?

(A) 204 g (B) 8.17 g (C) 306 g (D) $1.84 \times 10^3$ g (E) 167 g

*Solution* g O → moles O → moles $KClO_3$ → g $KClO_3$

$$\left(80.0 \cancel{\text{g O}}\right)\left(\frac{1 \cancel{\text{mole O}}}{16.0 \cancel{\text{g O}}}\right)\left(\frac{1 \cancel{\text{mole KClO}_3}}{3 \cancel{\text{mole O}}}\right)\left(\frac{122.5 \text{ g KClO}_3}{1 \cancel{\text{mole KClO}_3}}\right) =$$

*Answer*

| | |
|---|---|
| Calculator | 204.1666667 |
| Significant figures required | 3 |
| Reported answer | 204 g $KClO_3$ |

[A]

---

mole data in the form of moles of a given element per mole of the element(s) present to the least extent (here, Cr). In many cases this gives directly the whole-number atom ratio (and hence the simplest formula), but at times a fraction is still left and it is necessary to multiply by a small integer to clear the data of fractions (*e.g.*, if it had been found that 1.333 moles of A and 1.500 moles of B were present per 1 mole of C, these numbers would have to be multiplied through by 6 to give 8 moles of A, 9 moles of B, and 6 moles of C, *i.e.*, $A_8B_9C_6$). The data in such problems frequently are experimentally obtained % composition-by-weight data and are subject to small experimental errors. From such data it usually will be necessary to round off the last significant figure to obtain the integer required for a formula, *e.g.*, 1.997 moles of N would have to be rounded to 2. Analytical data should be quite accurate and, although it may be necessary to round off a third or fourth significant figure, rounding off too many figures leads to erroneous results (*e.g.*, 1.840 should not be rounded off to 2; this number is close to 1-5/6 and the data probably should be multiplied through by 6).

### 2-J. Weight Percent of a Component to Number of Atoms of the Component in the Molecule

The compound *equilin,* a hormone found in the urine of pregnant mares, has a molecular weight of 268.3 and contains 80.6% carbon by weight. How many carbon atoms are in each *equilin* molecule?

(A) 7 (B) 9 (C) 18 (D) 22 (E) 28

#### *Solution*

Since the mole is a fixed number, the ratio "moles of carbon/mole of equilin" is numerically equal to the desired ratio, "atoms of carbon/molecule of equilin." That is, **a ratio by mole is equal to a ratio by number of individual units.** Thus, by knowing that we need to find "moles of carbon/mole of equilin" and realizing that the given molecular weight carries with it the ". . . per mole of equilin" part of this ratio already, we can follow the route (where "E" = equilin):

$$\frac{\text{g E}}{\text{mole E}} \rightarrow \frac{\text{g C}}{\text{mole E}} \rightarrow \frac{\text{mole C}}{\text{mole E}} \quad \left(\text{which equals} \quad \frac{\text{atoms of C}}{\text{molecule of E}}\right)$$

$$\left(\frac{268.3\ \cancel{\text{g E}}}{\text{mole E}}\right)\left(\frac{80.6\ \cancel{\text{g C}}}{100\ \cancel{\text{g E}}}\right)\left(\frac{1\ \text{mole C}}{12.011\ \cancel{\text{g C}}}\right) =$$

***Calculator*** 18.00431271 mole C/1 mole Equilin [C]

#### *Answer*

Exactly 18! The calculation shows 18.0 moles of carbon per mole of equilin, which is limited to 3 significant figures by the % C given. The exactness of the eighteen is due to belief in the Law of Definite Composition, which states that in a discrete molecule there will be an *integral* number of combined atoms. Another point (which might be obvious): The ratio, "18.0 mole C/mole E" means "18.0 mole C/1 mole E," and multiplication of this ratio by $N_0/N_0 = 1$, where $N_0$ = Avogadro's number, gives the atom-to-molecule ratio explicitly:

$$\left(\frac{18.0\ \cancel{\text{mole C}}}{1\ \cancel{\text{mole E}}}\right)\left(\frac{N_0\ \text{atoms C}/\cancel{\text{mole C}}}{N_0\ \text{molecules E}/\cancel{\text{mole E}}}\right) = \left(\frac{18\ \text{atoms of C}}{1\ \text{molecule of E}}\right)$$

## Exercises on The Mole Concept—Applications to Pure Substances

11. The stoichiometric link that would be the multiplier to convert "moles of Cu" to "moles of $Cu_2(OH)_2CO_3$" is:

    (A) 2/1 (B) 2/2 (C) 3/2 (D) 1/2 (E) none of these

12. What mass of $Cu_2(OH)_2CO_3$ (FW = 221) contains 100. g of copper?

    (A) 358 g (B) 174 g (C) 57.5 g (D) 130 g (E) 112 g

13. A 1.0-mole sample of $C_3H_6$

    (A) contains 3.0 moles of carbon atoms (B) has a mass of 42 g
    (C) contains $6.0 \times 10^{23}$ $C_3H_6$ molecules
    (D) contains 6.0 g of hydrogen (E) has all of the above characteristics

14. Which of the following contains the greatest mass of chlorine?

    (A) 10.0 g $Cl_2$ (B) 50.0 g $KClO_3$ (C) 0.100 mole NaCl
    (D) 10.1 g NaCl (E) 0.100 mole $Cl_2$

15. The % nitrogen by weight in urea, $(H_2N)_2CO$, is: ($FW_{urea}$ = 60.0)

    (A) 23.3% (B) 31.3% (C) 38.0% (D) 46.7% (E) 60.9%

16. You are given a sample of sodium hydroxide (NaOH) containing $1.2 \times 10^{23}$ atoms of H. How many grams of sodium hydroxide do you have? (AW's: Na = 23, O = 16, H = 1.0)

    (A) 20 g (B) 2.0 g (C) 4.0 g (D) 8.0 g (E) 2.7 g

17. Which of the following is an empirical formula?

    (A) $C_2H_2$ (B) $C_2H_6O$ (C) $Au_2Cl_6$ (D) $Hg_2(NO_3)_2$
    (E) All of these are empirical formulas.

18. A compound having the empirical formula $C_2H_3Cl$ has a molecular weight of about 190. Its molecular formula is:

    (A) $C_2H_3Cl$ (B) $C_6H_{12}Cl_3$ (C) $C_6H_9Cl_3$ (D) $C_9H_{11}Cl_2$
    (E) $C_4H_6Cl_2$

19. Given a sample of $Mg(ClO_3)_2$ (FW = 191) containing 1.20 moles of oxygen atoms. How many grams of chlorine does the sample contain? (AW's: Mg = 24.3, Cl = 35.5, O = 16.0)

    (A) 14.2 g (B) 128 g (C) 38.2 g (D) 42.6 g (E) 85.2 g

20. A compound $X_3O_4$ contains 72.03% X and 27.97% oxygen by weight. What is the atomic weight of X?

    (A) 30.98 (B) 9.013 (C) 19.00 (D) 54.94 (E) 50.95

21. A 1.61-g sample of an oxide of chromium contains 1.00 g of chromium. Calculate the simplest formula of the compound. (AW's: Cr = 52.0, O = 16.0)

    (A) CrO (B) $CrO_2$ (C) $Cr_3O_2$ (D) $Cr_2O_2$ (E) $Cr_2O_3$

22. Pure vitamin E is a compound containing 29 carbon atoms per molecule and is 80.87% C by weight. What is its molecular weight?

    (A) 348.3 (B) 430.7 (C) 281.7 (D) 390.8 (E) $5.7 \times 10^{-22}$

## DRILL ON COMPOUND COMPOSITION

***Student's Name*** ______________________

Fill in all of the rectangles in each row that correspond to one quantity given in that row for the compound $P_2O_3Cl_4$, F.W. = 252. Your instructor may ask that you carefully tear this sheet out and turn it in. X is either Cl or O (assigned by the instructor).

| g $P_2O_3Cl_4$ | mole $P_2O_3Cl_4$ | molecules $P_2O_3Cl_4$ | mole P | atoms P | g P | mole X | atoms X | g X |
|---|---|---|---|---|---|---|---|---|
| 12.6 | | | | | | | | |
| | 0.300 | | | | | | | |
| | | $1.00 \times 10^{18}$ | | | | | | |
| | | | 1.00 | | | | | |
| | | | | ——— | | | | |
| | | | | | 5.16 | | | |
| | | | | | | ——— | | |
| | | | | | | | 12 | |

## THE MOLE CONCEPT–REACTION STOICHIOMETRY

In the previous section, the *stoichiometric link* within a compound was used to calculate weight, mole, and number relationships between the compound and its constituent parts. A similar *stoichiometric link* exists in a chemical reaction, *i.e.*, the coefficients in the balanced equation of the substance for which information is given and of the substance for which information is sought. The basis of the calculation of the weight, moles, or number of molecules of a reactant or product, when given the weight, moles, or number of molecules of another substance involved in the reaction, is this *stoichiometric link*. The setup for such problems is the same as that given under category (1) on p. 38. Inherent in these calculations are the assumptions that (a) there are sufficient reactants available to prepare the desired quantity of product or to react with the given quantity of reactant, and (b) the reaction proceeds 100% to completion.

In situations where data are given for two or more reactants it is necessary to determine which of the given reactants would be used up first, the *limiting reactant*, before carrying out calculations regarding the *theoretical (maximum) yield* of a product. The calculation then must be based upon the quantity of this reactant; data with respect to other reactants are no longer of interest unless it is desired to calculate the quantities of these substances remaining upon completion of the reaction.

### Example Problems Involving the Mole Concept–Reaction Stoichiometry

#### 2-K. Mass of Substance A to Mass of Substance B Required

Boron carbide, $B_4C$, is prepared from the reaction of $B_2O_3$ and carbon in an electric furnace. CO is produced as a by-product. What minimum weight of carbon is needed to prepare 150. g of $B_4C$? (FW of $B_4C$ = 55.25; FW of $B_2O_3$ = 69.6)

(A) 228 g (B) 196 g (C) 31.0 g (D) 181 g (E) 79.6 g

*Solution*

(1) Write and balance the equation

(2) g $B_4C$ → moles $B_4C \xrightarrow{a}$ moles C → g C
The stoichiometric link is used in the step marked "*a*."

(1) $2\,B_2O_3 + 7\,C \rightarrow B_4C + 6\,CO$

(2) $$\left(150\text{ g }B_4C\right)\left(\frac{1\text{ }\cancel{\text{mole }B_4C}}{55.25\text{ }\cancel{\text{g }B_4C}}\right)\left(\frac{7\text{ }\cancel{\text{moles C}}}{1\text{ }\cancel{\text{mole }B_4C}}\right)\left(\frac{12.011\text{ g C}}{1\text{ }\cancel{\text{mole C}}}\right) =$$

*Answer*

| | | |
|---|---|---|
| Calculator | 228.2633484 | |
| Significant figures required | 3 | |
| Reported Answer | 228 g of carbon | [A] |

## 2-L. Mass of Substance A to Number of Molecules of B Required

Ammonia ($NH_3$) reacts with $O_2$ to give NO and water. How many $O_2$ molecules are required to react with 68 g of ammonia?

(A) $1.9 \times 10^{24}$ (B) $1.2 \times 10^{24}$ (C) $1.9 \times 10^{23}$ (D) $9.5 \times 10^{22}$
(E) $3.0 \times 10^{24}$

*Solution*

(1) Write and balance the equation
(2) g $NH_3$ → mole $NH_3$ → mole $O_2$ → molecules $O_2$

(1) $4\ NH_3 + 5\ O_2 \rightarrow 4\ NO + 6\ H_2O$

(2) $$\left(68\ \cancel{g\ NH_3}\right)\left(\frac{1\ \cancel{mole\ NH_3}}{17\ \cancel{g\ NH_3}}\right)\left(\frac{5\ \cancel{moles\ O_2}}{4\ \cancel{moles\ NH_3}}\right)\left(\frac{6.02 \times 10^{23}\ \text{molecules}\ O_2}{1\ \cancel{mole\ O_2}}\right) =$$

*Answer*

| | | |
|---|---|---|
| Calculator | $3.01 \times 10^{24}$ | |
| Significant figures required | 2 | |
| Reported answer | $3.0 \times 10^{24}$ molecules of $O_2$ | [E] |

## 2-M. Limiting Reactant

When $Ca_3P_2$ is treated with water, the products are $Ca(OH)_2$ and $PH_3$. What is the maximum weight of $PH_3$ that can be prepared from 2.00 g of $Ca_3P_2$ (FW = 182.2) and 1.00 g $H_2O$ (FW = 18.0)?

(A) 0.0185 g (B) 0.187 g (C) 0.630 g (D) 0.748 g (E) 5.67 g

*Solution*

(1) Write and balance the equation
(2) g each reactant → moles each reactant → moles of **limiting** reactant → moles $PH_3$ → g $PH_3$

(1) $Ca_3P_{2\,(s)} + 6\ H_2O_{(\ell)} \rightarrow 3\ Ca(OH)_{2\,(s)} + 2\ PH_{3\,(g)}$

(2) $\frac{2.00}{182.2} = 0.0110$ mole $Ca_3P_2$ ; $\frac{1.00}{18.0} = 0.0556$ mole $H_2O$

Since 0.0110 mole of $Ca_3P_2$ would require 0.0660 mole of $H_2O$, the latter is deficient and $H_2O$ is therefore the *limiting reactant.* Alternatively, since

0.0556 mole of $H_2O$ can react with no more than 0.00927 mole of $Ca_3P_2$, the $Ca_3P_2$ is **in excess** and (again) we find that $H_2O$ is the **limiting reactant**. The number of **moles** of the **limiting reactant must be used** in this calculation.

$$\left(0.0556 \text{ mole } H_2O\right)\left(\frac{2 \text{ mole } PH_3}{6 \text{ mole } H_2O}\right)\left(\frac{34.0 \text{ g } PH_3}{1 \text{ mole } PH_3}\right) =$$

*Answer*

| | | |
|---|---|---|
| Calculator | 0.6301333333 | |
| Significant figures required | 3 | |
| Reported answer | 0.630 g $PH_3$ | [C] |

## 2-N. Mass of Product A to Mass of Product B—Pound Mole

White phosphorus ($P_4$) is prepared industrially by heating a mixture of phosphate rock, $Ca_3(PO_4)_2$, sand, $SiO_2$, and coke, C, in an electric furnace. Carbon monoxide and calcium silicate, $CaSiO_3$, are also produced. What weight of $CaSiO_3$ is produced from a run in which 225 lb of phosphorus is produced? Give the answer in pounds.

(A) $1.26 \times 10^3$ (B) $3.84 \times 10^2$ (C) $2.10 \times 10^2$ (D) 63.9 (E) 35.1

*Solution*

(1) Write and balance the equation
(2) lb $P_4$ → lb-mole $P_4$ → lb-mole $CaSiO_3$ → lb $CaSiO_3$

(1) $2\ Ca_3(PO_4)_2 + 6\ SiO_2 + 10\ C \rightarrow P_4 + 10\ CO + 6\ CaSiO_3$

(2) $$\left(225 \text{ lb } P_4\right)\left(\frac{1 \text{ lb-mole } P_4}{124 \text{ lb } P_4}\right)\left(\frac{6 \text{ lb-mole } CaSiO_3}{1 \text{ lb-mole } P_4}\right)\left(\frac{116 \text{ lb } CaSiO_3}{1 \text{ lb-mole } CaSiO_3}\right) =$$

*Answer*

| | | |
|---|---|---|
| Calculator | 1262.903226 | |
| Significant figures required | 3 | |
| Reported answer | $1.26 \times 10^3$ lb $CaSiO_3$ | [A] |

## Exercises on the Mole Concept—Reaction Stoichiometry

23. In the reaction between magnesium and nitrogen to form magnesium nitride, $Mg_3N_2$, 0.6 mole of magnesium would require:

    (A) 0.2 mole of $N_2$ molecules (B) 0.4 mole of $N_2$ molecules
    (C) 0.2 mole of N atoms (D) 0.6 mole of N atoms
    (E) $2.4 \times 10^{23}$ molecules of $N_2$

24. How much iron can be converted into $Fe_3O_4$ by 1000 g of 80% oxygen?

    (A) about 2 kg (B) about 1 lb (C) about 10 lb (D) about 10 kg
    (E) about 100 g

25. What mass of carbon dioxide is produced by the complete combustion of 52 g of $C_2H_2$?

    (A) 36 g (B) $1.8 \times 10^2$ g (C) 88 g (D) $3.5 \times 10^2$ g (E) 44 g

26. $FeS_2$ (FW = 120) reacts with oxygen to form $Fe_2O_3$ (FW = 160) and $SO_2$ (FW = 64). The mass of oxygen required to convert 40. g of $FeS_2$ into $Fe_2O_3$ is:

    (A) 40. g (B) 23 g (C) 2.6 g (D) 29 g (E) none of these

27. The compound $NH_4V_3O_8$ is to be prepared from the following stepwise reactions:

    *Step 1:* $N_2 + 3\,H_2 \rightarrow 2\,NH_3$
    *Step 2:* $2\,NH_3 + V_2O_5 + H_2O \rightarrow 2\,NH_4VO_3$
    *Step 3:* $3\,NH_4VO_3 + 2\,HCl \rightarrow NH_4V_3O_8 + 2\,NH_4Cl + H_2O$

    Assuming an abundance of the other reagents, what is the maximum number of moles of $NH_4V_3O_8$ that could be prepared from one mole of $N_2$ **and** one mole of $H_2$?

    (A) 1/3 mole (0.33 mole) (B) 2/3 mole (0.67 mole) (C) 1 mole
    (D) 2/9 mole (0.22 mole) (E) 2 moles

28. Ethane, $C_2H_6$, burns in the presence of oxygen to form $CO_2$ and $H_2O$ (*combustion*). If 8.00 g of oxygen are used, how many moles of $CO_2$ will be produced?

    (A) 0.143 (B) 0.286 (C) 0.438 (D) 1.00 (E) 4.00

29. The combustion reaction used in Question 28 was carried out in a closed container containing 10.0 g of ethane and 10.0 g of oxygen. What is the maximum number of moles of water that can be formed?

    (A) 0.111 (B) 0.268 (C) 0.365 (D) 0.534 (E) 1.00

30. What is the maximum weight of $P_2I_4$ that can be prepared from 5.00 g of $P_4O_6$ and 8.00 g of iodine by the reaction:

$$5\,P_4O_6 + 8\,I_2 \rightarrow 4\,P_2I_4 + 3\,P_4O_{10}$$

    (A) 13.00 g (B) 10.34 g (C) 8.98 g (D) 5.17 g (E) 4.99 g

31. For the thermite reaction between chromium(III) oxide ($Cr_2O_3$) and aluminum metal, producing chromium metal and aluminum oxide ($Al_2O_3$), what is the maximum weight of metallic Cr that can be prepared upon reaction of 38.0 g of $Cr_2O_3$ with 9.00 g of aluminum? (AW's: Cr = 52.0, O = 16.0, Al = 27.0)

## DRILL ON REACTANTS & PRODUCTS

***Student's Name*** ____________________

Fill in all of the rectangles in each row that correspond to the one quantity given in that row for the reaction:

____ $NH_3$ + ____ $O_2$ → ____ NO + ____ $H_2O$. You may be asked to turn in this sheet.

| ← REACTING | | | → | ← FORMING | | | | → |
|---|---|---|---|---|---|---|---|---|
| moles $NH_3$ | g $NH_3$ | moles $O_2$ | g $O_2$ | moles NO | g NO | moles $H_2O$ | g $H_2O$ | molecules $H_2O$ |
| 1.00 | | | | | | | | |
| | 85.0 | | | | | | | |
| | | 0.600 | | | | | | |
| | | | ____ | | | | | |
| | | | | 0.320 | | | | |
| | | | | | 11.5 | | | |
| | | | | | | ____ | | |
| | | | | | | | | $9.0 \times 10^{24}$ |

## DRILL ON LIMITING REACTANT, EXCESS REACTANT & YIELD

***Student's Name*** ________________

Fill in all rectangles in each row that correspond to the quantities given in that row for the reaction:

$$\_\_\_ S_2Cl_2 + \_\_\_ NH_3 \rightarrow \_\_\_ N_4S_4 + \_\_\_ NH_4Cl + \_\_\_ S_8$$

| F.W. | $S_2Cl_2$ | $NH_3$ | $N_4S_4$ | $NH_4Cl$ | $S_8$ |
|---|---|---|---|---|---|
| | 135 | 17.0 | 184 | 53.5 | 256 |

Unless your instructor assigns otherwise, let X be $N_4S_4$. You may be asked to turn in this sheet.

| ← REACTION MIXTURE → | | | | | | | ← THEO. (MAX) YIELD → | | |
|---|---|---|---|---|---|---|---|---|---|
| g $S_2Cl_2$ | moles of $S_2Cl_2$ | g $NH_3$ | moles of $NH_3$ | Formula of Limiting Reactant | Formula of *Excess* Reactant | moles of *XS* Reactant Left | of X, moles | of X, grams | of $S_8$ molecules |
| 10.0 | | 5.00 | | | | | | | |
| | 1.00 | | 2.67 | | | | | | |
| 40.5 | | | 0.700 | | | | | | |
| | 0.750 | 40.0 | | | | | | | |
| ______ | | ______ | | | | | | | |
| | ______ | | ______ | | | | | | |

(A) 47.0 g (B) 26.0 g (C) 8.67 g (D) 200. g (E) 17.3 g

32. A student prepared bromobenzene, $C_6H_5Br$, by the reaction of benzene, $C_6H_6$, with bromine,

$$C_6H_6 + Br_2 \rightarrow C_6H_5Br + HBr$$

The student weighed out 20.0 g of benzene (FW = 78.0) and 50.0 g of bromine (FW = 160.) and obtained 28.0 g of bromobenzene (FW = 157). What is the % yield of $C_6H_5Br$?

(A) 69.7% (B) 57.0% (C) 40.0% (D) 75.6% (E) 36.4%

33. It is desired to produce 325 tons of $FeCl_3$ by the reaction of $FeS_2$ with $Cl_2$. The by-product of the reaction is $S_2Cl_2$. What minimum weight of $Cl_2$ is needed? (FW's: $FeS_2$, 120.; $Cl_2$, 70.9; $FeCl_3$, 162; $S_2Cl_2$, 135)

(A) 356 tons (B) 142 tons (C) 56.9 tons (D) 813 tons
(E) 435 tons

---

## THE USE OF THE MOLE CONCEPT IN ANALYSIS PROBLEMS

Chemical analysis involves the conversion of one or more components in a sample into a stoichiometrically-related species whose quantity can be measured. For example, the silver in an ore sample may actually exist in the chemical form $Ag_2S$. By appropriate chemical treatment, the weighed mixture (ore) may be dissolved and all of the silver it contains converted to a precipitate of AgCl which can be filtered, dried, and weighed. Then,

$$(\text{mole AgCl obtained})\left(\frac{1 \text{ mole Ag}}{1 \text{ mole AgCl}}\right)\left(\frac{108 \text{ g Ag}}{1 \text{ mole Ag}}\right) = \text{mass of Ag in ore sample analyzed}$$

$$\text{and \% Ag in ore} = \left(\frac{\text{mass of Ag}}{\text{mass of ore sample analyzed}}\right)(100).$$

Or if we wanted the % $Ag_2S$ (FW = 248) in the ore,

$$(\text{mole AgCl obtained})\left(\frac{1 \text{ mole Ag}}{1 \text{ mole AgCl}}\right)\left(\frac{1 \text{ mole } Ag_2S}{2 \text{ mole Ag}}\right)\left(\frac{248 \text{ g } Ag_2S}{1 \text{ mole } Ag_2S}\right)$$

$$\text{and \% } Ag_2S \text{ in the ore} = \left(\frac{\text{mass of } Ag_2S}{\text{mass of ore sample analyzed}}\right)(100).$$

## Example Problems Involving the Use of the Mole Concept in Analysis Problems

### 2-O. Purity of a Sample by Analysis of Iron

Commercial, crude *copperas* is a mixture that contains $FeSO_4 \cdot 7H_2O$ (FW = 278.0) as the sole source of iron. A 1.000-gram sample of *copperas* was dissolved in water and treated with $NH_3$ solution. The iron was completely converted into a precipitate of $Fe(OH)_3$ that was filtered, ignited, and weighed as $Fe_2O_3$ (FW = 159.7). The $Fe_2O_3$ obtained had a mass of 0.2671 g. What is the percent $FeSO_4 \cdot 7H_2O$ in the *copperas* sample?

(A) 92.99% (B) 83.01% (C) 99.44% (D) 77.77% (E) 96.00%

#### *Solution*

All of the Fe in the $FeSO_4 \cdot 7H_2O$ component of the *copperas* is completely converted into $Fe(OH)_3$ which, in turn, is dehydrated by heating and converted to $Fe_2O_3$. The chemical equivalence, shown by ⇳, is that:

$$2 \text{ moles } FeSO_4 \cdot 7H_2O \Leftrightarrow 2 \text{ moles } Fe(OH)_3 \Leftrightarrow 1 \text{ mole } Fe_2O_3.$$

This indicates that we may skip the intermediate $Fe(OH)_3$ and say that each mole of $Fe_2O_3$ that is finally obtained must correspond to two moles of $FeSO_4 \cdot 7H_2O$ in the mixture (conservation of iron atoms). Thus, the amount of $FeSO_4 \cdot 7H_2O$ in the sample may be calculated by:

$$\left(\frac{0.2671}{159.7} \text{ mole } Fe_2O_3\right)\left(\frac{2 \text{ mole } FeSO_4 \cdot 7H_2O}{1 \text{ mole } Fe_2O_3}\right)\left(\frac{278.0 \text{ g } FeSO_4 7H_2O}{1 \text{ mole } FeSO_4 \cdot 7H_2O}\right)$$
$$= 0.9299 \text{ g of } FeSO_4 \cdot 7H_2O.$$

Since the *copperas* sample weighed a total of 1.000 g, the % $FeSO_4 \cdot 7H_2O$ is:

$$\left(\frac{0.9299 \text{ g } FeSO_4 \cdot 7H_2O}{1.000 \text{ g ore sample}}\right)(100) = 92.99\% \ FeSO_4 \cdot 7H_2O.$$

#### *Answer*

92.99% $FeSO_4 \cdot 7H_2O$. The answer is good to four significant figures. Much of the equipment used in analytical chemistry is designed to yield four significant figures: the analytical balance, the burette, volumetric flasks, *etc.* Since considerable care has been taken in the analytical laboratory, the analyst, in particular is careful to report answers to the correct number of significant figures.

### 2-P. Empirical Formula from Reaction Products

A hydrocarbon, $C_xH_y$, is burned completely (*complete combustion* reaction) in excess oxygen, producing 344 mg of $CO_2$ and 56.2 mg of $H_2O$. The empirical formula of $C_xH_y$ is:

(A) $C_7H_{16}$ (B) CH (C) $C_2H_5$ (D) $C_5H_4$ (E) $C_7H_5$

*Solution*

(1) mg $CO_2$ → mmole $CO_2$ → mmole C ↘ C:H ratio
(2) mg $H_2O$ → mmole $H_2O$ → mmole H ↗

$$(1)\ \left(344\ \cancel{\text{mg } CO_2}\right)\left(\frac{1\ \cancel{\text{mmole } CO_2}}{44.0\ \cancel{\text{mg } CO_2}}\right)\left(\frac{1\ \text{mmole C}}{1\ \cancel{\text{mmole } CO_2}}\right) = 7.82\ \text{mmole C}$$

$$(2)\ \left(56.2\ \cancel{\text{mg } H_2O}\right)\left(\frac{1\ \cancel{\text{mmole } H_2O}}{18.0\ \cancel{\text{mg } H_2O}}\right)\left(\frac{2\ \text{mmole H}}{1\ \cancel{\text{mmole } H_2O}}\right) = 6.24\ \text{mmole H}$$

$$\frac{1.25\ \text{mmole C}}{1.00\ \text{mmole H}}$$

It is necessary to multiply by 4 to obtain integers.

*Answer* $C_5H_4$ [D]

## Exercises on the Use of the Mole Concept in Analysis Problems

34. If 2.19 g of compound X is burned, 7.40 g of carbon dioxide, $CO_2$, is formed. What % by weight of compound X is carbon?

 (A) 92.2% (B) 50.5% (C) 29.6% (D) 12.4% (E) 8.07%

35. If compound Y contains 2.98 g of carbon per gram of hydrogen, its empirical formula is:

 (A) $CH_2$ (B) CH (C) $CH_3$ (D) $C_2H$ (E) $CH_4$

36. A sample of unknown hydrocarbon, $C_xH_y$, was completely burned in an excess of oxygen. The weights of products formed were:

| | | |
|---|---|---|
| Carbon dioxide | $CO_2$ | 220 mg |
| Water | $H_2O$ | 45 mg |

 Based on these data, the empirical formula of the compound is:

 (A) $C_2H$ (B) $CH_2$ (C) $CH_3$ (D) $C_2H_5$ (E) CH

37. The complete combustion of a hydrocarbon in oxygen produced 176 mg of $CO_2$ and 108 mg of $H_2O$. Its empirical formula is:

 (A) CH (B) $CH_2$ (C) $C_2H_3$ (D) $CH_3$ (E) $C_3H_2$

38. A 25.0-g mixture containing *blue vitriol*, $CuSO_4 \cdot 5H_2O$ (FW = 249.6), was found to contain 5.00 g of copper. Assuming that all of the copper originally was present as $CuSO_4 \cdot 5H_2O$, what % of the impure material is actually $CuSO_4 \cdot 5H_2O$?

 (A) 5.09% (B) 20.0% (C) 78.6% (D) 100.% (E) none of these

39. If 0.560 g of a metal (M) reacts with excess $HCl_{(aq)}$ to form 5.00 × $10^{-3}$ moles of $H_2$ and a solution of $MCl_{2\,(aq)}$, the atomic weight of the metal is:

(A) 14.0 (B) 28.0 (C) 7.00 (D) 112 (E) all of these

40. A mixture of $Na_2O$ and BaO that weighs 6.50 g is dissolved in water and this solution is then treated with dilute sulfuric acid ($H_2SO_4$). $BaSO_4$ precipitates from the solution, but $Na_2SO_4$ is soluble and remains in solution. The $BaSO_4$ is collected by filtration and, when dried, is found to weigh 7.61 g. What percentage of the original sample of mixed oxides is BaO? (FW of BaO = 153.3, $BaSO_4$ = 233.4, $Na_2O$ = 62.0)

(A) 23.1% (B) 65.2% (C) 76.9% (D) 50.0% (E) 82.3%

41.*A 100.0-mg sample of a compound that contains only carbon, hydrogen, and oxygen was burned in oxygen, yielding $CO_2$ and water. 199.8 mg of $CO_2$ and 81.8 mg of $H_2O$ were collected. What is the empirical formula of $C_xH_yO_z$?

(A) $C_2H_4O$ (B) $C_2H_2O$ (C) $C_4H_8O$ (D) $C_5H_2O_2$
(E) none of these

42.*A 4.000-g sample of $M_2S_3$ was ignited in air yielding 3.658 g of $MO_2$. What is element M?

(A) Al (B) Ga (C) Ce (D) Eu (E) Pr

43. Given the equation: $MCl_2 + 2\ AgNO_3 \rightarrow 2\ AgCl + M(NO_3)_2$. If 5.00 g of $MCl_2$ is reacted with excess $AgNO_3$ and 11.0 g of AgCl is formed, what is the best estimate of the atomic weight of M? (FW's: AgCl = 143, $AgNO_3$ = 170)

(A) 60 (B) 30 (C) 24 (D) 40 (E) 9

---

## THE MOLE CONCEPT–MOLAR CONCENTRATION

A *solution* is a homogeneous mixture of two or more components. The major component is referred to as the *solvent;* the other component(s) as the *solute(s).* Although, in general terms, the solvent and solute may be gas, liquid, or solid, the type of solution most frequently encountered involves a liquid solvent. (Indeed, water is the most commonly used solvent.)

Solutions are important in chemistry for several reasons: (a) reactions frequently occur more readily when the reactants are in solution, (b) the rate of a reaction can be readily controlled by changing the quantity of dissolved reactant, and (c) controlled quantities of reactants are more easily measured and added when the reactant is in solution. There are many ways of expressing how much solute is dissolved in the solution, *i.e.*, the *concentration* of the solution, but the way most frequently used in the chemistry laboratory is the number of moles of solute per liter of solution, *i.e.*, *molarity*, abbreviated as $M$;

$$\text{molarity}, (M) = \frac{\text{moles of solute}}{\text{liter of solution}} = \frac{\text{mmoles of solute}}{\text{mL of solution}}$$

By knowing the volume of solution and its molar concentration the number of moles of solute can be readily ascertained and then used in other stoichiometry calculations. A solution of known molarity may be diluted and the concentration of the diluted solution calculated, since the number of moles of solute remains constant, *i.e.*,

$$\begin{array}{ccc} \text{Moles solute}_{init.} & = & \text{Moles solute}_{final} \\ \| & & \| \\ (M_{init.})(V_{init.}) & = & (M_{final})(V_{final}) \end{array}$$

Concentration may be expressed simply in terms of the weight of solute per volume of solution, *e.g.*, g/L, mg/L, g/100 mL. Many commercial solutions, *e.g.*, hydrochloric acid, ammonia, nitric acid, have their concentrations given in weight %, *i.e.*, g of solute per 100 g of solution. To convert to molar concentration one must know the density of the solution and then it follows that

$$M = \left(\frac{\cancel{\text{g solute}}}{100 \text{ g soln}}\right)\left(\frac{\cancel{\text{g soln}}}{1\ \cancel{\text{mL soln}}}\right)\left(\frac{1000\ \cancel{\text{mL}}}{\text{L}}\right)\left(\frac{1 \text{ mole solute}}{\text{FW}\ \cancel{\text{g solute}}}\right)$$

For solutes present only to a very small extent, the concentration frequently is given in **parts per million (ppm)** or in **parts per billion (ppb).** As noted in Chapter 1, a part per million would refer to 1 gram of solute in 1,000,000 grams of solution. The density of such a solution is essentially the same as that of the solvent; in the case of water as a solvent this would be ~1 g/mL. Therefore, for ***aqueous*** solutions:

$$1 \text{ ppm} = \frac{1 \text{ g solute}}{1{,}000{,}000 \text{ mL solution}} = \frac{1 \text{ g solute}}{1{,}000 \text{ L solution}} = \frac{1 \text{ mg solute}}{1 \text{ L solution}}$$

## Examples of the Mole Concept—Molar Concentration

### 2-Q. Molar Concentration from Solution Make-Up

A 1.00-g sample of $Na_2CO_3 \cdot 10H_2O$ was dissolved in 20.0 mL of distilled water. Additional water was added so as to give 250. mL of solution. What is the molar concentration of $Na_2CO_3$?

(A) 0.0377 $M$ (B) $1.39 \times 10^{-8}$ $M$ (C) $8.74 \times 10^{-4}$ $M$ (D) 0.0140 $M$
(E) 1.14 $M$

***Solution***

$$\frac{\text{g cpd}}{\text{mL soln}} \rightarrow \frac{\text{g cpd}}{\text{L soln}} \rightarrow \frac{\text{moles cpd}}{\text{L soln}} \equiv M$$

$$\left(\frac{1.00 \cancel{\text{g cpd}}}{250 \cancel{\text{mL soln}}}\right)\left(\frac{10^3 \cancel{\text{mL soln}}}{\text{L soln}}\right)\left(\frac{1 \text{ mole cpd}}{286 \cancel{\text{g cpd}}}\right) =$$

***Answer***

| | | |
|---|---|---|
| Calculator | 0.0139860139 | |
| Significant figures required | 3 | |
| Reported answer | 0.0140 $M$ $Na_2CO_3$ | [D] |

## 2-R. Grams of Solute Required to Prepare a Particular Solution

How many grams of solid $NaC_2H_3O_2$ are needed to prepare 300. mL of a 0.060 molar solution?

(A) 1.5 g (B) 0.016 g (C) $4.1 \times 10^2$ g (D) 8.2 g (E) 16 g

***Solution***

L soln, molarity → moles solute → g solute

$$\left(0.300 \cancel{\text{L}}\right)\left(0.060 \frac{\cancel{\text{mole NaC}_2\text{H}_3\text{O}_2}}{\cancel{\text{L}}}\right)\left(82 \frac{\text{g NaC}_2\text{H}_3\text{O}_2}{\cancel{\text{mole NaC}_2\text{H}_3\text{O}_2}}\right) =$$

***Answer***

| | | |
|---|---|---|
| Calculator | 1.476····g | |
| Significant figures required | 2 | |
| Reported answer | 1.5 g $NaC_2H_3O_2$ | [A] |

## 2-S. Dilution

If 25.0 mL of 2.50 $M$ $CuSO_4$ is diluted with water to a final volume of 450. mL, what is the molarity of solute in the resulting solution?

(A) 0.139 $M$ (B) 0.132 $M$ (C) 0.0222 $M$ (D) 0.0211 $M$ (E) 7.20 $M$

***Solution***

$$\frac{\text{mL}_{init.}}{\text{mL}_{final}} \times M_{init.} \rightarrow M_{final}$$

$$\left(\frac{25.0 \cancel{\text{mL}}}{450. \cancel{\text{mL}}}\right)\left(2.50 \frac{\text{mmole CuSO}_4}{\text{mL}}\right) =$$

***Answer***

| | | |
|---|---|---|
| Calculator | 0.1388888889 | |
| Significant figures required | 3 | |
| Answer reported | 0.139 $M$ $CuSO_4$ | [A] |

## 2-T. Volume Required to Deliver a Given Mass of Solute

A nitric acid solution has a density of 1.249 g/mL and is 40.% $HNO_3$ by weight. How many milliliters of this solution should be delivered to provide 10. g of $HNO_3$?

(A) 5.0 mL (B) 20. mL (C) 31 mL (D) 28 mL (E) 3.2 mL

### *Solution*

$$\text{g } HNO_3 \rightarrow \text{g soln} \rightarrow \text{mL soln}$$

$$\left(10.\ \cancel{\text{g } HNO_3}\right)\left(\frac{100\ \cancel{\text{g soln}}}{40\ \cancel{\text{g } HNO_3}}\right)\left(\frac{1\ \text{mL soln}}{1.249\ \cancel{\text{g soln}}}\right) =$$

### *Answer*

| | | |
|---|---|---|
| Calculator | 20.01601281 | |
| Significant figures required | 2 | |
| Answer reported | 20. mL of 40% $HNO_3$ | [B] |

## 2-U. Volume of A Required to React with a Given Volume of B

Assuming a quantitative reaction, what minimum volume of 0.150 *M* $AgNO_3$ would be needed in order to precipitate (as $Ag_2CrO_4$) all of the chromate from 25.0 mL of 0.100 *M* $K_2CrO_4$?

(A) 8.33 mL (B) 16.7 mL (C) 33.3 mL (D) 75.0 mL (E) 18.8 mL

### *Solution*

(1) Write and balance the equation
(2) mL, *M* $K_2CrO_4$ → mmole $K_2CrO_4$ → mmole $AgNO_3$ → mL$AgNO_3$

$$K_2CrO_{4\,(aq)} + 2\ AgNO_{3\,(aq)} \rightarrow Ag_2CrO_{4\,(s)} + 2\ KNO_{3\,(aq)}$$

$$\left(25.0\ \cancel{\text{mL } K_2CrO_4}\right)\left(0.100\ \frac{\cancel{\text{mmole } K_2CrO_4}}{\cancel{\text{mL } K_2CrO_4}}\right)\left(\frac{2\ \cancel{\text{mmole } AgNO_3}}{1\ \cancel{\text{mmole } K_2CrO_4}}\right) \times$$

$$\left(\frac{1\ \text{mL } AgNO_3}{0.15\ \cancel{\text{mmole } AgNO_3}}\right) =$$

### *Answer*

| | | |
|---|---|---|
| Calculator | 33.33333333 | |
| Significant figures required | 3 | |
| Answer reported | 33.3 mL of 0.150 *M* $AgNO_3$ | [C] |

## Exercises on the Mole Concept—Molar Concentration

44. Given the following solutions: 1.00 L of 6.0 *M* HCl; 2.00 L of 3.0 *M* HCl; and 3.00 L of 2.0 *M* HCl, how many **total** moles of HCl do these solutions contain?

    (A) 2.3 (B) 11 (C) 8.2 (D) 18 (E) 7.3

45. Which contains the greatest quantity of $KMnO_4$ (FW = 158.0)?

    (A) 158 mg $KMnO_4$ (B) 100. mL of 0.100 *M* $KMnO_4$
    (C) 0.100 L of 0.0100 *M* $KMnO_4$ (D) 1.00 mL of 1.00 *M* $KMnO_4$
    (E) All of the above contain the same quantity of $KMnO_4$.

46. How many grams of $AgNO_3$ (FW = 170.) are required to make 0.200 L of a 0.100 *M* solution?

    (A) 3400 g (B) 85.0 g (C) 17.0 g (D) 3.40 g (E) 0.850 g

47. How many milliliters of water should be **added to** 25.0 mL of 12.0 *M* HCl to make 5.00 *M* HCl? (Assume that the volumes are additive.)

    (A) 60.0 mL (B) 35.0 mL (C) 10.4 mL (D) 300. mL (E) 275 mL

48. What weight of $HNO_3$ (FW = 63.0) is present in 13.0 mL of 0.0872 *M* $HNO_3$?

    (A) 71.4 mg (B) 9.39 g (C) 422 mg (D) 0.422 g (E) 0.0180 g

49. The volume of 0.200 molar $K_2CO_3$ (FW = 138) solution that contains 69.0 g of $K_2CO_3$ is:

    (A) 0.400 L (B) 200. mL (C) 1.60 L (D) 500. mL (E) 2.50 L

50. A 25.0% by weight ammonia solution has a density of 0.910 g/cm$^3$ What is the molarity of the solution?

    (A) 13.4 *M* (B) 6.50 *M* (C) 14.7 *M* (D) 7.14 *M* (E) 1.47 *M*

51. A solution has a density of 1.46 g/mL and contains 655 g of KOH (FW = 56.1) per liter of solution. What is the percentage, by weight of KOH?

    (A) 10.5% (B) 25.0% (C) 44.9% (D) 65.5% (E) 38.4%

52. It is desired to convert 0.10 mole of $GeO_2$ into $GeCl_4$ by reaction with concentrated (12 *M*) HCl, producing $H_2O$ as a by-product. What minimum volume of the concentrated acid is required?

    (A) 7.5 mL (B) 4.8 mL (C) 1.2 mL (D) 33 mL (E) 12 mL

53. When $AgNO_3$ solution is added to a solution of KCl, AgCl is quantitatively precipitated, leaving a solution of $KNO_3$. How many milliliters of 0.50 *M* $AgNO_3$ would be required to precipitate all of the chloride from 10. mL of 0.40 *M* KCl?

    (A) 2.0 mL (B) 4.0 mL (C) 5.0 mL (D) 8.0 mL (E) 6.0 mL

54. A sodium chloride solution is prepared by mixing 3.65 L of 0.105 *M* NaCl with 5.11 L of 0.162 *M* NaCl to give 8.76 L of the new solution.

## DRILL ON MOLARITY AND DILUTION

***Student's Name*** ______________

Data is given for an initial silver nitrate solution ($AgNO_3$, F.W. = 169.87). Calculate all quantities not filled in for this initial solution. In some cases additional substance(s) are added to the initial solution. Assume that any solid added dissolves and does not change the volume of the solution and that if a liquid is added that the final volume is the sum of the two starting volumes. You may be asked to turn in this sheet.

| INITIAL SOLUTION | | | | FINAL SOLUTION | | |
|---|---|---|---|---|---|---|
| g $AgNO_3$ | moles $AgNO_3$ | mL of solution | M, molarity of $AgNO_3$ | added | M $AgNO_3$ in final sol'n | g $AgNO_3$ present |
| 10.0 | | 75.00 | | X | | X |
| | 0.0025 | | 0.0015 | X | | X |
| 17.0 | | 250.0 | | 100.0 mL water | | |
| | | 450.0 | 0.1500 | 125.0 mL 0.2500 M $AgNO_3$ | | |
| | | | 0.100 | 75.0 mL water | | 5.00 |
| | 0.125 | | | 20.0 mL 0.333 M $AgNO_3$ | 0.280 | |

How many grams of NaCl are contained in 1.00 L of the new solution?

(A) 22.4 g (B) 8.08 g (C) 1.21 g (D) 0.138 g (E) 48.4 g

55. It is desired to precipitate $Ag_2CrO_4$ by mixing solutions of $AgNO_3$ and $K_2CrO_4$. How many milliliters of 0.30 $M$ $AgNO_3$ would be needed to react with 25 mL of 0.20 $M$ $K_2CrO_4$?

(A) 33 mL (B) 25 mL (C) 17 mL (D) 19 mL (E) 38 mL

---

## GENERAL EXERCISES ON CHAPTER 2–THE MOLE CONCEPT AND ITS USE

These exercises are divided into two series of 46 questions. Each series consists of a random selection of questions over material taken from the entire chapter.

### SERIES 1.

56. The number of $C_2H_6$ molecules in 15 g of $C_2H_6$ is (molecular weight of $C_2H_6$ = 30):

(A) $6.0 \times 10^{23}$ (B) $1.2 \times 10^{24}$ (C) 0.50 (D) $3.0 \times 10^{23}$
(E) $2.4 \times 10^{24}$

57. The number of $C_2H_6$ molecules that you have in 0.050 mole of $C_2H_6$ is:

(A) $2.4 \times 10^{23}$ (B) $3.0 \times 10^{22}$ (C) $8.3 \times 10^{-26}$ (D) $3.0 \times 10^{-2}$
(E) none of these

58. The number of moles of sulfur atoms in 2.5 moles of $Au_2(SO_4)_3$ is:

(A) 7.5 (B) 10. (C) 30. (D) 5.0 (E) $4.5 \times 10^{24}$

59. Given 2.5 g of carbon (AW of C = 12.0). You want a sample of neodymium (AW of Nd = 144) containing the same **number** of atoms as contained in 2.5 g of carbon. You should take ______ g of neodymium.

(A) 2.5 (B) 15 (C) 50 (D) 30 (E) $3.6 \times 10^2$

60. A one-gram sample of carbon contains $x$ carbon atoms. You wish to obtain a sample of sodium that contains $2x$ sodium atoms. The mass of sodium that should be taken is about:

(A) 3.8 g (B) 1.7 g (C) 2.0 g (D) 2.3 g (E) 0.95 g

61. One **molecule** of water contains:

(A) 1 mole (g-atom) of O (B) 2.016 grams of hydrogen
(C) 32.0 grams of $O_2$ (D) 1 atom of oxygen (E) 1/2 mole of $O_2$

62. A compound of boron and hydrogen contains 18.9% hydrogen and 81.1% boron. The empirical formula for the compound is:

(A) $BeH_2$ (B) $Be_2H$ (C) $B_4H$ (D) $B_2H_5$ (E) none of these

63. Metallic copper has a density of 8.9 g/cc. **One mole** of copper occupies a volume of:

    (A) 560 mL (B) 0.0071 L (C) 89 cc (D) 0.56 L (E) 140 mL

64. The formula weight (FW) of urea, $(NH_2)_2CO$, is:

    (A) 44 (B) 46 (C) 58 (D) 43 (E) 60

65. The weight of a millimole of $(NH_4)_2HPO_4$ is: (AW's: H = 1.01, N = 14.0, O = 16.0, P = 31.0)

    (A) 132 g (B) 114 g (C) $1.14 \times 10^{-3}$ g (D) 0.132 g
    (E) $6.02 \times 10^{20}$ g

66. A mixture of 4.00 moles of $H_2$ and 3.00 moles of $O_2$ is ignited, forming water. What is the composition of the system by weight after the reaction is completed?

    (A) 16.0 g of oxygen and 72.0 g of water
    (B) 2.02 g of hydrogen and 36.9 g of water
    (C) 1.00 g of oxygen and 4.00 g of water
    (D) 32.0 g of oxygen and 36.0 g of water
    (E) 32.0 g of oxygen and 72.0 g of water

67. What is the maximum weight of water that can be obtained by igniting a mixture of 6.05 g of $H_2$ and 44.0 g of $O_2$?

    (A) 49.5 g (B) 54.0 g (C) 27.0 g (D) 24.8 g (E) 56.0 g

68. The weight of pure ammonium dichromate that contains 16 g of oxygen is:

    (A) 36 g (B) 13 g (C) 25 g (D) 18 g (E) 176 g

69. 100. mL of a solution contains 24.5 g of $H_2SO_4$ (FW = 98.0). What is the molar concentration (molarity) of $H_2SO_4$?

    (A) 2.50 *M* (B) 0.245 *M* (C) 0.400 *M* (D) 0.00250 *M*
    (E) none of these

70. The complete combustion of acetylene ($C_2H_2$) in oxygen ($O_2$) produces carbon dioxide ($CO_2$) and water ($H_2O$). How many moles of oxygen are needed to react with 4.0 moles of acetylene?

    (A) 5.0 (B) 4.0 (C) 10. (D) 2.0 (E) 7.0

71. What weight of $Fe_2O_3$ (FW = 160.) contains 100. g of oxygen?

    (A) 333 g (B) $1.00 \times 10^3$ g (C) $3.00 \times 10^3$ g (D) 666 g
    (E) none of these

72. Which of the following pertains to the statement: "one-half mole of hydrogen peroxide, $H_2O_2$"?

    (A) 1 mole of H atoms (B) $6.0 \times 10^{23}$ oxygen atoms

(C) $3.0 \times 10^{23}$ $H_2O_2$ molecules (D) 17 grams of $H_2O_2$
(E) **all** of the above

73. What would be the concentration of a solution prepared by dissolving 15.8 g of $KMnO_4$ (FW = 158.0) and diluting this to 250. mL?

    (A) 0.000400 *M* (B) 0.0632 *M* (C) 0.100 *M* (D) 0.400 *M*
    (E) 15.8 *M*

74. 100 mL of a solution of NaCl (0.10 molar) is left in a graduated cylinder until evaporation of solvent reduces its volume to 80.0 mL. What is the molarity of NaCl in the resulting solution?

    (A) 0.080 *M* (B) 0.80 *M* (C) 0.18 *M* (D) 0.13 *M* (E) 0.15 *M*

75. If 15.0 mL of 2.50 M HCl is diluted to 50.0 mL, the concentration of HCl in the final solution will be:

    (A) 2.50 *M* (B) 1.50 *M* (C) 1.00 *M* (D) 0.750 *M* (E) none of these

76. One-half **mole** of sulfuric acid ($H_2SO_4$) contains:

    (A) 32 g of oxygen (B) 2.0 moles of $H_2$ (C) 1 atom of hydrogen
    (D) 1 mole (g-atom) of sulfur (E) $6.02 \times 10^{23}$ atoms of sulfur

77. Given the balanced equation:

    $$2\ NaCl_{(s)} + H_2SO_{4\,(\ell)} \rightarrow 2\ HCl_{(g)} + Na_2SO_{4\,(s)}$$

    The number of grams of HCl that can be prepared from 14.6 g of NaCl and 49.0 g of $H_2SO_4$ is: (FW's: NaCl = 58.4, $H_2SO_4$ = 98.0, HCl = 36.5, $Na_2SO_4$ = 142)

    (A) 9.12 g (B) 190. g (C) 63.6 g (D) 30.4 g (E) 69.2 g

78. A sample of $C_{12}H_{22}O_{11}$ that contains 72 g of carbon also contains how many grams of oxygen?

    (A) 88 g (B) 44 g (C) $1.8 \times 10^2$ g (D) 11 g (E) 72 g

79. The % aluminum in $Al_2O_3$ by weight is: (AW's: Al = 27.0, O = 16.0)

    (A) 47.1% (B) 26.5% (C) 52.9% (D) 40.0% (E) 60.0%

80. What weight of 75.0% NaOH (by weight) must be dissolved in water to prepare 250. mL of 1.50 *M* NaOH (FW = 40.0)?

    (A) 20.0 g (B) 15.0 g (C) 11.2 g (D) 5.33 g (E) 4.00 g

81. 20.0 g of an unknown metal chloride MCl is dissolved in 100. mL of water. If 357 mL of 0.750 *M* $AgNO_3$ are required to precipitate (as AgCl) all the chloride in the solution, what is the identity of element M?

    (A) Ag (B) Na (C) Li (D) Tl (E) K

82. How many grams of NaOH would be present in 200. mL of 2.00 *M* NaOH?

(A) 0.100 g (B) 0.400 g (C) 1.00 g (D) 16.0 g (E) 10.0 g

83. A solution contains 200 g of solute X per liter of solution. If the solution contains 18% X by weight, what is the density of the solution?

    (A) 1.2 g/cc (B) 1.1 g/cc (C) 0.036 g/cc (D) 3.6 g/cc (E) 0.90 g/cc

84. How many total **atoms** are present in 265 mg of $KAuCl_4$ (FW = 378)?

    (A) $2.53 \times 10^{21}$ (B) $2.22 \times 10^{20}$ (C) $2.53 \times 10^{24}$
    (D) $4.22 \times 10^{23}$ (E) $9.57 \times 10^{23}$

85. What is the weight of one atom of sulfur? (AW of S = 32.06)

    (A) $1.90 \times 10^{23}$ g (B) $6.02 \times 10^{-23}$ g (C) 32.06 g (D) 32.06 mg
    (E) $5.33 \times 10^{-23}$ g

86. The products of the combustion of methanol in oxygen are indicated below:

$$CH_3OH + O_2 \rightarrow CO_2 + H_2O.$$

When properly balanced, the equation indicates that ______ mole(s) of $O_2$ are required for each mole of $CH_3OH$ reacting.

    (A) 1 (B) 1-1/2 (C) 2 (D) 2-1/2 (E) 3

87. An oxide of lead contains 90.65% Pb. Its formula is:

    (A) Pb (B) PbO (C) $Pb_2O_3$ (D) $Pb_3O_4$ (E) $PbO_2$

88. A mole of a compound is composed of $6.02 \times 10^{23}$ atoms of hydrogen, 35.45 grams of chlorine, and 64.0 grams of oxygen. The formula of the compound is:

    (A) $HClO_2$ (B) HOCl (C) $HClO_3$ (D) $H(ClO)_2$ (E) $HClO_4$

89. When $CaSO_4 \cdot nH_2O$ is heated, all of the water is driven off. If 34.0 g of $CaSO_4$ (FW = 136) is formed from 43.0 g of $CaSO_4 \cdot nH_2O$, what is the value of $n$?

    (A) 1 (B) 2 (C) 3 (D) 4 (E) 5

90. 0.10 mole of $Al(NO_3)_3$ contains ______ nitrogen **atoms.**

    (A) $6.0 \times 10^{23}$ (B) $2.0 \times 10^{23}$ (C) $1.8 \times 10^{23}$ (D) 0.10
    (E) $1.4 \times 10^{22}$

91. A student mixes 1.50 L of 0.300 *M* NaCl with 2.50 L of 0.700 *M* NaCl, producing 4.00 L of a new NaCl solution. What is the molarity of the new solution?

    (A) 1.00 *M* (B) 0.550 *M* (C) 0.500 *M* (D) 0.197 *M* (E) 0.700 *M*

92. A solution of density 2.00 $g/cm^3$ contains solute X (FW = 80.0). The solution analyzes 60.0% X by weight. What is its molarity?

    (A) 24.0 *M* X (B) 12.5 *M* X (C) 15.0 *M* X (D) 12.0 *M* X
    (E) none of these

93. Commercial vinegar is an aqueous solution of acetic acid, $C_2H_4O_2$ (FW = 60.0). Titration analysis of a sample showed the solution to be 0.640 *M* in acetic acid. How many grams of $C_2H_4O_2$ would be present in 1 pint (473 mL) of the vinegar?

   (A) 81.2 (B) 44.3 (C) 18.2 (D) 93.8 (E) 28.4

94. What is the concentration of a solution made by dissolving 4.76 g of $MgCl_2$ in water and making the solution up to 1500 mL?

   (A) 0.025 *M* (B) 0.033 *M* (C) 0.011 *M* (D) 0.050 *M* (E) 0.075 *M*

95. If 4.55 g of an oxide $X_2O_5$ contains 2.55 g of element X, what is the atomic weight of X?

   (A) 91.0 (B) 31.4 (C) 20.4 (D) 51.0 (E) 101

96. One mmole of $H_2O_2$ weighs: (AW's: H = 1.0, O = 16.0)

   (A) 34.0 mg (B) 34.0 g (C) 1.00 mg (D) 0.0180 g
   (E) $1.0 \times 10^{-1}$ g

97. A molecular compound contains 92.3% carbon and 7.7% hydrogen. If the molecular weight of the compound is 78.1, what is its molecular formula? (AW's: C = 12.0, H = 1.0)

   (A) CH (B) $C_6H_6$ (C) $C_4H_8O_2$ (D) $C_5H_{12}$ (E) $C_5H_{18}$

98. Vitamin $B_1$ contains 16.6% N by weight. If 1 molecule of $B_1$ contains 4 nitrogen atoms, what is the molecular weight of $B_1$?

   (A) 337 (B) 930 (C) 58.1 (D) 211 (E) 474

99. Cochineal powder is obtained by pulverizing the dried bodies of the females of the insect *Coccus cacti.* This powder contains 10.% of the compound "neutral red" (FW = 492.38), a red dye used as food coloring. If 150,000 insects are required to produce 1 kg of the powder, approximately how many "neutral red" molecules are present in each insect?

   (A) $\sim 6 \times 10^{23}$ molecules/insect (B) $\sim 8 \times 10^{17}$ molecules/insect
   (C) $\sim 2 \times 10^{5}$ molecules/insect (D) $\sim 5 \times 10^{22}$ molecules/insect
   (E) $\sim 3 \times 10^{13}$ molecules/insect

100. The % carbon in $H_4C_2O_2$ by weight is: (AW's: C = 12.0, H = 1.0, O = 16.0)

   (A) 40.0% (B) 25.0% (C) 60.0% (D) 75.0% (E) 20.0%

101. Given a sample of pure ammonium dichromate, $(NH_4)_2Cr_2O_7$, which contains $1.0 \times 10^{23}$ oxygen **atoms**. The mass in grams of the sample is:

   (A) 42 g (B) $2.9 \times 10^2$ g (C) 0.38 g (D) 6.0 g (E) none of these

SERIES 2.

102. What weight of $CaCO_3$ (FW = 100.) contains the same weight of oxygen that 171 g of $Al_2(SO_4)_3$ (FW = 342) contains?

(A) 50.0 g (B) 100. g (C) 171 g (D) 25.0 g (E) 200. g

103. The hemoglobin from the red corpuscles of most mammals contains 0.33% iron by weight. Physical measurements indicate that hemoglobin is a macromolecule with a molecular weight of 68,000. How many iron atoms are in **one** hemoglobin molecule?

(A) one (B) two (C) three (D) four (E) five

104. What is the mass of 1.00 lb-mole of ethane, $C_2H_6$?

(A) 30.0 g (B) 60.0 g (C) 1.00 lb (D) 30.0 lb (E) 60.0 lb

105. What weight of water is produced by the reaction of the oxygen in 25.0 g of air (20.0% oxygen by weight) with $C_3H_8$? The only other product formed is carbon dioxide.

(A) 3.62 g (B) 2.25 g (C) 11.2 g (D) 56.2 g (E) 25.0 g

106. Consider the reaction between aqueous $ScCl_3$ and aqueous $AgNO_3$, precipitating AgCl and leaving $Sc(NO_3)_3$ in solution. If 2.50 liters of 0.300 *M* $AgNO_3$ is added to 500. mL of 0.500 *M* $ScCl_3$ and the reaction goes to completion, what will be the molarity of the remaining $Sc(NO_3)_3$? (Assume additive volumes.)

(A) 0.100 *M* (B) 0.300 *M* (C) 0.0833 *M* (D) 0.0167 *M*
(E) 0.0233 *M*

107. $N_2$ reacts with $H_2$ to form ammonia, $NH_3$. From the balanced equation for this reaction, one can say that the consumption of 0.150 mole of $N_2$ would require:

(A) 0.450 mole $H_2$ (B) 0.150 mole $H_2$ (C) 0.150 mole $NH_3$
(D) 0.300 mole $H_2$ (E) 3.00 moles $H_2$

108. 0.100 mole of $(NH_4)_2Cr_2O_7$ contains ______ g of nitrogen (N). (AW's: N = 14.0, H = 1.0, O = 16.0, Cr = 52.0)

(A) 1.40 (B) 2.80 (C) 0.700 (D) 0.100 (E) 0.200

109.* Given two solutions: 0.125 *M* NaOH and 0.275 *M* NaOH. In what **proportion** by volume should these solutions be mixed in order to prepare a 0.250 *M* solution of NaOH? (Assume the volumes are additive; express your answer in volumes of 0.275 *M* NaOH per unit volume of 0.125 *M* NaOH.)

(A) 1.5/1 (B) 5/1 (C) 2.75/1 (D) 6/1 (E) 1/1

110. Balance the following expression: $FeS_2 + O_2 \rightarrow FeSO_4 + SO_3$. The coefficients of $FeS_2$ and $O_2$ in the balanced equation are:

(A) $4\ FeS_2 + 5\ O_2$ (B) $2\ FeS_2 + 5\ O_2$ (C) $2\ FeS_2 + 7\ O_2$
(D) $4\ FeS_2 + 11\ O_2$ (E) $4\ FeS_2 + 8\ O_2$

111. Given the reaction: $2\ S_2Cl_2 + 2\ H_2O \rightarrow SO_2 + 4\ HCl + 3\ S$. How many grams of $SO_2$ could be formed from the reaction of 67.5 g of $S_2Cl_2$ (FW = 135) with 10.0 g of water?

(A) 16.0 g (B) 32.0 g (C) 64.0 g (D) 128 g (E) 17.7 g

112. Concentrated HCl is 37% by weight HCl and has a density of 1.19 g/mL. What is the molarity of the concentrated HCl?

(A) 31 *M* (B) 12 *M* (C) 10 *M* (D) 8.5 *M* (E) 3.2 *M*

113. A compound has the following percent composition by weight:

| | |
|---|---|
| Na, | 36.5% |
| H, | 0.8% |
| P, | 24.6% |
| O, | 38.1% |

The simplest formula for the compound is:

(A) $Na_2HPO_3$ (B) $NaH_2PO_4$ (C) $NaPO_3 \cdot H_2O$ (D) $Na_2P_2O_3 \cdot 5H_2O$
(E) $Na_3PO_4 \cdot 5H_2O$

114. A compound contains by weight, 40.0% carbon, 6.7% hydrogen, and 53.3% oxygen. A 0.10-mole sample of this compound weighs 6.0 g. The molecular formula of the compound is:

(A) $C_3H_8O$ (B) $C_2H_4O_2$ (C) $C_4H_4O$ (D) $H_2CO_3$ (E) $H_2CO_2$

115. A 257.0-mg sample of hydrocarbon $C_xH_y$ gave, on complete combustion in oxygen, 880.2 mg of $CO_2$ (MW = 44.01) and 150.1 mg of $H_2O$ (MW = 18.02). The simplest formula of the hydrocarbon is:

(A) $C_2H_3$ (B) $C_2H_5$ (C) $C_4H_3$ (D) $C_{12}H_5$ (E) $C_6H_5$

116. A molecular compound has the empirical formula $C_3H_4OCl$ (FW = 91.52). The molecular weight of the compound could be:

(A) 45.76 (B) 137.28 (C) 183.04 (D) any of these
(E) none of these.

117. Consider the mineral *tschermigite*, essentially pure $NH_4Al(SO_4)_2 \cdot 12H_2O$ (FW = 453). The % oxygen by weight in this substance is about:

(A) 71% (B) 14% (C) 56% (D) 28% (E) 42%

118. In 1.00 mole of $Na_2CO_3$ there would be:

(A) 28.0 g of nitrogen (B) 16.0 g of oxygen (C) 3.00 g of oxygen
(D) 22.0 g of sodium (E) 46.0 g of sodium

119. Aqueous $(NH_4)_2Cr_2O_7$, when mixed with a solution of NaOH, reacts quantitatively to form $NH_{3\,(aq)}$, $Na_2CrO_{4\,(aq)}$, and water. If 200. mL of 0.100 *M* $(NH_4)_2Cr_2O_7$ is mixed with 100. mL of 0.800 *M* sodium hy-

droxide and the resulting solution diluted to 500. mL with water, the molar concentration of $Na_2CrO_4$ would be:

(A) 0.200 (B) 0.0400 (C) 0.100 (D) 0.320 (E) 0.0800

120. Consider the reaction described in Question 119 (above). A 12.6-g sample of $(NH_4)_2Cr_2O_7$ was dissolved in a minimum volume of water. To this solution was added 200. mL of 0.200 *M* NaOH. The final solution volume was 250. mL. What is the molar concentration of $Na_2CrO_4$?

(A) 0.400 (B) 0.200 (C) 0.160 (D) 0.320 (E) 0.0800

121. Upon mixing solutions of $Pb(NO_3)_2$ and KI the following reaction occurs:

$$Pb(NO_3)_{2\,(aq)} + 2\,KI_{(aq)} \rightarrow PbI_{2\,(s)} + 2\,KNO_{3\,(aq)}$$

Mixing 25.0 mL of 0.0500 *M* $Pb(NO_3)_2$ with 50.0 mL of 0.0250 *M* KI gave 0.187 g of recovered $PbI_2$ (FW = 461). The percentage yield of $PbI_2$ from the experiment was:

(A) 32.5% (B) 64.9% (C) 16.2% (D) 81.0% (E) 97.5%

122. What is the maximum weight of $Ca_3(PO_4)_2$ that can be obtained by mixing 2.00 liters of 1.00 *M* $CaCl_2$ with 3.00 liters of 0.500 *M* $Na_3PO_4$? (FW of $Ca_3(PO_4)_2$ = 310)

$$3\,CaCl_{2\,(aq)} + 2\,Na_3PO_{4\,(aq)} \rightarrow Ca_3(PO_4)_{2\,(s)} + 6\,NaCl_{(aq)}$$

(A) 207 g (B) 232 g (C) 0.667 g (D) 0.750 g (E) 610. g

123. 25.0 mL of 0.400 *M* $Co(NO_3)_2$ is diluted with water to yield exactly 1.00 liter of solution. The resulting solution is ______ *M* in $Co(NO_3)_2$.

(A) 0.100 *M* (B) 16.0 *M* (C) 0.0100 *M* (D) 0.160 *M* (E) 0.0400 *M*

124.* Consider the reaction, $Ba(OH)_{2\,(aq)} + H_3PO_{4\,(aq)}$, which forms water and quantitatively precipitates $Ba_3(PO_4)_2$. If 50.0 mL of 0.0300 *M* $H_3PO_4$ and 25.0 mL of 0.100 *M* $Ba(OH)_2$ are mixed, what is the resulting molar concentration of the **excess reactant**?

(A) $2.0 \times 10^{-2}$ *M* (B) $2.3 \times 10^{-3}$ *M* (C) $3.3 \times 10^{-3}$ *M*
(D) $4.3 \times 10^{-2}$ *M* (E) none of these

125. A sample of ammonium dichromate, $(NH_4)_2Cr_2O_7$, contains, $1.81 \times 10^{24}$ combined hydrogen atoms. How many **grams** of combined nitrogen are there in the sample?

(A) 2.80 g (B) 14.0 g (C) 17.0 g (D) 10.5 g (E) 5.60 g

126. A sample of ammonium dichromate that contains 13.0 g of chromium (AW = 52.0) also contains how many grams of oxygen?

(A) 4.00 (B) 14.0 (C) 1.14 (D) 13.0 (E) 27.0

127. Given 0.75 mole of ammonium dichromate (FW = 252). The number of chromium atoms contained in this sample is:

(A) $4.5 \times 10^{23}$ (B) $6.0 \times 10^{23}$ (C) $3.0 \times 10^{23}$ (D) $2.2 \times 10^{23}$ (E) $9.0 \times 10^{23}$

128. How many hydrogen **atoms** are present in 42 g of $(NH_4)_2Cr_2O_7$?

(A) $1.0 \times 10^{23}$ (B) $8.0 \times 10^{23}$ (C) $4.0 \times 10^{23}$ (D) $6.0 \times 10^{23}$ (E) $2.0 \times 10^{23}$

129. Given the following reaction (equation "unbalanced"):

$$\_\_\_\_ V_2O_5 + \_\_\_\_ C + \_\_\_\_ Cl_2 \rightarrow \_\_\_\_ VOCl_3 + \_\_\_\_ COCl_2.$$

When properly balanced, the equation shows that for every one **mole** of $VOCl_3$ formed, ______ mole(s) of carbon are required.

(A) 1 (B) 1-1/2 (C) 2 (D) 2-1/2 (E) 3

130. Consider the reaction given in Question 129. Starting with one mole of **each** of the reactants, the maximum number of **moles** of $VOCl_3$ that could be prepared is:

(A) 1 (B) 1/2 (C) 1/3 (D) 1/6 (E) 1/8

131. Which of the following $KMnO_4$ (FW = 158.0) solutions would be the most concentrated?

(A) 0.100 *M* $KMnO_4$ (B) 1.00 g $KMnO_4$/L (C) 100. mg $KMnO_4$/mL
(D) 10.0 g of $KMnO_4$ dissolved in water and made up to 1.000 L
(E) All of the above solutions have the same concentration.

132. How many grams of $KMnO_4$ would be required to prepare 200. mL of a 0.200 *M* solution?

(A) 6.32 (B) 31.6 (C) 158 (D) 398 (E) 632

133. What is the molar concentration of potassium permanganate ($KMnO_4$, FW = 158) in a solution prepared by dissolving 47.4 g of $KMnO_4$ in about 500 mL of water and then finally diluting the solution to 5.00 L?

(A) $6.00 \times 10^{-2}$ *M* (B) $6.67 \times 10^{-1}$ *M* (C) $6.00 \times 10^{-1}$ *M*
(D) $5.45 \times 10^{-2}$ *M* (E) $6.00 \times 10^{-4}$ *M*

134. Vitamin $K_5$ contains 76.27% carbon by weight and has a molecular weight of 173.21. How many C atoms are in one $K_5$ molecule?

(A) 2 (B) 11 (C) 19 (D) 27 (E) 5

135. A sample of element X weighing 156.0 mg combines completely with silicon, forming 268.3 mg of pure $X_3Si_4$. If we take the atomic weight of silicon to be 28.086, the atomic weight of X must be:

(A) 52.02 (B) 54.95 (C) 47.95 (D) 55.85 (E) 58.71

136. Balance the following expression: $NH_3 + O_2 \rightarrow NO_2 + H_2O$. The **balanced** equation shows that 1.00 mole of $NH_3$ requires ______ mole(s) of $O_2$.

(A) 1.25 (B) 1.33 (C) 2.67 (D) 1.75 (E) 0.57

137. Element Z forms an oxide $Z_2O_3$ that is 34.8% oxygen by weight. The atomic weight of Z must be:

(A) 65.2 (B) 45.0 (C) 56.2 (D) 95.0 (E) 27.0

138. How many grams of NaOH (FW = 40.) must be dissolved in water to prepare 500 mL of a 0.20 *M* solution?

(A) 40. g (B) 4.0 g (C) 8.0 g (D) 10. g (E) 20. g

139. How many moles of NaOH are in 22.0 mL of 0.150 *M* NaOH?

(A) 3.30 (B) 1.48 (C) 0.00330 (D) 0.00221 (E) 0.00148

140. How many milliliters of a 0.100 *M* solution of NaOH (FW = 40.0) would be needed to contain 2.00 g of NaOH?

(A) 0.500 mL (B) 8.00 mL (C) 50.0 mL (D) 80.0 mL (E) 500. mL

141. How many grams of $Al_2O_3$ can be produced from 9.0 g of Al and 9.0 g of oxygen?

(A) 17 g (B) 18 g (C) 19 g (D) 34 g (E) 68 g

142.* Given 1.00 L **each** of 0.195 *M* and 0.395 *M* HCl. Using **only** these solutions and assuming that volumes are additive, what is the **maximum** volume of 0.275 *M* HCl that can be prepared?

(A) 1.67 L (B) 2.00 L (C) 0.69 L (D) 1.18 L (E) 1.93 L

143. The simplest formula for a compound that contains 32.79% Na, 13.02% Al, and 54.19% F would be:

(A) $Na_3AlF_6$ (B) $Na_2AlF_5$ (C) NaAlF (D) $Na_2AlF_6$
(E) $Na_2Al_2F_4$

144. A compound with a molecular weight of 80 contains 50% by weight of a hypothetical element X (AW = 10). The remainder of the compound is made up of a hypothetical element Z (AW = 20). What is the molecular formula?

(A) $ZX_6$ (B) $Z_3X_2$ (C) $Z_2X_4$ (D) $Z_2X$ (E) $Z_6X_3$

145. A certain compound used for fertilizer analyzes 60.0% oxygen, 5.0% hydrogen, and 35.0% nitrogen, by weight. Its simplest formula is:

(A) $NOH_2$ (B) $N_2OH_4$ (C) $N_2O_3H_4$ (D) $NO_2H_3$ (E) $N_3O_2H_3$

146.* Phenolphthalein ($C_{20}H_{14}O_4$, FW = 318.31) is formed by reaction of 1 mole of phthalic anhydride ($C_8H_4O_3$, FW = 148.11) with 2 moles of phenol ($C_6H_5OH$, FW = 94.11). A particular process gives a 79.0% yield based upon the phenol used when a 10.0% excess of phthalic anhydride is used in the reaction. What quantities of reactants should be used if it is desired to produce 2.00 tons of phenolphthalein?

(A) 1.18 tons phthalic anhydride + 1.50 tons phenol
(B) 1.30 tons phthalic anhydride + 1.50 tons phenol

(C) 0.931 tons phthalic anhydride + 1.18 tons phenol
(D) 1.02 tons phthalic anhydride + 1.18 tons phenol
(E) 1.18 tons phthalic anhydride + 1.65 tons phenol

147* A martini is accurately prepared by mixing 6.0 volumes of gin with 1.0 volume of extra dry vermouth. The density of the gin is 95% that of the vermouth. The martini contains what percentage (by weight) of gin?

(A) 85% (B) 86% (C) 87% (D) 90% (E) 95%

148. Maggie Mitchell ordered a 22-ton car load of guano fertilizer. This particular batch analyzed 9.0% N, 6.0% P, and 2.0% K. Assuming all of the phosphorus is in the form of $Ca_3(PO_4)_2$, FW = 310, and that no other sources of calcium are present, what is the % Ca in this lot of guano?

(A) 6.0% (B) 12% (C) 18% (D) 36% (E) 40%

149. What weight of potassium dichromate, $K_2Cr_2O_7$ (F.W. = 294), should be taken in order to prepare 500 mL of an aqueous solution containing 10. ppm chromium?

(A) .0020 g (B) 5.0 mg (C) 14 mg (D) 28 mg (E) 0.52 g

150. For the solution in Question 149, what would be the concentration of potassium in ppm?

(A) 10. ppm (B) 7.5 ppm (C) 13 ppm (D) 0.50 ppm
(E) None of these

151.*You are to prepare 1.00 L of a solution that contains 2000 ppm potassium ***and*** 1000 ppm phosphorus. Using only $KH_2PO_4$ (F.W. = 136) and $K_2HPO_4$ (F.W. = 174) as sources of these elements, what weight of each compound should be taken?

(A) 3.47 g $KH_2PO_4$ and 2.05 g $K_2HPO_4$
(B) 5.32 g $KH_2PO_4$ and 4.29 g $K_2HPO_4$
(C) 0.956 g $KH_2PO_4$ and 4.09 g $K_2HPO_4$
(D) 1.82 g $KH_2PO_4$ and 3.29 g $K_2HPO_4$
(E) 5.32 g $KH_2PO_4$ and 2.38 g $K_2HPO_4$

## ANSWERS TO CHAPTER 2 PROBLEMS

| | | | | | |
|---|---|---|---|---|---|
| 1. C | 9. A | 17. B | 25. B | 33. A | 41. A |
| 2. A | 10. D | 18. C | 26. D | 34. A | 42. C |
| 3. C | 11. D | 19. A | 27. D | 35. E | 43. A |
| 4. D | 12. B | 20. D | 28. A | 36. E | 44. D |
| 5. D | 13. E | 21. B | 29. B | 37. D | 45. B |
| 6. C | 14. B | 22. B | 30. C | 38. C | 46. D |
| 7. E | 15. D | 23. A | 31. E | 39. D | 47. B |
| 8. D | 16. D | 24. A | 32. A | 40. C | 48. A |

49. E
50. A
51. C
52. D
53. D
54. B
55. A
56. D
57. B
58. A
59. D
60. A
61. D
62. D
63. B
64. E
65. D
66. E
67. A
68. A
69. A
70. C
71. A
72. E
73. D
74. D
75. D
76. A
77. A
78. A
79. C
80. A
81. E
82. D
83. B
84. A
85. E
86. B
86. D
88. E
89. B
90. C
91. B
92. C
93. C
94. B
95. D
96. A
97. B
98. A
99. B
100. A
101. D
102. E
103. D
104. D
105. B
106. C
107. A
108. B
109. B
110. C
111. A
112. B
113. A
114. B
115. E
116. C
117. A
118. E
119. E
120. E
121. B
122. A
123. C
124. C
125. D
126. B
127. E
128. B
129. B
130. C
131. C
132. A
133. A
134. B
135. A
136. D
137. B
138. B
139. C
140. E
141. A
142. A
143. A
144. C
145. C
146. B
147. A
148. B
149. C
150. B
151. D

# 3
# *Atomic Structure and the Periodic Table*

## FUNDAMENTAL PARTICLES AND THE ATOMIC NUCLEUS

The fundamental particles of which individual atoms are composed are *protons, neutrons,* and *electrons:*

| Particle | Symbol | Mass, amu | Charge, Electronic |
|---|---|---|---|
| Proton | $p$ | 1.007274 | +1 |
| Neutron | $n$ | 1.008665 | 0 |
| Electron | $e$ | 0.000549 | -1 |

Essentially the entire atomic mass (neutrons and protons) is concentrated in a tiny, but enormously dense *nucleus.* Surrounding the nucleus is a diffuse, negatively charged, electrical cloud composed of sufficient electrons to render the atom electrically neutral. It is this electron cloud that determines the volume of the atom and that is most directly related to the chemical properties of the atom.

The number of protons (the *atomic number, Z*) determines the identity of the element. Atoms of a particular element may vary in mass because of differing numbers of neutrons in the nucleus (*isotopes*). Particular isotopes are identified by the symbol of the element with the atomic number at the lower-left corner and the *mass number* (sum of the number of protons and neutrons) at the upper left corner, *e.g.*, ${}^{108}_{48}Cd$ identifies the isotope of cadmium ($Z = 48$) which has 60 neutrons in the nucleus ($108 - 48 = 60$). The natural distribution of these isotopes of an element determines the *atomic weight, i.e.*, the weighted-average of the masses of the naturally occurring isotopes. An atom is electrically neutral, but atoms may gain or lose electrons to form electrically charged species called *ions.* Thus, the symbol $^{34}_{16}S^{2-}$ (or equivalently $^{34}_{16}S^{=}$) designates an *anion* formed by a sulfur atom having 16 protons and 18 neutrons in its nucleus; two electrons have been accepted by this sulfur atom to give a total of 18 electrons. Similarly, positive ions, *cations,* are formed by the loss of one or more electrons; the symbol $K^+$ refers to an average potassium atom (19 protons and an undesignated number of neutrons) which has lost one electron to give a total of 18 electrons.

## Example Problems Involving Fundamental Particles and the Atomic Nucleus

### 3-A. Atomic Weight from % Abundances of Naturally Occurring Isotopes

There are three naturally occurring isotopes of silicon:

$^{28}_{14}Si$, natural abundance 92.21%, mass = 27.97693 amu
$^{29}_{14}Si$, natural abundance 4.70%, mass = 28.97649 amu
$^{30}_{14}Si$, natural abundance 3.09%, mass = 29.97376 amu

Calculate the atomic weight of silicon from these data.

(A) 28 (B) 28.98 (C) 28.08561 (D) 28.09 (E) 28.1

*Solution*

For each isotope: % abundance → fractional abundance → weighted contribution →

Sum the weighted contributions of all three isotopes to obtain the atomic weight.

$$\left(\frac{92.21}{100}\right)(27.97693 \text{ amu}) + \left(\frac{4.70}{100}\right)(28.97649 \text{ amu}) + \left(\frac{3.09}{100}\right)(29.97376 \text{ amu}) =$$

*Answer*

| | |
|---|---|
| Calculator | 28.08561137 |
| Significant figures required | 4 |
| Reported answer | 28.09 amu |

[D]

### 3-B. % Abundance of an Isotope from Atomic Weight

Natural silver consists of two isotopes: $^{107}Ag$ (mass = 106.9041 amu) and

$^{109}Ag$ (mass = 108.9047 amu). The periodic table gives the atomic weight of silver as 107.868 amu. Find the % abundance of the lighter isotope of silver.

(A) 48.17% (B) 51.82% (C) 50.00% (D) 0.52% (E) none of these

### *Solution*

Let $f$ = fraction of the lighter isotope; then $(1 - f)$ = fraction of the heavier isotope. The atomic weight then is the sum of the individual weights times their respective fractional contritions, *i.e.*,

$$107.868 \text{ amu} = (106.9041 \text{ amu})f + (108.9047 \text{ amu})(1 - f)$$

### *Answer*

| | | |
|---|---|---|
| Calculator | 51.81945416 | |
| Significant figures required | 4 | |
| Reported answer | 51.82% $^{107}Ag$ | [B] |

## Exercises on Fundamental Particles and the Atomic Nucleus

1. Which of the following has **no net** electrical charge?

   (A) an electron (B) a proton (C) an atom (D) a nucleus
   (E) none of the above

2. Which of the following contains the greatest number of neutrons?

   (A) $^{112}_{48}Cd$ (B) $^{112}_{49}In$ (C) $^{112}_{47}Ag$ (D) $^{114}_{47}Ag$ (E) $^{114}_{48}Cd$

3. A **nucleus** of $^{56}Co$ contains:

   (A) 27 protons, 29 neutrons, and 27 electrons
   (B) 29 protons, 27 neutrons, and 29 electrons
   (C) 29 protons and 27 neutrons
   (D) 27 protons and 29 neutrons
   (E) 27 protons, 29 neutrons, and 25 electrons

4. A specific isotope has an atomic number of 18 and a mass number of 35. How many electrons are in the neutral atom?

   (A) 8 (B) 17 (C) 18 (D) 35 (E) 53

5. In question 4, how many neutrons are in the nucleus?

   (A) 8 (B) 17 (C) 18 (D) 35 (E) 53

6. Which of the following has 16 protons and 18 electrons?

   (A) $S^{2+}$ (B) $Ar^{2-}$ (C) $Cl^-$ (D) $K^+$ (E) none of these

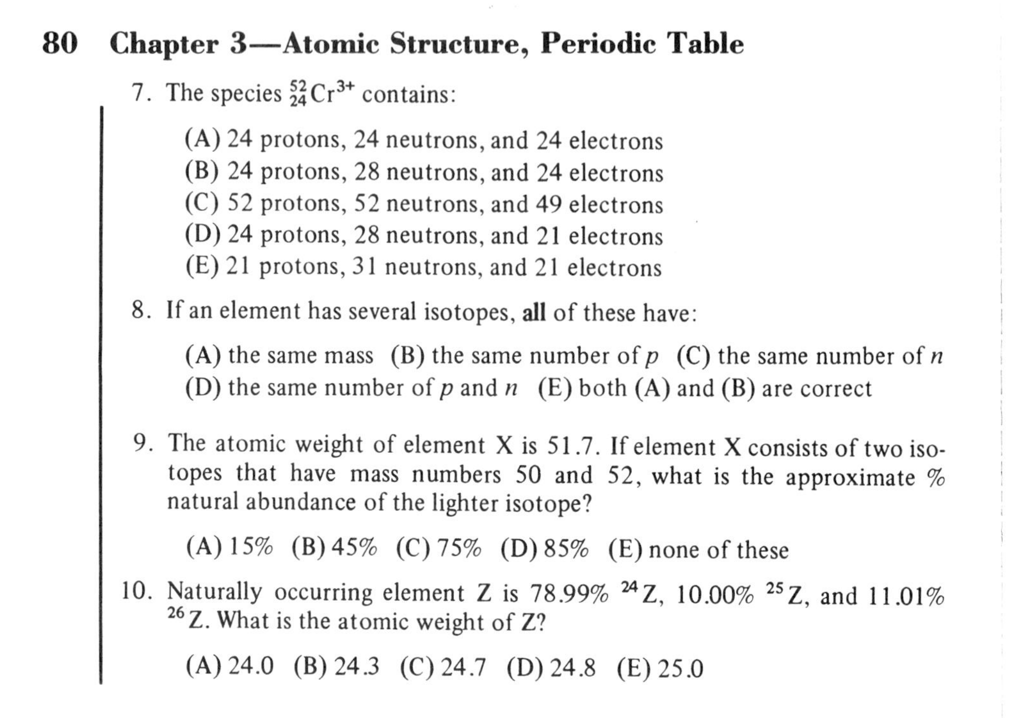

7. The species $^{52}_{24}Cr^{3+}$ contains:

   (A) 24 protons, 24 neutrons, and 24 electrons
   (B) 24 protons, 28 neutrons, and 24 electrons
   (C) 52 protons, 52 neutrons, and 49 electrons
   (D) 24 protons, 28 neutrons, and 21 electrons
   (E) 21 protons, 31 neutrons, and 21 electrons

8. If an element has several isotopes, **all** of these have:

   (A) the same mass (B) the same number of $p$ (C) the same number of $n$
   (D) the same number of $p$ and $n$ (E) both (A) and (B) are correct

9. The atomic weight of element X is 51.7. If element X consists of two isotopes that have mass numbers 50 and 52, what is the approximate % natural abundance of the lighter isotope?

   (A) 15% (B) 45% (C) 75% (D) 85% (E) none of these

10. Naturally occurring element Z is 78.99% $^{24}Z$, 10.00% $^{25}Z$, and 11.01% $^{26}Z$. What is the atomic weight of Z?

    (A) 24.0 (B) 24.3 (C) 24.7 (D) 24.8 (E) 25.0

---

## AN INTRODUCTION TO THE PERIODIC TABLE

The approximately 100 known elements have been arranged in a very important tabular form known as the *periodic table*. The elements are placed in the table in order of increasing atomic number. A new row is started when it is noted that the chemical and physical properties of an element are very similar to those of an element already placed in the table. Rows do not necessarily have the same number of elements; indeed, as one goes down the table, the number of elements in a row tends to increase. The rows are usually referred to as *periods* and the vertical columns as *families* (since the elements in a column will have similar chemical and physical properties). The first period consists of only 2 elements, H and He, the second period has 8 elements, and the 6th period has 32. Note that in the usual forms of the table, 14 of the 6th period elements have been moved to the bottom of the table; this is also the case for the 7th period. This is done merely to keep the table from being particularly wide.

The elements are divided into two very general groups, the *metals* (elements placed on the left and in the center of the table) and the *nonmetals* (elements located on the right, particularly the upper right, part of the table). If you try to draw a line that separates the metals from the nonmetals (roughly from B to Po), one finds that elements on and near this line have some properties that are more or less metallic, others that are

## DRILL ON PROTONS, NEUTRONS, AND ELECTRONS

***Student's Name*** ____________________

Complete the table by entering the symbol and the number of protons, neutrons, and electrons for each species described. If an entry is undefined, enter "?". You may be asked to turn in this sheet.

| Description | Symbol | Number of protons | Number of neutrons | Number of Electrons |
|---|---|---|---|---|
| the nucleus of a calcium atom | | | | |
| an atom of silver-108 | | | | |
| | $^{43}_{21}Sc^{3+}$ | | | |
| | | 56 | 82 | 54 |
| | $Cl^{-}$ | | | |
| phosphide ion | $P^{3-}$ | | | |
| a fluorine-19 atom | | | | |
| ____________________ | | | | |

rather nonmetallic. These elements are frequently referred to as *metalloids.*

**1. The Metals.** Metals make-up about 75% of the elements. The metal portion of the periodic table is shown below.

| IA | | IIIB | IVB | VB | VIB | VIIB | VIII | VIII | VIII | IB | IIB | | | | |
|---|---|---|---|---|---|---|---|---|---|---|---|---|---|---|---|
| 3 Li 6.941† | 4 Be 9.01218 | | | | | | | | | | | | | | |
| 11 Na 22.98977 | 12 Mg 24.305 | | | | | | | | | | | 13 Al 26.98154 | | | |
| 19 K 39.0983 | 20 Ca 40.08 | 21 Sc 44.9559 | 22 Ti 47.88 | 23 V 50.9415 | 24 Cr 51.996 | 25 Mn 54.9380 | 26 Fe 55.847 | 27 Co 58.9332 | 28 Ni 58.69 | 29 Cu 63.546 | 30 Zn 65.38 | 31 Ga 69.72 | | | |
| 37 Rb 85.4678 | 38 Sr 87.62 | 39 Y 88.9059 | 40 Zr 91.22 | 41 Nb 92.9064 | 42 Mo 95.94 | 43 Tc (98) | 44 Ru 101.07 | 45 Rh 102.9055 | 46 Pd 106.42 | 47 Ag 107.8682 | 48 Cd 112.41 | 49 In 114.82 | 50 Sn 118.69 | | |
| 55 Cs 132.9054 | 56 Ba 137.33 | 57 *La 138.9055 | 72 Hf 178.49 | 73 Ta 180.9479 | 74 W 183.85 | 75 Re 186.207 | 76 Os 190.2 | 77 Ir 192.22 | 78 Pt 195.08 | 79 Au 196.9665 | 80 Hg 200.59 | 81 Tl 204.383 | 82 Pb 207.2 | 83 Bi 208.9804 | 84 Po (209) |
| 87 Fr (223) | 88 Ra 226.0254 | 89 † Ac 227.0278 | 104 Unq§ (261) | 105 Unp§ (262) | 106 Unh§ (263) | | | | | | | | | | |

| 58 Ce 140.12 | 59 Pr 140.9077 | 60 Nd 144.24 | 61 Pm (145) | 62 Sm 150.36 | 63 Eu 151.96 | 64 Gd 157.25 | 65 Tb 158.9254 | 66 Dy 162.50 | 67 Ho 164.9304 | 68 Er 167.26 | 69 Tm 168.9342 | 70 Yb 173.04 | 71 Lu 174.967 |
|---|---|---|---|---|---|---|---|---|---|---|---|---|---|

*Lanthanides

| 90 Th 232.0381 | 91 Pa 231.0359 | 92 U 238.0289 | 93 Np 237.0482 | 94 Pu (244) | 95 Am (243) | 96 Cm (247) | 97 Bk (247) | 98 Cr (251) | 99 Es (252) | 100 Fm (257) | 101 Md (258) | 102 No (259) | 103 Lr (260) |
|---|---|---|---|---|---|---|---|---|---|---|---|---|---|

†Actinides

**FIG. 1. The Metals**

Except for mercury, metals are solids (although on a hot day gallium and cesium melt). Metallic solids consist of layers of closely packed, essentially spherical atoms. Each atom contributes one or more electrons to the overall solid that can freely move throughout the 'atomic' array. It is these *delocalized electrons* at the surface of a metallic solid that interact with light rays to produce the typical *metallic luster*. Although metals vary widely in hardness, the atoms can be forced to slide one over the other; thus, metals can be pounded into sheets (*malleability*) or drawn into wires (*ductility*). A very important chemical property of metals is that metal atoms give up one or more electrons to other species in the formation of compounds. The metal, now part of the compound, is present as a positive ion. Metals are thus said to be *electropositive*. Metallic character, and hence electropositivity, increases as you go down a family and as you go to the left in a period. Chemically, the best metal is the little known element francium.

**2. The Nonmetals.** The electrons of the nonmetallic atoms are strongly attracted to the atom. Nonmetallic solids lack metallic luster and are brittle rather than malleable. They are poor conductors of heat and with the exception of graphite, a form of carbon, are electrical insulators. The nonmetals can be split into two groups, the *'noble'* or *inert gases* and the *reactive nonmetals*. These are shown in the partial periodic table below.

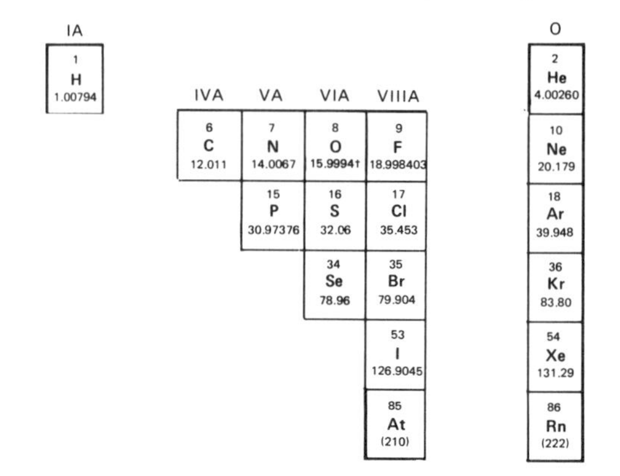


**FIG. 2. Reactive Nonmetals Inert Gases**

The noble gases are the elements at the ends of each period. The term 'noble' refers to the fact that these elements show but little tendency to combine with themselves or with other elements. As free elements they exist as single atoms even when cooled enough to liquefy or to solidify. A few xenon compounds are known, *e.g.* $XeO_3$, $XeF_4$, $XeF_6$. No compounds of helium, neon, and argon are known and evidence for compound formation by krypton and radon is meager.

As the elements, the reactive nonmetals exist either as small molecular clusters of atoms or as *macromolecules* (chains, layers, or 3-dimensional networks of an extremely large number of atoms). Several of them occur at room temperature as diatomic molecules (*i.e.* $H_2$, $N_2$, $O_2$, $F_2$, $Cl_2$, $Br_2$, $I_2$). Examples of slightly larger clusters include $O_3$, $P_4$ and $S_8$. The reactive nonmetals are noted for their ability to *catenate* (like atoms linking together). In some cases, one of these elements can link atoms together in two or more distinctly different ways to form substances having different structures and properties, but which are made up only of atoms of this element, *allotropic forms*. The most common form of the element oxygen is $O_2$; an allotropic form is $O_3$ which is given the special name ozone. White phosphorus consists of discrete tetraatomic molecules, $P_4$. White phosphorus actually is a yellowish wax-like solid that is readily soluble in carbon disulfide ($CS_2$) and melts at 44.1°C in absence of oxygen. In the presence of oxygen it spontaneously ignites. Red phosphorus is an allotropic modification consisting of long chains of -$P_4$- units. It is reddish-black in color, stable in air at room temperature, insoluble in carbon disulfide, and melts about 600°C. Both of these are forms of the element phosphorus, but they differ in how the phosphorus atoms are linked and hence, differ in chemical and physical properties.

Hydrogen is a unique reactive nonmetal in that it has some properties that slightly resemble the alkali metals, others that slightly resemble the halogens. For these reasons some periodic tables place hydrogen twice on the table, *i.e.* in both of these families. Most of the properties of hydrogen differ markedly from properties of either of these families and thus, we have put hydrogen in the first period, but associated with no family.

In forming compounds atoms of the reactive non-metals either gain electrons to form negative ions (*e.g.* $Cl^-$, $S^{2-}$, $N^{3-}$) or to bind to one another by sharing electrons to form molecules (*e.g.* $CO_2$, $P_4O_{10}$, $C_{27}H_{44}O_3$) or polyatomic ions (*e.g.*, $OH^-$, $SO_4^{2-}$, $NH_4^+$, $N_3^-$). Since these nonmetals tend to gain electrons they are said to be *electronegative*.

**3. Metalloids.** There is not a sharp dividing line between metals and nonmetals. Elements that fall roughly along a stair-step division between the metals and the nonmetals, the heavy line in Fig. 3, have some properties that are metallic, others that are nonmetallic. The elements are given the special name metalloids (or sometimes, *semimetals*). There is disagreement as to whether certain of these elements should be with the metals or with the metalloids or whether certain others should be placed with the nonmetals or with the metalloids. At times boron and selenium are categorized as nonmetals. One allotrope of tin definitely is not metallic and metalloid in character. Here, astatine has been included both as a nonmetal and as a metalloid.

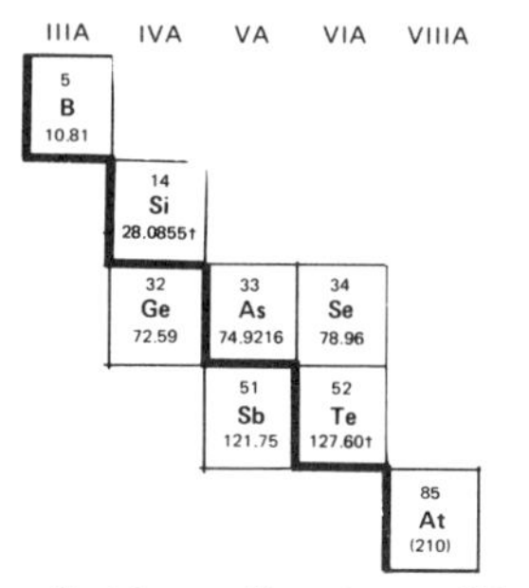


**FIG. 3. Metalloids or Semimetallic Elements**

Metalloids are poor conductors of electricity, but their electrical conductivity increases with increasing temperature (unlike metals, whose conductivity decreases with increasing temperature). Many of these elements are important themselves, or in combination with trace impurities of other elements, in solid state electronics.

## Chemical Families

The elements in any of the vertical columns of the periodic table are referred to as *chemical families* or *groups*. Members of a family differ, but have a number of distinct similiarities. The 'power' of the periodic table is that if I know something about one member of a family, I can predict properties of other members of the family based on this knowledge of the one member. Each family is labeled with a Roman numeral and either the letter 'A' or 'B'. The families that carry the letter 'A' in their name make up the *representative elements*. As you proceed down any one of these families, the metallic chararacter of the elements increases. The center portion of the table is made up of families with a 'B' label. These are collectively called *transition metals*. The two series at the bottom of the table, elements following lanthanum (the *lanthanides*) and elements following actinium (the *actinides*) are called *inner transition elements*. There are some similarities between corresponding 'A' and 'B' Groups, particularly in the formulas of compounds. Thus, the IVA element tin forms oxides of formula SnO and $SnO_2$; the IVB element titanium forms oxides TiO and $TiO_2$.

Families or groups can always be named using the Roman numeral designation. They also are named by identifying the first member of the family. Group VIA is also the oxygen family; Group VB, the vanadium family. Some of the more important families have yet another distinctive name which needs to be remembered. The IA elements are called *alkali metals*; the IIA Group, *alkaline earth metals*; the VIIA family, *halogens*, and we have already noted that VIIIA is called the noble gases as well as the inert gases (this family at times is referred to as *Group 0* to emphasize the unreactive nature of the members).

### Exercises on an Introduction to the Periodic Table

11. An important class of solid state electronic materials involves the combination of a Group III representative element with a Group V representative element. An example of such of a pair of elements is:

    (A) Sc and V (B) Y and As (C) A1 and Ga
    (D) Ga and As (E) Nd and Eu

12. From the answers given pick the metalloid that has the greatest atomic weight.

    (A) boron (B) uranium (C) iodine (D) potassium (E) silver

13. Which of the following symbols represents a member of the inert gas family?

    (A) N (B) Ne (C) H (D) Cl (E) Ce

14. At room temperature, which of the following elements would be the most ductile?

    (A) Sb (B) B (C) S (D) Ag (E) I

## DRILL ON THE PERIODIC TABLE

***Student's Name*** ____________

Fill in the symbol of the element which fits the description given and has (a) the highest atomic weight and (b) the lowest atomic weight. You may be asked to turn in this sheet.

| Description | Symbol of the Element with the | |
|---|---|---|
| | Highest Atomic Weight | Lowest Atomic Weight |
| The alkali metal | | |
| The non-metal | | |
| The metal | | |
| The metalloid | | |
| The Group IVB element | | |
| The 4th period transition metal | | |
| The non-metal having an atomic weight lighter than arsenic | | |
| The element, occurring at room temperature as a diatomic molecule | | |
| ____________ | | |

15. Pick the set of answers that includes symbols for elements in the halogen family.

    (A) Fl, Cl, and Br (B) N, O, and F (C) Mn, Tc, and Re
    (D) I, Br, and Cl (E) I, At, and Rn

16. The Group I transition metals are sometimes called the coinage metals. An example of a coinage metal would be:

    (A) nickel (B) rubidium (C) zinc (D) tin (E) gold

17. Which of the following pairs of elements would be chemically most similar?

    (A) H and O (B) H and He (C) H and Li (D) O and S
    (E) O and F

18. Which of the following transition elements has the largest atomic weight?

    (A) Sn (B) Cr (C) Fe (D) Co (E) Ni

19. Experiments with the cathode-ray tube (discharge tube) have shown:

    (A) that all nuclei contain protons
    (B) that all forms of matter contain electrons
    (C) that all positive rays were actually protons
    (D) that alpha particles are heavier than protons
    (E) none of the above

20. The emission of radiation from the **nuclei** of certain atoms is known as:

    (A) the photoelectric effect (B) atomic spectra (C) thermal emission
    (D) the periodic law (E) none of these

21. After Rutherford observed that some alpha particles were scattered at large angles by gold foil, he concluded that:

    (A) most of the mass of the atom is in the nucleus
    (B) the nucleus is positively charged (C) the nucleus is very small
    (D) the nucleus is very dense (E) all of the above

22. The particle with the greatest mass is the:

    (A) alpha particle (B) proton (C) neutron (D) electron (E) photon

23. The nucleus of an atom of $^{238}U$ contains:

    (A) 92 electrons and 92 protons (B) 92 neutrons and 238 protons
    (C) 146 neutrons and 92 protons (D) 146 protons and 92 neutrons
    (E) 238 protons

24. The atomic number of fluorine is:

    (A) 19 (B) 26 (C) 38 (D) 9 (E) 10

25. Which of the following elements would be more nonmetallic in character than arsenic?

(A) Sb (B) W (C) Pb (D) Al (E) S

26. Elements whose atoms lose electrons easily are said to function as reducing agents. Which of the following should be the best 'reducing agent'?

(A) F (B) Cu (C) Ba (D) Kr (E) O

27. Elements whose atoms gain electrons readily are said to function as 'oxidizing agents'. Which of the following should be the best 'oxidizing agent'?

(A) F (B) Cu (C) Ba (D) Kr (E) O

28. The species that contains 24 protons, 26 neutrons and 22 electrons would be represented by the symbol:

(A) $^{50}_{24}V^{3+}$ (B) $^{26}_{24}Cr^{2+}$ (C) $^{50}_{24}Cr^{2+}$ (D) $^{50}_{22}Mn^{2-}$ (E) none of these

29. The electrical charge on the *nucleus* of the magnesium ion ($Mg^{2+}$) is:

(A) 0 (B) -2 (C) -12 (D) +12 (E) +10

30. Which of the following contains more electrons than neutrons?

(A) $^{25}_{12}X^{2+}$ (B) $^{26}_{13}X$ (C) $^{75}_{33}X^{3-}$ (D) $^{33}_{16}X^{2-}$ (E) none of these

31. The yet-undiscovered element $Z = 114$ should fall in ______ of the periodic table.

(A) the sixth period (B) the actinide series
(C) the fourth transition series (D) Group IVA
(E) the alkali metal family

32. One of the naturally occurring isotopes of xenon consists of atoms with a mass of 135.90722 amu. How many neutrons are present in each of these atoms?

(A) 81 (B) 81.90722 (C) 82 (D) 135.90722 (E) 136

33. Aluminum forms a compound with oxygen that has the formula $Al_2O_3$. Which of the following pairs of elements would form a compound having the same general type formula ($M_2X_3$)?

(A) In and S (B) B and F (C) Si and O (D) Zn and N
(E) Ba and S

34. A metallic, representative element, M, forms oxides $M_2O$ and $M_2O_3$. Metal M must belong to:

(A) The alkali metals
(B) Group IIIA
(C) The halogen family

(D) Family VIIIB
(E) The carbon family

35. Which chemical family contains the greatest number of nonmetallic atoms?

(A) the halogens
(B) the oxygen family
(C) the inert gas family
(D) the representative elements
(E) the fourth period transition elements

36 A representative (A-Group) element that is distinctly metallic is:

(A) Fe (B) Cu (C) Ba (D) Si (E) none of these

37.*In a mass spectrometer, atoms of element X form ions ($X^+$) of masses 49.95, 51.94, 52.94, and 53.94 amu. Analysis reveals that these occur in the proportions 184:3540:402:100 by number, respectively. What is the percentage natural abundance of the most predominant isotope?

(A) 83.77% (B) 24.88% (C) 75.12% (D) 80.62% (E) 79.88%

38. The number of electrons in a neutral atom of phosphorus is:

(A) 5 (B) 16 (C) 15 (D) 18 (E) 31

39. The most metallic element in the third period would be:

(A) boron (B) aluminum (C) thallium (D) sodium (E) none of these

40. Which of the following contains the same total number of electrons as a krypton atom?

(A) Ar (B) $Se^{2-}$ (C) $Se^{2+}$ (D) $Br_2$ (E) $Sr^{2-}$

41. The particle with the *smallest* mass is the

(A) alpha particle (B) proton (C) neutron (D) electron
(E) all have the same mass

42. If a helium atom loses two electrons, the resulting particle is:

(A) electrically neutral (B) a hydrogen atom (C) an alpha particle
(D) a proton (E) a cathode ray

43. Consider the species $^{60}Co$, $^{59}Fe$ and $^{62}Cu$. These species have:

(A) the same mass number (B) the same nuclear charge
(C) the same number of electrons. (D) the same number of neutrons
(E) the same number of protons *plus* neutrons

44. The species $^{78}_{34}Se^{2-}$ contains:

(A) 44 protons, 34 neutrons and 42 electrons
(B) 34 protons, 44 neutrons and 36 electrons

(C) 78 protons, 34 neutrons and 80 electrons
(D) 34 protons, 44 neutrons and 34 electrons
(E) 34 protons, 112 neutrons and 36 electrons

45. The symbol $^{31}_{15}P^{+}$ refers to a species that is an *isotope* of which of the following?

| | protons | neutrons | electrons |
|---|---|---|---|
| (A) | 28 | 13 | 28 |
| (B) | 27 | 15 | 26 |
| (C) | 15 | 13 | 13 |
| (D) | 14 | 13 | 13 |
| (E) | 13 | 13 | 13 |

46. The most metallic element of Group VA is:

(A) vanadium (B) lead (C) rubidium (D) bismuth (E) phosphorus

47. Which has the same number of electrons as a germanium atom?

(A) Zn (B) $Zn^{2+}$ (C) Se (D) $Se^{2-}$ (E) None of these

---

## ANSWERS TO CHAPTER 3 PROBLEMS

| | | | | |
|---|---|---|---|---|
| 1. C | 11. D | 21. E | 31. D | 41. D |
| 2. D | 12. A | 22. A | 32. C | 42. C |
| 3. D | 13. B | 23. C | 33. A | 43. D |
| 4. C | 14. D | 24. D | 34. B | 44. B |
| 5. B | 15. D | 25. E | 35. C | 45. C |
| 6. E | 16. E | 26. C | 36. C | 46. D |
| 7. D | 17. D | 27. A | 37. A | 47. E |
| 8. B | 18. D | 28. C | 38. C | |
| 9. A | 19. B | 29. D | 39. D | |
| 10. B | 20. E | 30. D | 40. B | |

# 4
# *Electronic Structure and the Periodic Table*

## ELECTRONS IN ATOMS

Electrons in atoms are *quantized,* which means that electrons in atoms can have **only** certain allowed energies (*i.e.* occupy only allowed energy states). These states are indexed by $\Psi_{n\ell m}$ where $\Psi$ represents an *atomic orbital* capable of holding **two electrons** and $n$, $\ell$, and $m$ are *quantum numbers* indicating the energy and spatial arrangement of the electron cloud. The energy of an electron in an atom is determined mainly by the value of $n$ (the *principal* quantum number) and secondarily by $\ell$ (the *azimuthal quantum number*). The principal quantum number defines a region in space (a *shell*) at some average distance from the nucleus capable of holding several electrons having about the same energy. The values of $n$ can be any positive integer with the lowest energy electrons having $n = 1$; usually only values up to $n = 7$ will be of interest. Shells with different values of $n$ may have electrons with energies and average distances from the nucleus that overlap.

The shells consist of one or more *sub-shells* indexed by the azimuthal quantum number, $\ell$. Allowed values of $\ell$ include any positive integer from 0 to $(n - 1)$. Usually instead of being expressed as these integers, values of $\ell$ are assigned the following set of letters (the first 4 of which will be important):

| 0 | 1 | 2 | 3 | 4 | 5 | 6 | 7 |
|---|---|---|---|---|---|---|---|
| *s* | *p* | *d* | *f* | *g* | *h* | *i* | *j* |

The energy of an electron can be designated by merely indicating the principal and azimuthal quantum numbers, *e.g.* 1*s*, 4*f*, 2*p*. Not only does the azimuthal quantum number indicate differences in energy of electrons within a shell, but it also indicates the shape of the electron cloud. *s*-electron clouds are spherically shaped; *p*-electron clouds have two lobes, one on either side of the nucleus (placing one electron in a *p*-subshell **partially fills both** lobes).

Each sub-shell consists of one or more *degenerate* (having the same energy) orbitals indexed by the *magnetic* quantum number, *m*. Permitted values of *m* are any integer starting with $-\ell$ up to $+\ell$; if $\ell = 1$, *m* must be $-1$, $0$, or $+1$. These all have **exactly the same energy**, but lie at right angles to one another along x-, y-, and z-coordinates having an origin at the nucleus:

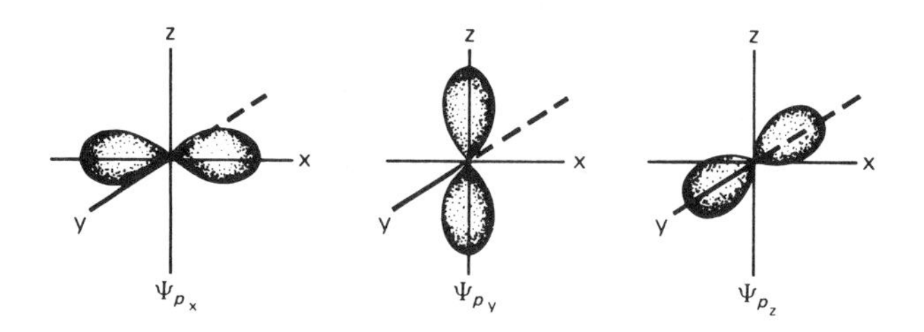


Electrons behave as if they were spinning. It is found that the 2 electrons that occupy a single orbital **must** 'spin' in opposite directions; this electron spin is indexed by a *spin quantum number, s,* which takes on values of $\pm 1/2$. Two electrons in an orbital are called an *electron pair* and are indicated in an orbital diagram by a set of opposed arrows:

⇅

This limitation is summarized by the *Pauli Exclusion Principle* which states that in an atom no two electrons may have all four quantum numbers identical.

The order of the shells and sub-shells in an atom with respect to energy and distance from the nucleus is summarized in the orbital diagram at the top of page 95. Assignment of electrons to appropriate shells and sub-shells, the *electronic configuration,* of any atom in its *ground* (most stable, or lowest energy) *state* is frequently referred to as the *aufbau* (building up) procedure. This is accomplished by adding two electrons per orbital (per box) until the number of added electrons equals the number of protons in the nucleus (making the atoms electrically neutral).

When electrons are placed in a sub-shell consisting of more than one orbital, it makes no difference which orbital is occupied by the first electron nor whether the spin is $+1/2$ or $-1/2$ (up or down). It is found, however, that since electrons repel one another, the second electron will go

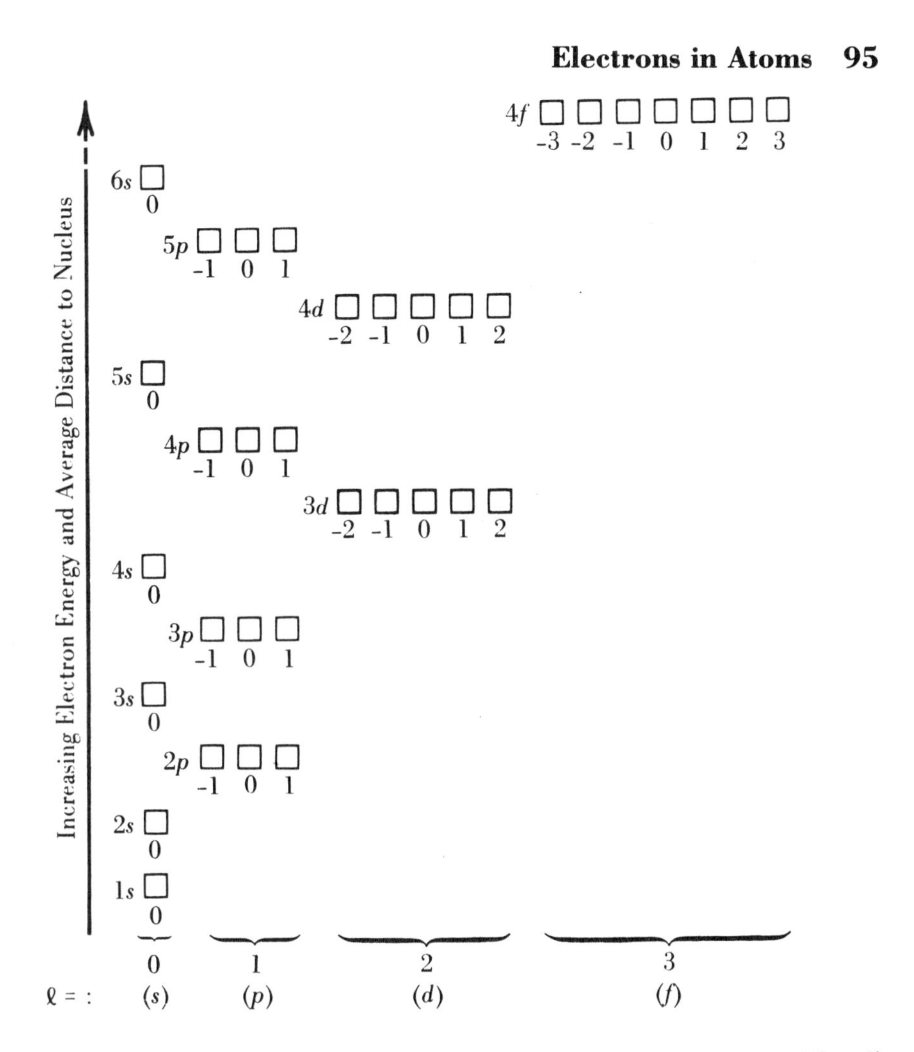


into a vacant orbital rather than pair up with the first electron (*Hund's Rule*) and, further, that its spin direction will be the same as that of the first electron (*spins parallel*). For example:

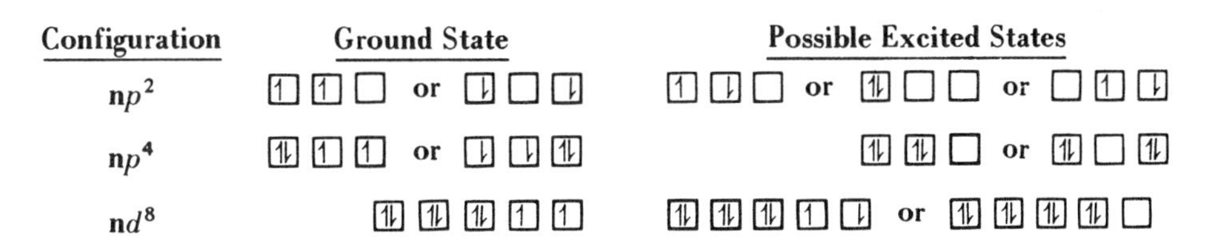


A single electron in an orbital is called an *unpaired electron* and gives rise to a magnetic property whereby the species is attracted into a magnetic field, *paramagnetism.* If all electrons are paired, the species is repelled from a magnetic field and is said to be *diamagnetic.*

The ground state configuration of an atom indicates the sub-shells that contain one or more electrons along with the number of electrons in each. Hence for scandium $Z = 21$), the configuration can be shown by writing each of the sub-shells with a superscript to indicate the number of electrons in the sub-shell; *i.e.* $1s^2$, $2s^2$, $2p^6$, $3s^2$, $3p^6$, $4s^2$, $3d^1$; Note that the sum of the superscripts is equal to 21, the number of electrons in the species, and also note that the scandium atom is paramagnetic since it has an unpaired electron. This notation can be abbreviated by [Ar], $4s^2$, $3d^1$, which indicates that Sc contains the same set of electrons as does the rare gas argon plus a pair of 4*s* electrons and a 3*d* electron. Particularly stable electronic arrangements occur when all of the orbitals in a *p*-, *d*-, or *f*-sub-shell are either half-filled or completely filled. This causes two notable exceptions to the predicted configurations in the first third of the periodic table–chromium and copper. Thus, one would anticipate for Cr ($Z = 24$), [Ar], $4s^2$, $3d^4$, but actually the 3*d* sub-shell is half-filled at expense of the 4*s* to give [Ar], $4s^1$, $3d^5$. In the case of the next atom, Mn ($Z = 25$), the expected configuration is attained–[Ar], $4s^2$, $3d^5$. Copper is similar in that $3d^{10}$ is attained again at expense to the 4*s*. The energy levels in the heavier elements (the second half of the periodic table) become more closely spaced and there are a number of exceptions that occur from the predicted ordering.

A mnemonic device for getting ground state electronic configurations is to write columns of *s*, *p*, *d*, and *f* but start the *p*'s in the second row, the *d*'s in the third, and the *f*'s in the fourth. Draw diagonals through each to give the order of filling:

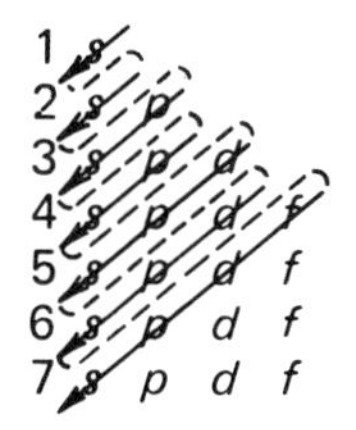


Thus, reading in order and remembering the maximum number of electrons in each sub-shell, $1s^2$, $2s^2$, $2p^6$, $3s^2$, $3p^6$, $4s^2$, $3d^{10}$ can be written for Zn ($Z = 30$). Configurations can also readily be obtained from a *periodic table.* The table sets together elements having similar arrangements of outer electrons (the *valence electrons*) and chemical properties.

## Example Problem Involving Electrons in Atoms

### 4-A. Electron Configuration

Which of the following atoms would have a ground-state electron configuration fulfilling all of the following criteria?

**Three** completely filled shells; **one or more** electrons in a total of **eleven sub-shells** and **27 orbitals**; a valence shell consisting of two filled orbitals as well as 2 unpaired electrons

(A) Ar (B) Co (C) Zn (D) Zr (E) Te

### *Solution*

The fact that 3 shells are completely filled tells us that the build-up must include 28 electrons to fill the 1*s*, 2*s*, 2*p*, 3*s*, 3*p*, and 3*d* sub-shells. From a knowledge of the *aufbau* order we may then write:

$$1s^2\,2s^2\,2p^6\,3s^2\,3p^6\,4s^2\,3d^{10}\cdots$$

At this point, 7 populated sub-shells and 15 orbitals have been used. Adding electrons to 4 **additional** sub-shells would allow us to continue,

$$\cdots 4p^6\,5s^2\,4d^{10}\,5p^?$$

This would mean that the valence shell is the fifth ($n$ = 5) shell:

5*s* [↑↓]

Now, so as to have the valence shell contain 2 filled orbitals and 2 unpaired electrons, we must progressively add (following Hund's rule), 4 electrons to the 5*p* sub-shell, leading to the valence shell,

5*p* [↑↓] [↑] [↑]

5*s* [↑↓]

This leads to the final, completed configuration,

$$1s^2\,2s^2\,2p^6\,3s^2\,3p^6\,4s^2\,3d^{10}\,4p^6\,5s^2\,4d^{10}\,5p^4 \quad \text{or} \quad [\text{Kr}]\,4d^{10}\,5s^2\,5p^4$$

### *Answer*

The total number of electrons is 52, so that the atom must have an atomic number, Z, equal to 52, *i.e.* tellurium, Te. It should be noted that this answer was arrived at by merely working upward through 11 sub-shells and 27 orbitals on the orbital diagram (the *aufbau* order), following the Pauli principle and Hund's rule. **[E]**

## Exercises on Electrons in Atoms

1. An atom of Si ($Z$ = 14) has a **total** of ______ *p*-electrons.

   (A) 2 (B) 4 (C) 6 (D) 8 (E) 20

2. What is the maximum number of electrons for an $n = 3$ shell?

   (A) 18 (B) 6 (C) 10 (D) 8 (E) 2

3. The number of orbitals in a *d*-subshell ($\ell = 2$) is:

   (A) 1 (B) 3 (C) 5 (D) 7 (E) 10

4. In a cobalt atom (Co, $Z = 27$) in its ground state, the **total** number of shells populated by one or more electrons is:

   (A) 1 (B) 2 (C) 3 (D) 4 (E) 5

5. Consider the cobalt atom in Question 14. The **total** number of **orbitals** populated by one or more electrons is:

   (A) 15 (B) 12 (C) 9 (D) 6 (E) none of these

6. Again consider the cobalt atom in Question 14. The **total** number of **subshells** populated by one or more electrons is:

   (A) 9 (B) 7 (C) 5 (D) 4 (E) none of these

7. In the ground state of a cobalt atom there are ______ **unpaired** electrons and the atom is ______.

   (A) 3, paramagnetic (B) 5, paramagnetic (C) 2, diamagnetic
   (D) 0, diamagnetic (E) 2, paramagnetic

8. The electronic configuration of Ti is:

   (A) $1s^2 2s^2 2p^6 3s^2 3p^6 4s^2 3d^2$ (B) $[Ar] 3d^4$ (C) $[Ar] 4s^2 4p^2$
   (D) $[Kr] 5s^2 4d^{10} 5p^2$ (E) $1s^2 2s^2 2p^6 3s^2 3p^6$

9. A possible set of quantum numbers for the last electron added to a gallium atom ($Z = 31$) in its ground state is:

|  | $n$ | $\ell$ | $m$ | $s$ |
|---|---|---|---|---|
| (A) | 4 | 1 | -1 | -1/2 |
| (B) | 4 | 0 | 0 | -1/2 |
| (C) | 3 | 2 | +2 | +1/2 |
| (D) | 3 | 1 | +1 | +1/2 |
| (E) | 3 | 0 | 0 | -1/2 |

10. A given orbital is labeled by the magnetic quantum number as $m = -1$. This orbital **could NOT** be a:

    (A) *s*-orbital (B) *p*-orbital (C) *d*-orbital (D) *f*-orbital (E) *g*-orbital

---

## PERIODIC TABLE AND PERIODICITY

The periodic table is an arrangement of the elements in order of increasing atomic number ($Z$). Elements having similar properties are placed in verti-

## DRILL ON ELECTRONS IN ATOMS

***Student's Name*** ____________________

Complete the following table. An occupied shell, sub-shell, or orbital contains at least one electron, but is not necessarily filled. You may be asked to turn in this sheet.

| Atomic Symbol | Total Number of Electrons in the Atom | Ground State Electronic Configuration | Number of Occupied | | | Diamagnetic or Paramagnetic |
|---|---|---|---|---|---|---|
| | | | Shells | Sub-shells | Orbitals | |
| F | | | | | | |
| | | $1s^2, 2s^2, 2p^6, 3s^2, 3p^6, 4s^1$ | | | | |
| V | | | | | | |
| Cu | | | | | | |
| | | $[Kr], 5s^2, 4d^{10}, 5p^2$ | | | | |
| | | | | | 41 | |
| ________ | | | | | | |

cal columns called *groups* (or *families*); horizontal rows of elements are called *periods.* Atoms in a given group have similar electronic configurations. Electronically, it is convenient to view the table in terms of "blocks" of elements (with $n$ being the period number):

**s-block**: Groups 1A and IIA, with configurations $ns^1$ and $ns^2$, respectively.
**p-block**: Groups IIIA through VIIA plus the 0 Group with valence shells of $ns^2$, $np^1$ through $ns^2$, $np^6$. The 0 Group atoms (*noble gases*) have stable, "closed-shell" configurations which, with the exception of helium ($1s^2$), can be represented by $ns^2$, $np^6$. Groups IIIA through VIIA have a total number of valence $s$- and $p$-electrons equal to the group number.
**d-block**: *Transition elements,* Groups IIIB through IB, typically have configurations $(n\text{-}1)d^{1-10}$, $ns^2$ (some are irregular, particularly in Groups VIB and IB).
**f-block**: *Inner transition metals* (*lanthanides and actinides*) typically (but with irregularities) have configurations $(n\text{-}2)f^{1-14}$, $(n\text{-}1)d^{10}$, $ns^2$.

The general organization of the Periodic Table was known 36 years prior to the discovery of the electron–indeed, a triumph of the quantized, shell-theory of electrons in atoms was that this theory was able to generate the order that is known as the Periodic Table.

Periodicity refers to the variations of properties within groups and between families. Since all members of a family have similar valence shell electronic configurations, they form similar series of compounds. Thus, in the alkali metal family (IA), since sodium forms a compound $Na_2CO_3$, one would predict that the other members of the family would form similar compounds, $M_2CO_3$; this is helpful in making predictions even across *sub-groups*–if $BaSO_4$ exists, then one should not be surprised to find a compound $ZnCrO_4$. It must be remembered, however, that in families there are gradations caused principally by size differences (which lead to differences in other properties). Very marked differences in chemical character are to be expected and are observed going across a period (different valence electron configurations).

Some important physical properties that exhibit periodic behavior are (1) the relative sizes (*radii*) of atoms and ions, (2) the tightness with which electrons are bound to the atom (*ionization energy*), and (3) the degree to which isolated atoms attract and acquire additional electrons (*electron affinity*). The magnitudes of all of these are direct reflections of the attractions felt by outer shell electrons. The force of attraction for a single electron is given by Coulomb's Law:

$$f = \frac{\text{(positive attraction felt by the electron)(charge on the electron)}}{\text{(distance from the nucleus)}^2}$$

where the "positive charge felt by the electron" or effective nuclear charge, $Z_{eff}$, is equal to the atomic number ($Z$) minus a screening factor that is approximated as equal to the number of electrons in sub-shells between the nucleus and the subshell containing the electron of interest,

and the distance from the nucleus squared is proportional to the number of the shell (principal quantum number) containing the electron.

Thus, for an outer shell electron,

$$\text{force of attraction} = f \propto \frac{Z_{eff}}{n_{max}}.$$

Hence, for an outer shell electron in nitrogen, $n = 2$, $Z_{eff} \simeq 7 - 2 = 5$, and $f \propto 5/2 = 2.5$; for an outer electron in phosphorus, $n = 3$, $Z_{eff} \simeq 15 - 10 = 5$, and $f \propto 5/3 = 1.7$.

Going across a period containing only representative elements, $Z_{eff}$ continually increases while $n$ remains constant. This means that the force binding the outer electrons as we go across the period increases. This results in the atoms becoming smaller, having higher ionization energies, and having greater attraction for an additional electron (higher electron affinities). Similar trends would be observed as we go up any family, but now this is due to $Z_{eff}$ remaining constant while $n$ decreases. When we have a period containing a transition of inner-transition metal series, it is found that within the series the value of $Z_{eff}/n_{max}$ remains almost constant due to the fact that the added electron as we go from one element in the series to the next is not added to the outermost shell, but to one of the inner shells. The properties' atomic size, ionization energy, and electron affinity do not change markedly within the series.

The above approximation does not take into consideration the fact that electrons within a shell do repel one another. Removal of an outer shell electron to form an ion reduces the forces of repulsion between the remaining outer shell electrons, resulting in a positive ion that is smaller, that holds the remaining outer shell electron more tightly, and that has a greater tendency to accept an electron than the atom. Similarly, addition of an electron to form a negative ion results in a species that is larger and easier to remove an electron from.

Some interesting exceptions to the general trends occur. (1) A IIIA element has a lower ionization energy than does the IIA element in the same period, and a VIA element has a lower ionization energy than does the corresponding VA element. Note that this correlates with changes in configuration $ns^2 \rightarrow ns^2\,np^1$ and $ns^2\,np^3 \rightarrow ns^2\,np^4$, corresponding to the higher energy associated with putting the first electron into a new sub-shell or the pairing up of electrons within a sub-shell. (2) Upon filling an *inner*-transition series the ratio, $Z_{eff}/n_{max}$, actually very slowly increases so that the first few atoms beyond an inner-transition metal series are very nearly the same size and have almost identical chemical properties as the respective elements just above them (the *lanthanide contraction*).

## Exercises Involving the Periodic Table and Periodicity

11. In which of the following pairs are the elements most similar chemically (because of similar electron configurations)?

    (A) Li, C (B) P, Al (C) F, C (D) S, Te (E) P, S

12. Given the following atoms: Ca, K, Be, Na, Li. The one with the **lowest** ionization potential (ionization energy) is:

    (A) Ca (B) Be (C) Na (D) Li (E) K

13. Atoms with high first ionization energies always have:

    (A) low electron affinities (B) large atomic radii
    (C) metallic properties (D) tightly bound valence electrons
    (E) none of the above

14. The yet-undiscovered element $Z = 117$ should:

    (A) be a halogen (B) have seven valence electrons
    (C) have the valence-shell configuration $7s^2, 7p^5$
    (D) be in the seventh period (E) have all of the above properties

15. The atom with which of the following electronic configurations has the largest first ionization potential, *i.e.*, is most reluctant to lose an electron?

    (A) [Ne], $3s^2, 3p^1$ (B) [Ne], $3s^2, 3p^2$ (C) [Ne], $3s^2, 3p^3$
    (D) [Ne], $3s^2, 3p^4$ (E) [Ar], $3d^{10}, 4s^2, 4p^3$

16. Which of the following has the largest radius?

    (A) N (B) Cl (C) S (D) F (E) O

17. Which of the series of elements listed below would have most nearly the same atomic radius?

    (A) F, Cl, Br, I (B) Li, Be, B, C (C) He, Ne, Ar, Kr
    (D) B, Si, As, Te (E) Mn, Fe, Co, Ni

18. The atom having the valence-shell configuration $6s^2, 6p^2$ would be in:
    (A) Group VIA and Period 4 (B) Group IVB and Period 6
    (C) Group IVA and Period 4 (D) Group VIB and Period 6
    (E) Group IVA and Period 6

19.*Which of the following pairs of elements are most nearly the same in size (radius)?

## DRILL ON THE PERIODIC TABLE AND PERIODICITY

***Student's Name*** ______________________

In each case from the four given elements, pick the element that best fits the specified property. You may be asked to turn in this sheet.

| Elements | Most Metallic | Largest Radius | Highest 1st Ionization Potential | Lowest Electron Affinity |
|---|---|---|---|---|
| Li, Be, B, C | | | | |
| Be, Mg, Ca, Sr | | | | |
| Sn, As, S, F | | | | |
| Ga, Ge, As, Se | | | | |
| Se, Br, Te, I | | | | |
| ______________ | | | | |

(A) K, Ca (B) Ti, Zr (C) K, Rb (D) Ne, Kr (E) Hf, Zr

20. In order of increasing atomic size (left-to-right), the atoms aluminum, boron, carbon, potassium, and sodium should be arranged:

    (A) B, C, Na, Al, K (B) B, C, Al, K, Na (C) C, B, Al, Na, K
    (D) C, B, Na, Al, K (E) C, B, Na, K, Al

---

## GENERAL EXERCISES ON CHAPTER 4—ELECTRONIC STRUCTURE AND THE PERIODIC TABLE

21. Atomic line spectra give a direct measure of:

    (A) the number of protons in the nucleus
    (B) the absolute energy of an electronic energy level
    (C) the number of electrons in an atom
    (D) the energy difference between two energy levels
    (E) the radius of the electronic orbit–Bohr relationship

22. The following electronic transitions occur when lithium atoms are sprayed into a hot flame ("$4s \rightarrow 3p$" means an electronic transition from a $4s$ to a $3p$ orbital). The various steps are numbered for identification.

$$2s \xrightarrow{\text{I}} 2p \xrightarrow{\text{II}} 3d \xrightarrow{\text{III}} 3p \xrightarrow{\text{IV}} 4s \xrightarrow{\text{V}} 2p$$

which of these would result in **emission** of light?

    (A) all of the steps (B) I, II, and V (C) III and V (D) III, IV, and V
    (E) only step III

23. How many electrons are in the innermost shell of the atom that has $Z = 18$ and $A = 35$?

    (A) 0 (B) 2 (C) 8 (D) 18 (E) none of these

24. A titanium atom in its ground state has one or more electrons in how many electronic orbitals?

    (A) 4 (B) 7 (C) 11 (D) 12 (E) 22

25. The electronic configuration of a $^{31}_{15}P$ atom is:

    (A) $1s^2, 2s^2, 2p^6, 3s^2, 3p^4$ (B) $1s^2, 2s^2, 2p^6, 3s^2, 3p^3$
    (C) $1s^2, 2s^2, 2p^6, 3s^2, 3p^2, 3d^1$ (D) $1s^2, 2s^2, 2p^6, 3s^1, 2d^4$
    (E) $1s^2, 2s^2, 2p^6, 3s^2, 3p^6, 4s^2, 3d^{10}, 4p^1$

26. If a phosphorus atom were to gain three additional electrons, the resulting species would be:

    (A) negatively charged and isoelectronic with argon
    (B) negatively charged and isoelectronic with neon
    (C) positively charged and isoelectronic with argon
    (D) positively charged and isoelectronic with magnesium
    (E) smaller than a phosphorus atom

27. The maximum number of electrons that can be accommodated in a subshell for which $\ell = 3$ is:

    (A) 2 (B) 10 (C) 6 (D) 14 (E) 8

28. Which of the following subshells has room for 10 electrons?

    (A) 5*s* (B) 4*p* (C) 2*p* (D) 3*d* (E) 3*p*

29. Naturally occurring rubidium consists of just two isotopes. One of the isotopes consists of atoms having a mass of 84.912 amu; the other of 86.901 amu. What is the percent natural abundance of the **heavier** isotope? (AW of Rb = 85.4678)

    (A) 15% (B) 28% (C) 37% (D) 72% (E) 85%

30. An electron having the following set of quantum numbers, $n = 4$, $\ell = 2$, $m = 0$, $s = 1/2$, would be classified as a:

    (A) 3*d* electron (B) 4*d* electron (C) 3*p* electron (D) 4*p* electron (E) 4*s* electron

31. Which of the following atoms has the greatest number of **unpaired** electrons?

    (A) Zn (B) P (C) Ti (D) Ag (E) I

32. All electrons in a *d*-subshell must have the quantum number:

    (A) $n = 3$ (B) $m = 2$ (C) $\ell = 2$ (D) $n = 4$ (E) $s = -1/2$

33. Which one of the following atoms has the lowest electron affinity?

    (A) O (B) F (C) Cl (D) S (E) P

34. The lightest atom with a filled 3*d*-subshell in the ground state is:

    (A) zinc (B) gallium (C) krypton (D) copper (E) rubidium

35. In the periodic table, the lightest atom having the ground state electron configuration $(n\text{-}1)d^8, ns^2$ would be found in:

    (A) Group IIA (B) Group IIB (C) Period 8 (D) Period 4 (E) Period 6

36. The orbital diagram that best describes a carbon atom in its ground state is:

|  | 1s | 2s | 2p |  |  |
|---|---|---|---|---|---|
| (A) | ↑↓ | ↑↓ | ↑ | ↑ |  |
| (B) | ↑↓ | ↑↓ | ↑↓ |  |  |
| (C) | ↑↓ | ↑ | ↑ | ↑ | ↑ |
| (D) | ↑↓ | ↑ | ↑↓ | ↑ |  |
| (E) | ↑ | ↑ | ↑↓ | ↑ | ↑ |

37. The reason why the atomic weight of Cl is 35.453 rather than almost

exactly 35 is because:

(A) every chlorine atom contains 17 protons
(B) all chlorine atoms have identical chemical properties
(C) there are at least two naturally occurring isotopes of Cl
(D) protons and neutrons do not have exactly a mass of 1 amu
(E) every chlorine atom has a mass of 35.453 amu

38. In which one of the following pairs does **each** ion have a noble-gas (inert-gas) configuration?

(A) $Br^-$, $Ge^{3-}$ (B) $Se^{2+}$, $O^{2-}$ (C) $Be^{2+}$, $As^{3-}$ (D) $Fe^{2+}$, $Cl^-$
(E) $Te^{2-}$, $O^-$

39. The nucleus of $^{19}_{9}F$ contains:

(A) 9 protons + 10 neutrons + 9 electrons
(B) 9 protons + 19 neutrons
(C) 9 protons + 10 neutrons + 19 electrons
(D) 9 protons + 10 neutrons
(E) 10 protons + 9 neutrons

40. Which of the following is a valid set of quantum numbers for an electron in the ground state of a carbon atom?

| | $n$ | $\ell$ | $m$ | $s$ |
|---|---|---|---|---|
| (A) | 1 | 0 | 1 | 1/2 |
| (B) | 3 | 1 | -1 | 1/2 |
| (C) | 2 | 2 | -1 | -1/2 |
| (D) | 1 | 1 | 0 | 1/2 |
| (E) | 2 | 0 | 0 | -1/2 |

41. An atom has the ground state configuration $1s^2 2s^2 2p^6 3s^2 3p^6 4s^2 3d^3$ How many **orbitals** are populated with one or more electrons?

(A) 4 (B) 7 (C) 13 (D) 23 (E) 5

42. The **total** number of $p$-electrons in the ground state of a gallium atom is:

(A) 6 (B) 3 (C) 13 (D) 1 (E) none of these

43. Which of the following has the largest radius?

(A) $K^+$ (B) $Cl^-$ (C) $S^{2-}$ (D) $F^-$ (E) $O^{2-}$

44. Which of the following has the **smallest** radius?

(A) $Be^{2+}$ (B) Li (C) Be (D) $O^{2-}$ (E) $F^-$

45. Transition elements are those whose valence electrons may come from:

(A) only a $d$-subshell
(B) subshells that differ in principal quantum number
(C) different subshells of the same principal quantum number
(D) only $d$- and $f$-subshells
(E) only $s$-subshells

46. The number of valence electrons in a chlorine atom is:

(A) 7 (B) 5 (C) 3 (D) 1 (E) 0

47. The atom that has a $4s^2$, $4p^2$ valence-shell configuration is:

(A) Ti (B) Si (C) Ca (D) Kr (E) Ge

48. If first ionization potentials are plotted *vs.* $Z$, a graph results with

(A) transition elements at the maxima, nonmetals at the minima
(B) noble gases at the maxima, the halogens at the minima
(C) alkali metals at the maxima, halogens at the minima
(D) noble gases at the maxima, the alkali metals at the minima
(E) Group VA elements at the maxima, Group VB elements at the minima

49.*Which of the following elements has the largest first ionization potential?

(A) Al (B) Si (C) P (D) S (E) As

50. Of the ions $B^{3-}$, $C^{4-}$, $Al^{3+}$, $Na^+$, $K^+$, the one with the smallest radius is:

(A) $B^{3-}$ (B) $C^{4-}$ (C) $Al^{3+}$ (D) $Na^+$ (E) $K^+$

51. When arranged (left-to-right) in order of increasing ionization potential, the atoms carbon, lithium, neon, and silicon would be:

(A) Li, Si, C, Ne (B) C, Si, Li, Ne (C) Li, Si, Ne, C
(D) Li, C, Si, Ne (E) Ne, C, Si, Li

52. Which of the following has the highest first ionization energy (ionization potential)?

(A) P (B) S (C) Cl (D) As (E) Se

53. In order of increasing ionization energy (left-to-right) the atoms aluminum, boron, carbon, potassium and sodium are:

(A) C, B, K, Al, Na (B) B, C, Al, K, Na (C) C, B, Na, K, Al
(D) B, C, Na, Al, K (E) K, Na, Al, B, C

54. For any permitted value of $\ell$ (the angular momentum quantum number) the number of orbitals will be:

(A) $2n^2$ (B) $2\ell + 1$ (C) $2\ell - 1$ (D) $n - 1$ (E) $(\ell - 1)^{1/2}$

55. Which of the following quantum numbers is often designated by the letters *s, p, d,* and *f,* instead of numbers?

(A) $n$ (B) $m$ (C) $\ell$ (D) $s$ (E) none of these

56. An orbital is labeled by the magnetic quantum number as $m = 1$. This could **not** be a:

(A) *s*-orbital (B) *p*-orbital (C) *d*-orbital (D) *f*-orbital (E) *g*-orbital

57. Which of the following would be paramagnetic?

(A) $Ti^{4+}$ (B) Mg (C) $Fe^{2+}$ (D) $Mg^{2+}$ (E) $Sc^{3+}$

58. Which of the following isoelectronic species would lose an electron most easily?

(A) $S^{2-}$ (B) $Cl^{-}$ (C) Ar (D) $K^{+}$ (E) $Ca^{2+}$

---

## ANSWERS TO CHAPTER 4 PROBLEMS

| | | | | | |
|---|---|---|---|---|---|
| 1. D | 11. D | 21. D | 31. B | 41. C | 51. A |
| 2. A | 12. E | 22. C | 32. C | 42. C | 52. C |
| 3. C | 13. D | 23. B | 33. E | 43. C | 53. E |
| 4. D | 14. E | 24. D | 34. D | 44. A | 54. B |
| 5. A | 15. C | 25. B | 35. D | 45. B | 55. C |
| 6. B | 16. C | 26. A | 36. A | 46. A | 56. A |
| 7. A | 17. E | 27. D | 37. C | 47. E | 57. C |
| 8. A | 18. E | 28. D | 38. C | 48. D | 58. A |
| 9. A | 19. E | 29. B | 39. D | 49. C | |
| 10. A | 20. C | 30. B | 40. E | 50. C | |

# 5
# General Concepts of Bonding and Chemical Nomenclature

All chemical bonding is electrostatic in origin, *bonds* (linkages) between atoms resulting when the **net** of the attractive and repulsive forces is one of attraction. Chemical bonds may be formed (a) by complete transfer of electrons from one atom to another (*ionic bond*) thereby forming ions, (b) by *sharing* electrons between two atoms (*covalent bond*) to form a *molecule*, a *polyatomic ion*, or an *extended network*, and (c) by sharing electrons among a group of atoms (*delocalized* or *metallic bond*). Here, we primarily are interested in ionic and covalent bonds. These terms, although long used to categorize bonds, represent two extremes and in actuality there is a continuous gradation of bonds between these two types. Thus, some bonds are readily classified as predominantly ionic, others as predominantly covalent, and still others occur in an undefined region that cannot readily be categorized.

## SIMPLE IONS AND IONIC BONDS

Typically, compounds of metals and nonmetals are made up of positive and negative ions. Positive ions are formed from atoms by removing the appropriate number of electrons in order of decreasing principal quantum number (**outermost electrons are removed first**). For outer-shell electrons in two or more sub-shells, electrons are removed in order of decreasing azimuthal quantum number. Thus the configuration of $Mn^{3+}$ is: $1s^2\,2s^2\,2p^6\,3s^2\,3p^6\,3d^4$. The configuration for simple (monoatomic) anions may be obtained by adding electrons to fill the outermost sub-shell; this makes the anion *isoelectronic* with the rare gas at the end of the period, a particularly stable electronic configuration. This means that the arsenic anion would be $As^{3-}$ with the configuration $[Ar]\,4s^2\,3d^{10}\,4p^6 = [Kr]$, since the As atom (valence shell, $4s^2\,4p^3$) had 3 electron vacancies in the $4p$ sub-shell.

Valence shell configurations for atoms and ions frequently are indicated by writing Lewis symbols. These consist merely of the symbol of the element along with a dot for each valence electron and, in the case of an ion, the charge on the ion. Thus, the Lewis symbol for the Li atom is Li˙; for the lithium cation, merely $Li^+$; for the sulfur atom, :Ṡ̤: and for the sulfur anion (sulfide ion), $:\ddot{\underset{..}{S}}:^{2-}$.

Formulas of binary (made up of only 2 elements) ionic compounds may readily be predicted by knowing the formulas of the respective ions and combining a sufficient number of each so as to make a combination that is electrically neutral. The empirical formula for the compound between $Li^+$ and $S^{2-}$ would be $Li_2S$, and that between $Fe^{3+}$ and $O^{2-}$, $Fe_2O_3$.

## Example Problem Involving Simple Ions

### 5-A. Electronic Configuration of an Ion

In the "ground state," the ion $V^{3+}$ has its electrons distributed over __W__ main energy levels (shells), __X__ sublevels (subshells), and __Y__ orbitals. Of these electrons, __Z__ are unpaired. The correct numbers needed to fill the blanks are:

| | W | X | Y | Z |
|---|---|---|---|---|
| (A) | 4 | 7 | 13 | 3 |
| (B) | 3 | 6 | 20 | 0 |
| (C) | 4 | 6 | 10 | 0 |
| (D) | 4 | 7 | 15 | 4 |
| (E) | 3 | 6 | 11 | 2 |

#### *Solution*

In order to derive the configuration of an ion, that of the parent atom should be considered first. The aufbau order for the vanadium atom is:

$$V^0\ (Z = 23):\ 1s^2, 2s^2, 2p^6, 3s^2, 3p^6, 4s^2, 3d^3$$

Three electrons must be **removed** to obtain $V^{3+}$ (with 23 protons, 20 electrons). These are **not** removed in the reverse of the order given above [which would give an incorrect answer (C)]. If we **first** remove the two electrons of the **highest principal quantum number** ($n = 4$), $V^{2+}$ would result, leaving $3s$, $3p$, and $3d$ as occupied subshells next highest in $n$. Removal of the additional electron from the subshell of **highest** $\ell$ ($3d$, $\ell = 2$) then gives $V^{3+}$ with the configuration:

$$V^{3+}:\ 1s^2, 2s^2, 2p^6, 3s^2, 3p^6, 3d^2$$

The use of three principal quantum numbers (1, 2, and 3) indicates that 3 shells are populated with electrons. The underlined entries above denote the 6 populated subshells ($1s$, $2s$, $2p$, $3s$, $3p$, and $3d$). Of these, the $1s$, $2s$, $2p$, $3s$, and $3p$ are filled, and consist of nine orbitals; by Hund's rule **two** of the five $3d$-orbitals are needed, giving a total of **11** orbitals used and **2** unpaired electrons.

***Answer***

3 shells
6 subshells
11 orbitals
2 unpaired electrons

**[E]**

## Exercises on Simple Ions and Ionic Bonds

1. The electronic structure of $Ti^{2+}$ is:

   (A) $1s^2, 2s^2, 2p^6, 3s^2, 3p^6, 4s^2, 3d^2$ (B) $1s^2, 2s^2, 2p^6, 3s^2, 3p^6, 4s^2$
   (C) $1s^2, 2s^2, 2p^6, 3s^2, 3p^6, 4s^2, 3d^4$ (D) $1s^2, 2s^2, 2p^6, 3s^2, 3p^6, 3d^2$
   (E) the same structure as the rare gas argon

2. Which of the following is the Lewis symbol that is most similar to that for a potassium ion?

   (A) $\cdot\ddot{\underset{\cdot}{X}}\cdot$ (B) $X\cdot$ (C) $X^+$ (D) $:\ddot{\underset{\cdot\cdot}{X}}:^{3-}$ (E) $\cdot\dot{\underset{\cdot}{X}}\cdot^+$

3. Which of the following is a correct Lewis symbol for a monoatomic ion derived from tellurium (Z = 52)?

   (A) $:\ddot{\underset{\cdot\cdot}{Te}}:^{2-}$ (B) $\cdot\underset{\cdot}{Te}\cdot^{2+}$ (C) $\cdot\ddot{Te}\cdot^{2-}$ (D) $:\ddot{Te}^{2-}$ (E) $:\ddot{Te}:^{2+}$

4. Manganese exhibits a variety of cations in its binary compounds with nonmetals:

   I. with fluorine, $Mn^{3+}$
   II. with silicon, $Mn^{2+}$
   III. with oxygen, $Mn^{4+}$
   IV. with arsenic, $Mn^{3+}$
   V. with nitrogen, $Mn^{2+}$

   The expected formulas of the compounds would be:

| | I | II | III | IV | V |
|---|---|---|---|---|---|
| (A) | $MnF_3$ | $MnSi$ | $MnO_4$ | $Mn_2As_3$ | $Mn_3N_2$ |
| (B) | $MnF_3$ | $Mn_2Si$ | $MnO_2$ | $MnAs$ | $Mn_3N_2$ |
| (C) | $Mn_3F$ | $MnSi_2$ | $MnO_4$ | $Mn_3As_2$ | $MnN_2$ |
| (D) | $MnF_3$ | $MnSi_2$ | $MnO$ | $MnAs$ | $Mn_2N$ |
| (E) | $MnF_3$ | $Mn_2Si_3$ | $MnO_2$ | $MnAs_3$ | $Mn_2N_3$ |

# DRILL ON THE ELECTRONIC CONFIGURATION OF SIMPLE IONS

***Student's Name*** ______________________

Complete the table for ground state configurations for each of the given ions. In some cases, the charge on the ion is given in the Formula column and you will have to determine the symbol.

| Formula | Configuration | Number of Occupied Shells | Number of Occupied sub-shells | Number of Occupied Orbitals | Number of Unpaired Electrons | Name of the Ion |
|---|---|---|---|---|---|---|
| $S^{2-}$ | | | | | | |
| | | | | | | vanadium(II) |
| $Cu^{2+}$ | | | | | | |
| $\_\_\_^{3+}$ | $1s^2,2s^2,2p^6,3s^2,3p^6,3d^6$ | | | | | |
| $\_\_\_^{-}$ | | | | 18 | | |
| $Sb^{3+}$ | | | | | | |
| | | | | | | |

5. Which of the following does **not** have a noble gas configuration?

   (A) $S^{2-}$ (B) Ar (C) $Al^{3+}$ (D) $Sb^{5+}$ (E) $Sc^{3+}$

6. Atom X forms $X^+$ with the electronic configuration $5d^{10}$, $6s^2$. Atom X is:

   (A) Ag (B) Tl (C) Cs (D) Au (E) At

---

## COVALENT BONDS

Bonds between nonmetals are usually formed by sharing electrons; hence, bonds holding molecules or polyatomic ions together are covalent bonds. Lewis structures are frequently written for molecules and polyatomic ions in which **all** of the electrons in the valence shells are shown as dots (for individual electrons) or lines (for electron pairs), *e.g.* H:H (or H–H) and :Ö::C::Ö: (or |$\bar{O}$=C=$\bar{O}$|). These structures are obtained by carefully counting all electrons in valence shells and, as a first approximation, distributing these as bond-pairs or lone-pairs so that each atom has a rare gas configuration (the *isoelectronic principle* or *octet rule*). There are numerous exceptions when dealing with atoms beyond the second period, *e.g.* $SF_4$ and $PF_6^-$. See Exercise 4-**B** or your text for a detailed method of writing Lewis structures.

Electrons in covalent bonds are rarely equally shared; the best example of equal sharing of electrons would be in a bond where the atoms are identical, *e.g.* H–H or in the C–C bond in H–C≡C–H. When two different atoms are covalently bonded, one usually has a greater attraction for the bonding electrons than does the other. Indeed, the electrons will be closer to the one atom, which results in that atom carrying a fractional negative charge, the other atom an equal fractional positive charge. The **bond** between the two atoms is referred to as a *polar, covalent bond.* A function, *electronegativity,* has been devised to semi-quantitatively express the relative ability to attract electrons in a covalent bond. In the most commonly used scale of electronegativity, fluorine, the most electronegative atom, is assigned the value 4.0; the least electronegative atom, cesium, a value of 0.7. Bonds formed between nonmetals with electronegativity differences of less than 1.7 are best described as covalent; bonds between atoms having electronegativity differences greater than 1.7, as ionic. In terms of electronegativity differences, bonds can be roughly categorized as follows:

Nonpolar covalent - - - - - - - → Polar covalent - - - - - - - - - - - - - - - → Ionic

0 1.7 3.3

Increasing difference in Electronegativity ⟶

## Example Problem Involving Covalent Bonds

### 5-B. Valence Electrons, Lone Pairs, Single Bonds, Double Bonds, and Bond Order

A Lewis formula for the formate ion, $HCO_2^-$, would show a total of __(V)__ valence electrons, with __(W)__ lone (unshared) pairs on the carbon atom. There would be depicted a total of __(X)__ single bond(s) and __(Y)__ double bond(s), and the carbon-oxygen bond order would be __(Z)__.

| | (V) | (W) | (X) | (Y) | (Z) |
|---|---|---|---|---|---|
| (A) | 17 | 0 | 3 | 0 | 1/2 |
| (B) | 18 | 0 | 2 | 1 | 1-1/2 |
| (C) | 16 | 0 | 3 | 0 | 2 |
| (D) | 18 | 0 | 3 | 1 | 1-1/2 |
| (E) | none of the above | | | | |

#### *Solution*

The total number of valence electrons will be the sum of those from the neutral atoms plus one additional electron due to the –1 ionic charge, giving [1 + 4 + 2(6) + 1] = 18 valence electrons. In order to apportion these electrons, it is necessary to **know** how the species is connected together. In many instances, the formula for a simple species is written so that the first nonhydrogen atom is the central atom with all other atoms (or groups in parentheses) attached only to that atom. Thus, here the carbon is the central atom with the hydrogen and the oxygens attached only to the carbon, H—C(O)(O). The number of electrons that must be shared is calculated assuming the isoelectronic principle (*i.e.*, each atom effectively attains an electronic "closed-shell" structure of the nearest noble gas). This means that the **separate**-atom, **closed**-shell requirement would be:

| | |
|---|---|
| One H-atom with 2 $e^-$, [He], requires | 2 $e^-$ |
| One C-atom with 8 $e^-$, [Ne], requires | 8 $e^-$ |
| Two O-atoms with 8 $e^-$, [Ne], requires | 16 $e^-$ |
| Total requirement: | 26 $e^-$ |

Therefore, the number of pairs that must be shared is (closed-shell requirement - electrons available) ÷ 2, or (26 - 18)/2 = 4 pairs shared. Since the C—H bond can be involved with only one of these pairs, we must have:

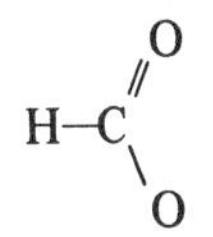

This gives the "closed-shell *via* sharing" to both C and H using 8 of the 18 electrons. Completing the oxygen "octets" with the remaining 5 pairs gives the Lewis formula,

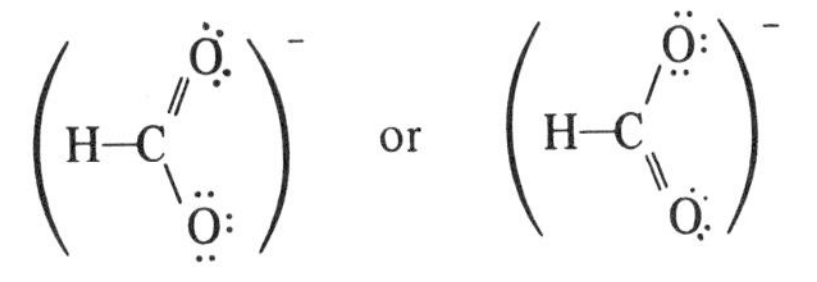

with 2 single bonds (C—H and C—O), 1 double bond (C=O), and no lone pairs on C. There are, however, two equivalent ways to write this formula, as shown above. These represent two equivalent resonance forms. Even transiently, the formate ion is NEITHER of these but is a "resonance hybrid" often shown as

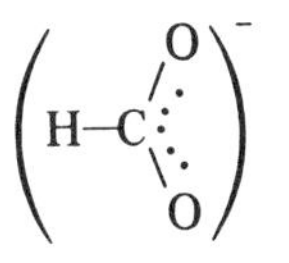

The two carbon-oxygen bonds are involved with three pairs of shared electrons; therefore the C—O bond order is

$$\frac{\text{Number of shared pairs}}{\text{Number of linkages}} = 3/2 \quad \text{or} \quad 1\text{-}1/2$$

### *Answer*

18 valence electrons
0 unshared electron pairs on carbon
2 single bonds
1 double bond
1-1/2 bond order for the C—O bonds

[E]

## Exercises on Covalent Bonds

7. Which of the following is **least** likely to show ionic bonding?

   (A) KF (B) $BaCl_2$ (C) $ZnCl_2$ (D) $S_2Cl_2$ (E) $ScCl_3$

8. Which of the following is the most electronegative element?

   (A) O (B) S (C) Se (D) Si (E) Ga

9. An atom with a high electronegativity generally has a

   (A) low electron affinity (B) small atomic number
   (C) large atomic radius (D) tendency to form positive ions
   (E) high ionization potential

10. In which of the following compounds do ALL atoms obey the octet rule?

    (A) NO (B) $PF_5$ (C) $PF_3$ (D) $SF_4$ (E) NaH

11. Which of the following formulas is a satisfactory representation of sulfur dioxide?

    (A) :Ö: S :Ö: (B) :Ö: S :Ö: (C) :Ö: S :Ö: (D) :O::S::O:

    (E) :O:::S::O:

12. Which of the following molecules contains **one** triple bond?

    (A) $NH_3$ (B) HCN (C) $SO_3$ (D) $H_2CCH_2$ (E) $H_2O$

13. The bond order in the carbonate ion is:

    (A) 2 (B) 4/3 (C) 1-1/2 (D) 1/2 (E) 0

14. The **total** number of valence electrons that must be shown in the Lewis formula for the phosphonium ion, $PH_4^+$, is:

    (A) 16 (B) 18 (C) 8 (D) 32 (E) 24

15. For which of the following are resonance structures important?

    (A) $NH_3$ (B) $CCl_4$ (C) $SO_4^{2-}$ (D) $CO_2$ (E) $NO_3^-$

16. The Lewis structure for the bromate ion, $BrO_3^-$, is:

    (A) (:Ö: / :Ö:Br::Ö:)⁻ (B) (:Ö: / :Ö::Br::Ö:)⁻ (C) (:Ö: / :Ö:Br:Ö:)⁻

    (D) (:Ö: / :Ö:Br:Ö:)⁻ (E) (:Ö: / :Ö:Br:Ö:)⁻

## DRILL ON LEWIS STRUCTURES

***Student's Name*** ____________________

Draw correct Lewis structures for each of the following species and complete the table. You may be asked to turn in this sheet.

| Formula | Lewis Structure | Bond Order of the X-O Bond | Formula | Lewis Structure | Bond Order of the N-X Bond |
|---|---|---|---|---|---|
| $SO_3^{2-}$ | | | $NH_4^+$ | | |
| $XeO_6^{4-}$ | | | $CN_2^{2-}$ | | |
| $CH_2O$ | | | $NO_2$ | | |

## OXIDATION NUMBERS

Atoms frequently are characterized by the assignment of *oxidation numbers*. These are the charges that the two atoms involved in a bond would assume if all bonding electrons were **arbitrarily** assigned to the more electronegative atom; *e.g.*, in $OF_2$ the oxidation number of F is −1 and that of 0 is +2. Oxidation numbers are useful in the systematic naming of compounds (*nomenclature*) and in the bookkeeping of electrons during chemical reactions. Reactions in which electrons are transferred are called *oxidation-reduction* reactions; these two processes must occur simultaneously and to the same extent.

**Oxidation** is: electron-loss, or an algebraic **increase** in oxidation number

**Reduction** is: electron-gain, or an algebraic **decrease** in oxidation number

In most cases oxidation numbers can easily be assigned by using the following rules **in order of priority given**:

**General Rules**: The oxidation number of an atom in an element is zero; in a monoatomic ion, it is the same as the charge on the ion. The sum of the oxidation numbers is equal to the charge on the overall species.
**Priority Rules**: An atom bonded to a different element is assigned an oxidation number in the order of precedence: F, −1; IA Metals, +1; IIA Metals, +2; H, +1; O, −2.
**Addendum**: In binary compounds, the oxidation number of the least metallic element can be predicted from the isoelectronic principle, that of the other atom by difference. Similarly, if one knows the charge on a polyatomic ion, the oxidation number of an atom(s) combined with it can be deduced since the sum must equal the net charge.

### Example Problem Involving Oxidation Numbers

#### 5-C. Determination of Oxidation Numbers from a Formula

The oxidation number of the phosphorus atom in the compound $Ba(H_2PO_2)_2$ is:

(A) +1 (B) +2 (C) +3 (D) +4 (E) +5

***Solution***

Since this is a neutral formula unit, the first rule that applies is that the sum of all of the individual oxidation numbers must equal zero. Second, if we wish to find the oxidation number for P, oxidation numbers must be assigned to all other atoms present. These assignments, **in order of decreasing priority**, are:

| Atom: | Ba (a Group IIA atom) | H | O | P |
|---|---|---|---|---|
| Oxidation Number: | +2 | +1 | -2 | ? |

Thus, rewriting the formula as $BaH_4P_2O_4$, and letting $x$ stand for the oxidation number of **each** phosphorus atom, we have:

$$(\text{Ox. No. Ba}) + 4(\text{Ox. No. H}) + 4(\text{Ox. No. O}) + 2(\text{Ox. No. P}) = 0$$
$$(+2) + 4(+1) + 4(-2) + 2(x) = 0$$

*Answer*

$x = +1$; **each** phosphorus atom in $Ba(H_2PO_2)_2$ would be assigned a +1 oxidation number (or *oxidation state*). [A]

**Exercises Involving Oxidation Numbers**

17. The oxidation number for nitrogen in $NH_2OH$ is:

    (A) -2 (B) -1 (C) 0 (D) +1 (E) +2

18. The oxidation number for nitrogen in $N_2O$ is:

    (A) -1 (B) 0 (C) +1 (D) +2 (E) +3

19. The oxidation number for nitrogen in $NO_3^-$ is:

    (A) -1 (B) +3 (C) +4 (D) +5 (E) +6

20. The oxidation number for nitrogen in $N_2H_5^+$ is:

    (A) -3 (B) -2 (C) +2 (D) +3 (E) +5

## FORMULAS AND CHEMICAL NOMENCLATURE

From your encounters with atomic structure and periodicity (Ch. 4) and the bonding concepts of this chapter, two characteristics concerning reactivity have emerged:

1. *Metallic elements,* which make up approximately 80% of the periodic table, have a pronounced tendency to form positive ions and exist as such in many compounds and in solution.†

---

†An aspect of the chemistry of metals that is not so obvious from the foregoing discussions of structure and bonding is that simple positive ions are seldom encountered

2. Hydrogen and the elements on the extreme right (particularly the upper right) of the periodic table are *nonmetals*. These elements have relatively high *ionization potentials*, and with the exception of the *noble gases*, have relatively high *electron affinities* and *electronegativities*. These properties are shown in two ways: (a) the formation of simple negative ions, following the *isoelectronic principle*, in **binary compounds with metals**; and (b) the formation of covalent bonds with themselves (*catenation*) or with other nonmetals to yield discrete molecules‡ (*e.g.*, $H_2S$, $P_4O_{10}$, and $SF_4$) and/or stable *polyatomic ions* (*e.g.*, $SO_4^{2-}$, $NH_4^+$, $O_2^{2-}$, and $C_2H_3O_2^-$), which also exist in compounds and in solution.

Although many of these may be predicted within the framework of the theories of structure and bonding, it is much easier, and certainly faster, if a minimum vocabulary of these species is **memorized**, *i.e.*, **known** by ***name, composition,*** and ***net charge***. Since most dealings with materials (including biology, medicine, engineering, and arts, as well as chemistry) require a knowledge of formulas and names of substances, the rapid accumulation of this vocabulary is of utmost importance.

## Formulas

All chemical compounds are **electrically neutral**, even though they are constituted from units that are electrical in nature. In writing formulas for **compounds** this net electrical neutrality must be preserved. Thus, the compound formed between $Ag^+$ and $CrO_4^{2-}$ requires two silver ions for every chromate ion, $Ag_2CrO_4$, to be electrically neutral. The binary compound formed between calcium and phosphorus can be considered as made up of ions having a noble gas configuration, $Ca^{2+}$ and $P^{3-}$, 2 $P^{3-}$

---

with charges exceeding +3; ions with charges of +4 and higher are exhibited only by very large atoms (*e.g.*, Th). The reason for this is that as the **oxidation state** of a metal increases, electronic charge is drawn away from neighboring electron-rich atoms or ions. The net result is that atoms in high oxidation states (even as low as +2 at times) do not exist as simple ions, but are covalently bonded to one or more electron-rich species. The result is the formation of independent, covalently bonded species such as $MnO_4^-$ and $CrO_4^{2-}$ that are observed both in solutions and in crystalline salts.

‡Very complex species involving an infinite number of atoms may be built up (*polymers*). Thus the formula for silicon carbide, SiC, looks very simple. In actuality all of the atoms in a crystal are covalently bonded to form a network of alternating Si and C so that the formula approaches $Si_\infty C_\infty$.

being required for 3 $Ca^{2+}$ to be electrically neutral, $Ca_3P_2$. Fill in the table below as practice in the writing of formulas of compounds.

| | $NO_3^-$ Nitrate | $O^{2-}$ Oxide | $SO_4^{2-}$ Sulfate | $PO_4^{3-}$ Phosphate | $Fe(CN)_6^{4-}$ Hexacycanoferrate(II) |
|---|---|---|---|---|---|
| $Ag^+$ | | | | | |
| $Mg^{2+}$ | | | | | |
| $Fe^{2+}$ Iron(II) | | | | | |
| $Fe^{3+}$ Iron(III) | | $Fe_2O_3$ | | | $Fe_4[Fe(CN)_6]_3$ |
| $Th^{4+}$ | | | | | |
| $NH_4^+$ Ammonium | | | | | |
| $H^+$ | $HNO_3$ | | | | |

## Nomenclature

The naming of compounds requires a considerable vocabulary. Included are the names of elements, ions, special names for certain compounds, prefixes, and special endings–these are **simply either known and understood** or you are unable to begin naming the vast majority of compounds. Names of some simple ions are given in Table I; names of some common polyatomic ions are given in Table II. We start our study of nomenclature with compounds containing only two different elements (*binary compounds*) along with some compounds that contain more than two elements but are named like binary compounds. Then we proceed to compounds containing polyatomic ions (including many common *acids*).

TABLE I. SOME COMMON SIMPLE IONS (IONS CONSISTING OF A SINGLE ATOM BEARING A CHARGE)

| Periodic Table Group | Ion Formula | Name |
|---|---|---|
| IA | $M^+$, *e.g.*, $K^+$ | *Potassium ion* |
| IIA | $M^{2+}$, *e.g.*, $Ra^{2+}$ | *Radium ion* |
| IIIA | $Al^{3+}$ | *Aluminum ion* |
| IVA | $Sn^{2+}$ | *Tin(II) ion* (old, *stannous)* |
| | $Sn^{4+}$ | *Tin(IV) ion* (old, *stannic)* |
| | $Pb^{2+}$ | *Lead(II) ion* (old, *plumbous)* |
| VA | $Bi^{3+}$ | *Bismuth(III) ion* |
| IB | $Cu^+$ | *Copper(I) ion* (old, *cuprous)* |
| | $Cu^{2+}$ | *Copper(II) ion* (old, *cupric)* |
| | $Ag^+$ | *Silver ion* |
| IIB | $Zn^{2+}$ | *Zinc ion* |
| | $Cd^{2+}$ | *Cadmium ion* |
| | $Hg^{2+}$ | *Mercury(II) ion* (old, *mercuric)* |
| IIIB | $M^{3+}$, *e.g.*, $La^{3+}$ | *Lanthanum ion* |
| VIB | $Cr^{3+}$ | *Chromium(III) ion* (old, *chromic)* |
| VIIB | $Mn^{2+}$ | *Manganese(II) ion* (old, *manganous)* |
| VIIIB | $Fe^{2+}$ | *Iron(II) ion* (old, *ferrous)* |
| | $Fe^{3+}$ | *Iron(III) ion* (old, *ferric)* |
| | $Co^{2+}$ | *Cobalt(II) ion* (old, *cobaltous)* |
| | $Ni^{2+}$ | *Nickel(II) ion* (old, *nickelous)* |
| VIIA | $X^-$, *e.g.*, $Br^-$ | *Bromide ion* |
| VIA | $X^{2-}$, *e.g.*, $S^{2-}$ | *Sulfide ion* |
| VA | $X^{3-}$, *e.g.*, $P^{3-}$ | *Phosphide ion* |

TABLE II. SOME COMMON POLYATOMIC IONS

| | Formula | Comments |
|---|---|---|
| *Groups VIIA and VIIB* | | |
| Perchlorate | $ClO_4^-$ | Analogously for Br, I (applies to $ClO_4^-$, $ClO_3^-$, $ClO_2^-$, $ClO^-$) |
| Chlorate | $ClO_3^-$ | |
| Chlorite | $ClO_2^-$ | |
| Hypochlorite | $ClO^-$ | |
| Permanganate | $MnO_4^-$ | Analogously for Tc, Re |
| *Groups VIA and VIB* | | |
| Peroxide | $O_2^{2-}$ | $(O{-}O)^{2-}$ |
| Hydroxide | $OH^-$ | |
| Hydronium | $H_3O^+$ | Abbreviated, $H^+_{(aq)}$ |
| Sulfate | $SO_4^{2-}$ | Analogously for Se (applies to sulfate through bisulfite) |
| Hydrogensulfate (bisulfate) | $HSO_4^-$ | $HOSO_3^-$ |
| Sulfite | $SO_3^{2-}$ | |
| Hydrogensulfite (bisulfite) | $HSO_3^-$ | $HOSO_2^-$ |
| Thiosulfate | $S_2O_3^{2-}$ | $S{-}SO_3^{2-}$ |
| Chromate | $CrO_4^{2-}$ | Analogously for Mo and W |
| Dichromate | $Cr_2O_7^{2-}$ | $(O_3Cr{-}O{-}CrO_3)^{2-}$ |
| *Group VA* | | |
| Nitrate | $NO_3^-$ | |
| Nitrite | $NO_2^-$ | |
| Phosphate | $PO_4^{3-}$ | Analogously for As (applies to phosphate through dihydrogenphosphate) |
| Hydrogenphosphate | $HPO_4^{2-}$ | $HOPO_3^{2-}$ |
| Dihydrogenphosphate | $H_2PO_4^-$ | $(HO)_2PO_2^-$ |
| Pyrophosphate | $P_2O_7^{4-}$ | $(O_3P{-}O{-}PO_3)^{4-}$ |
| Ammonium | $NH_4^+$ | |
| *Group IVA* | | |
| Cyanide | $CN^-$ | |
| Carbonate | $CO_3^{2-}$ | Analogously for Si |
| Hydrogencarbonate (bicarbonate) | $HCO_3^-$ | $HOCO_2^-$ |
| Formate | $CHO_2^-$ | $H{-}CO_2^-$ |
| Cyanate | $OCN^-$ | |
| Thiocyanate | $SCN^-$ | |
| Oxalate | $C_2O_4^{2-}$ | $(O_2C{-}CO_2)^{2-}$ |
| Acetate | $C_2H_3O_2^-$ | $H_3C{-}CO_2^-$ |
| Orthosilicate | $SiO_4^{4-}$ | |
| *Miscellaneous* | | |
| Hexacyanoferrate(III) (ferricyanide) | $Fe(CN)_6^{3-}$ | |
| Hexacyanoferrate(II) (ferrocyanide) | $Fe(CN)_6^{4-}$ | |
| Mercury(I) (mercurous) | $Hg_2^{2+}$ | $(Hg{-}Hg)^{2+}$ |

### Metal-Nonmetal Binary Compounds

The entire name of the more metallic species is written; if the metal can exist in more than one oxidation state,† the oxidation state is indicated by placing a Roman numeral, specifying the state, in parentheses directly after the name of the metal (no intervening space). A space is then left and the first portion (the *root*) of the nonmetallic element's name is written with the ending *ide* attached to it. The negative charge on the nonmetal is predicted by the isoelectronic principle. Examples are $Na_2S$, sodium sulfide; $Mg_3N_2$, magnesium nitride; $SnF_2$, tin(II) fluoride; $Cu_2O$, copper(I) oxide; CuO, copper(II) oxide.

The $NH_4^+$ (ammonium) ion is named in compounds as if it were a simple metallic ion; thus, $NH_4I$ is ammonium iodide, and $(NH_4)_2S$, ammonium sulfide. In addition, there are a number of polyatomic negative ions that have the *ide* ending. Most notable among these is $OH^-$, hydroxide—others are $CN^-$, cyanide; $O_2^{2-}$, peroxide, *etc.* Thus, $Fe(OH)_2$ is iron(II) hydroxide.

### Nonmetal-Nonmetal Binary Compounds

Many binary compounds of hydrogen bear trivial (unsystematic) names—*e.g.*, $H_2O$ (water), $NH_3$ (ammonia), and $CH_4$ (methane). Most of the other compounds are systematically named by the use of prefixes denoting the number of a specific atom in the formula of the compound. These prefixes are:

| | | | | |
|---|---|---|---|---|
| 1 *mono* | 2 *di* | 3 *tri* | 4 *tetra* | 5 *penta* |
| 6 *hexa* | 7 *hepta* | 8 *octa* | 9 *nona* | 10 *deca* |

The name is written by first writing the numerical prefix needed for the less electronegative element followed directly by the entire name of this element. A space is left and then the numerical prefix for the more electronegative element is written directly followed by the root of the element's name and the ending, *ide.* Often the numerical prefix *mono-* is not stated, but is understood if no prefix is given. Examples are: $OF_2$, oxygen difluoride; $H_2S$, dihydrogen sulfide (but more commonly just "hydrogen sulfide" since the 2:1 H-to-S ratio is clear); $P_4O_{10}$, tetraphosphorus decoxide; $N_2O_4$, dinitrogen teroxide; CO, carbon monoxide; *etc.*

---

†The Roman numeral frequently is omitted when one oxidation state is very common as compared to the others. Thus, although silver(I), silver(II), and silver(III) are all known, silver(I) is the commonly occurring state. In naming silver(I) compounds the Roman numeral is usually dropped; $AgNO_3$ is just silver nitrate, not silver(I) nitrate.

Aqueous solutions of binary hydrogen compounds that behave as acids, sources of $H^+$, have special names. These solutions are named by indicating hydrogen with the prefix *hydro,* which is attached directly to the root of the name of the other element and followed by the ending *ic,* a space, and the word *acid.* Thus, the pure compound HCl (hydrogen chloride) when dissolved in water gives a solution which is called hydrochloric acid; an aqueous solution of $H_2S$ likewise may be called hydrosulfuric acid. The compound $PH_3$, however, does not act as a source of $H^+$ in aqueous solution and hence would **never** be called hydrophosphoric acid; it goes by a trivial name, phosphine.

## Polyatomic Ions and Derivatives

Table II gives a list of formulas and names of many of the more common polyatomic ions. Most compounds containing these ions are named by naming the positive ion along with a Roman numeral where appropriate, followed by a space and then the name of the polyatomic ion.

Compounds made up by protonation (adding protons, $H^+$) of one of these polyatomic anions until an electrically neutral formula results (anion, $SO_4^{2-} \rightarrow$ acid, $H_2SO_4$) are usually classified as acids. The prevalent nomenclature for the derived "parent" acid of a polyatomic anion is as follows: If the negative ion has a name with the *-ate* suffix, the parent acid is called "*elementic acid*"; if the anion ends in *-ite,* the parent acid is called "*elementous acid.*" Thus, $CO_3^{2-}$ (carbon**ate** ion) leads to $H_2CO_3$ (carbon**ic acid**); and $ClO^-$ (hypochlor**ite** ion) leads to HClO (hypochlor**ous acid**). The parent acids, formally derived in this way, are generally referred to as *oxyacids;* the central atom is **covalently** bonded to one or more oxygen atoms and to one or more *hydroxy* (—OH) groups. The general form of an oxyacid is $(HO)_mXO_n$; specifically, $H_2SO_4$ is $(HO)_2SO_2$ which, upon the loss of one $H^+$, yields $HSO_4^-$ (hydrogensulfate ion, or bisulfate ion); upon the loss of the second proton, $SO_4^{2-}$ (sulfate ion) results.

The following are examples of formulas and names of some compounds derived from polyatomic anions:

| | | |
|---|---|---|
| $NH_4H_2PO_4$ | ammonium dihydrogenphosphate | |
| $Ca(HCO_3)_2$ | calcium hydrogencarbonate | (or calcium bicarbonate) |
| $FePO_4$ | iron(III) phosphate | (old name, ferric phosphate) |
| $CoS_2O_3$ | cobalt(II) thiosulfate | (old name, cobaltous thiosulfate) |
| $H_2SO_3$ | sulfurous acid | |
| $HC_2H_3O_2$ | acetic acid | |
| $Y_2(C_2O_4)_3$ | yttrium oxalate | |

Knowing these aspects of formula writing and nomenclature should allow one to be able to write and understand a wide variety of formulas and names. The exercises test your knowledge of a "minimum vocabulary." Also, write the name for each one of the formulas written in the grid on p. 126.

## Exercises on Formulas and Chemical Nomenclature

21. The formula for oxalic acid is:

    (A) $C_2O_4^{2-}$ (B) $HC_2O_4$ (C) $H_2CO_3$ (D) $HC_2H_3O_2$
    (E) none of these

22. Perrhenic acid has the formula $HReO_4$; the formula for the perrhenate ion must be:

    (A) $HReO_4^-$ (B) $ReO_4^-$ (C) $ReO_4^{2-}$ (D) $ReO_3^-$ (E) none of these

23. Chromium(III) sulfite would have the formula:

    (A) $CrSO_4$ (B) $CrSO_3$ (C) $Cr_2S_3$ (D) $Cr_2(SO_4)_3$ (E) $Cr_2(SO_3)_3$

24. If nitrite ions can be converted into nitrate ions using hydrogen peroxide and forming water, the nitrite ion and hydrogen peroxide should react in a ___:___ ratio, respectively.

    (A) 2:1 (B) 1:2 (C) 1:1 (D) 3:1 (E) 1:3

25. An 0.10 M solution of barium acetate is ______ *M* in acetate ions.

    (A) 0.30 (B) 0.15 (C) 0.050 (D) 0.10 (E) 0.20

26. A compound in many fertilizers is calcium dihydrogenphosphate. This compound would contain ______ oxygen atoms for every calcium atom.

    (A) 4 (B) 8/3 (C) 8 (D) 3 (E) 2

27. Consider a 35-0-0 fertilizer that is a pure compound with the first figure being the percent nitrogen by weight. Which compound could this be?

    (A) potassium nitrate (B) sodium nitrite (C) ammonia
    (D) ammonium nitrite (E) ammonium nitrate

28. In the reaction between sulfite and hypochlorite ions, an oxygen atom is transferred from the latter to the former. What are the products?

    (A) chlorite and sulfide (B) chloride and sulfate
    (C) chloride and sulfur dioxide (D) chlorate and sulfate
    (E) none of these

29. Suppose the following reaction occurs:

    1 mole hydrogen selenide + 3/2 mole $O_2$ → 1 mole "product"

The "product" is:

(A) selenium dioxide (B) selenious acid (C) selenic acid
(D) selenium trioxide (E) hydroselenic acid

30. The reaction of 1 mole of calcium oxide with 1 mole of dinitrogen trioxide yields 1 mole of a single product. The product is:

(A) calcium nitride (B) calcium nitrate (C) calcium nitrite
(D) calcium hyponitrite (E) none of these

31. Aluminum and sulfur may react violently once reaction is initiated. The maximum number of moles of the resulting compound obtainable from one mole **each** of Al and S is:

(A) 1/3 mole (B) 1/2 mole (C) 2/3 mole (D) 1 mole (E) 2 moles

32. The **total** number of **moles** of chloride ion contained in 1.00 liter **each** of 0.10 *M* lithium chloride, 0.10 *M* copper(II) chloride, 0.10 *M* iron(III) chloride, and 0.10 *M* magnesium perchlorate is:

(A) 0.40 (B) 0.60 (C) 0.80 (D) 1.0 (E) none of these

33. Lithium nitride, copper(II) nitrite, and vanadium(IV) sulfate have the formulas, respectively:

(A) $LiN_2$, $Cu_3N_2$, $VSO_4$ (B) $Li_3N$, $Co(NO_2)_2$, $V(SO_4)_2$
(C) $Li_2N_3$, $Co_3N_2$, $V_2(SO_4)_3$ (D) $Li_3N$, $Cu(NO_2)_2$, $V(SO_4)_2$
(E) LiN, $Cu(NO_3)_2$, $V_4SO_4$

34. One mole of iron(II) sulfite contains ___?___ g of oxygen.

(A) 192 (B) 64 (C) 48 (D) 32 (E) 16

35. The compound calcium phosphide is predominantly ______ with the formula ______.

(A) ionic; $Ca(PO_3)_2$ (B) polar; $Ca_3(PO_4)_2$ (C) ionic; $Ca_5P_2$
(D) ionic; $Ca_3P_2$ (E) covalent; CaP

## GENERAL EXERCISES ON CHAPTER 5—GENERAL CONCEPTS OF BONDING AND CHEMICAL NOMENCLATURE

36. Mixing a solution containing one mole of sodium carbonate with a solution containing one mole of sulfuric acid yields a solution containing 1 mole of sodium sulfate and 1 mole of X. What is X?

(A) sodium bicarbonate (B) carbonic acid (C) sodium bisulfate
(D) sodium dihydrogencarbonate (E) none of these

## DRILL ON NOMENCLATURE

***Student's Name*** ______________________

Give correct names for each of the indicated compounds where 'X' represents the symbol of various different elements. You may be asked to turn in this sheet.

| X | $XO_2$ | $XO_3$ | $XF_4$ | $XF_6$ |
|---|---|---|---|---|
| S | X | | | |
| Xe | | | | |
| Mo | | | | |

| X | $X_2O_3$ | $XO_2$ | $X_2O_7$ | $XF_3$ |
|---|---|---|---|---|
| Cl | | | | |
| P | here, actually $X_4O_6$ | here, actually $X_2O_4$ | X | |
| Mn | | | | |

# DRILL ON FORMULAS AND NAMES OF COMPOUNDS

***Student's Name*** ______________________

In each block write the formula and name of the compound formed from the given cation and anion. You may be asked to turn in this sheet.

| Cation \ Anion | | $NO_3^-$ | $SO_4^{2-}$ | $PO_4^{3-}$ | $Fe(CN)_6^{4-}$ |
|---|---|---|---|---|---|
| $Ag^+$ | F | | | | |
| | N | | | | |
| $Mg^{2+}$ | F | | $MgSO_4$ | | |
| | N | | magnesium sulfate | | |
| $Fe^{2+}$ | F | | | | |
| | N | | | | |
| $Fe^{3+}$ | F | | | | |
| | N | | | | |
| $Th^{4+}$ | F | | | | |
| | N | | | | |
| $NH_4^+$ | F | | | | |
| | N | | | | |
| $H^+$ | F | | | | |
| | N | | | | |

37. Mixing solutions of potassium chromate and silver nitrate yields a precipitate that does not contain potassium or nitrate ions. The precipitate is likely:

    (A) $Ag_2CrO_4$ (B) $AgCrO_4$ (C) $Ag(CrO_4)_2$ (D) $Ag_3(CrO_4)_2$
    (E) none of these

38. Zinc perbromate would be expected to have the formula:

    (A) $ZnBr_2$ (B) $Zn(BrO)_2$ (C) $ZnBrO_3$ (D) $Zn(BrO_3)_2$
    (E) $Zn(BrO_4)_2$

39. One mole of lanthanum selenate would be expected to contain ______ moles of oxygen **atoms.**

    (A) 0 (B) 3 (C) 8 (D) 4 (E) 12

40. Which of the following acids could **not** yield an "acid anion" of the form $HX^-$?

    (A) sulfuric acid (B) hydroselenic acid (C) perchloric acid
    (D) sulfurous acid (E) carbonic acid

41. From a knowledge of the possible oxidation states of the elements, which of the following is a peroxide?

    (A) $TiO_2$ (B) $BaO_2$ (C) $SnO_2$ (D) $SO_2$ (E) $SiO_2$

42. Hydrated sodium thiosulfate (photographer's *hypo*) contains sodium and sulfur atoms in a ___:___ ratio, respectively.

    (A) 1:1 (B) 2:1 (C) 1:2 (D) 2:3 (E) none of these

43. Aluminum atoms and aluminum cations differ in all of the following respects EXCEPT:

    (A) radius (B) number of electrons (C) formula
    (D) number of protons (E) net electrical charge

44. A $Ga^{3+}$ ion would have the electronic configuration:

    (A) [Ar], $4s^2$, $3d^8$ (B) [Ar], $3d^{10}$ (C) [Kr]
    (D) $1s^2$, $2s^2$, $2p^6$, $3s^2$, $3p^6$, $4s^2$, $3d^8$ (E) [Ar]

45. In order to draw a proper Lewis formula for the sulfuric acid molecule, a **total** of ______ electrons must be shown.

    (A) 24 (B) 26 (C) 28 (D) 30 (E) 32

46. The **total** number of **valence** electrons in the bromate ion, $BrO_3^-$, is:

    (A) 25 (B) 26 (C) 27 (D) 32 (E) 20

47. The **total** number of **valence** electrons in the sulfite ion is:

    (A) 24 (B) 22 (C) 26 (D) 32 (E) 34

48. The bonds, in order of **decreasing** ionic character, are:

(A) F—F, F—H, Cs—F (B) F—F, Cs—F, H—H (C) Cs—F, F—F, H—F
(D) Cs—F, H—F, F—F (E) H—F, F—F, Cs—F

49. Which of the following compounds is the most ionic?

(A) ICl (B) $BiCl_3$ (C) $AsCl_3$ (D) $CCl_4$ (E) $OSCl_2$

50. Of the pairs of elements listed, which would form the most ionic bond?

(A) B, N (B) H, Cl (C) K, Cl (D) C, O (E) F, Cl

51. The bonds between sodium and phosphate in $Na_3PO_4$ would be:

(A) ionic (B) covalent (C) metallic (D) hybrid
(E) none of the above

52. **Within the phosphate ion** in the compound $Na_3PO_4$, the bonds would be:

(A) ionic and nonpolar (B) covalent and polar
(C) covalent and nonpolar (D) triple bonds (E) none of the above

53. In the Lewis structure for the $CS_2$ molecule, the number of **lone** (unshared) **pairs** on the central atom (carbon) is:

(A) 0 (B) 1 (C) 2 (D) 3 (E) 4

54. The bonds in the $SO_3^{2-}$ anion are best described as:

(A) covalent, polar, and single (B) ionic, polar, and single
(C) covalent, nonpolar, and single (D) covalent, polar, and double
(E) ionic, nonpolar, and single

55. The reaction of 1 mole of barium hydroxide with 2/3 mole of arsenic acid yields 2 moles of water and 1/3 mole of a white product. The product is:

(A) barium dihydrogenarsenate (B) barium hydrogenarsenate
(C) barium arsenate (D) barium arsenide (E) diarsenic trioxide

56. One mole of aluminum sulfate could theoretically yield ______ mole(s) of aluminum permanganate. Assume an excess of other nonaluminum-containing reagents.

(A) 1 (B) 2 (C) 3 (D) 1/2 (E) 1/3

57. How many moles of carbon dioxide could be obtained by completely decomposing one mole of lanthanum carbonate? The by-product is $La_2O_3$.

(A) 1 (B) 1-1/2 (C) 2 (D) 2-1/2 (E) none of these

58. Which of the following exhibits **both** ionic **and** covalent bonding?

(A) $BaSO_4$ (B) $NH_4Cl$ (C) $(NH_4)_2SO_3$ (D) $Ca(NO_3)_2$
(E) all of these

59. The formula of the chloride formed by element X that has the electronic configuration $1s^2, 2s^2, 2p^1$ would be:

    (A) XCl (B) $XCl_2$ (C) $XCl_3$ (D) $X_2Cl$ (E) $XCl_4$

60. The number of **shared pairs** of electrons in the butane molecule, $C_4H_{10}$, is:

    (A) 8 (B) 26 (C) 16 (D) 15 (E) 13

61. Which of the following pure compounds contains **only** covalent bonds?

    (A) NaH (B) $Mg(OH)_2$ (C) $CaC_2$ (D) $HNO_3$ (E)$NH_4NO_2$

62. Which of the following always shows violation of the octet rule?

    (A) C (B) O (C) P (D) H (E) S

63. The electronic structure of the $SO_3$ molecule is best represented as a resonance hybrid of ______ equivalent structures.

    (A) 6 (B) 5 (C) 4 (D) 3 (E) 2

64. Which of the following contains the **most polar bond**?

    (A) $BeF_2$ (B) $BF_3$ (C) $CF_4$ (D) $NF_3$ (E) $F_2$

65. Given the species: $N_2$, CO, $CN^-$, and $NO^+$. Which of the following statements is **false**?

    (A) The species are all diatomic.
    (B) The atoms in each species are joined by a triple bond.
    (C) The species are all linear.
    (D) The bond in each case is polar.
    (E) The species are all isoelectronic.

66. The **central atom** in the chlorite ion is surrounded by:

    (A) two bonding and two unshared pairs of electrons
    (B) three bonding and one unshared pair of electrons
    (C) one bonding and three unshared pairs of electrons
    (D) two **double bonds** and **no** unshared pairs of electrons
    (E) four bonding and four lone pairs of electrons

67. The number of lone pairs of electrons on the **central atom** in the sulfite ion is:

    (A) 1 (B) 2 (C) 3 (D) 0 (E) none of these

68. An oxide of chlorine showing chlorine in its maximum permitted oxidation number (oxidation state) is:

(A) $ClO$ (B) $ClO_4$ (C) $ClO_7$ (D) $Cl_2O_7$ (E) $Cl_2O_5$

69. The oxidation number of sulfur in sodium sulfite is:

    (A) 0 (B) 1 (C) 2 (D) 3 (E) 4

70. Any one of the three equivalent resonance forms for the carbonate ion would show around the carbon atom:

    (A) three single bonds and one lone pair of electrons
    (B) three single bonds and a double bond
    (C) two single bonds and a double bond
    (D) four single bonds
    (E) none of the above

71. Transition metal X forms $X^{3+}$ . $X^{3+}$ has four unpaired electrons. X could be:

    (A) titanium (B) iron (C) gallium (D) manganese (E) vanadium

72. Which of the following represents an isoelectronic sequence?

    (A) Be, $Mg^{2+}$, Ca, $Sr^{2+}$ (B) $Na^+$, $Mn^{2+}$, $Al^{3+}$, $Si^{4+}$ (C) N, O, F, Ne
    (D) $N^{3-}$, $O^{2-}$, Na, Mg (E) $Cl^-$, Ar, $Ca^{2+}$, $Ti^{4+}$

73. Which of the following does **not** have a noble gas configuration?

    (A) $S^{2-}$ (B) $Ar^+$ (C) $Al^{3+}$ (D) $Sb^{3-}$ (E) $Sc^{3+}$

74. Based on chromium, if 1 mole of potassium dichromate is completely converted into chromium(III) sulfate, how many moles of the latter would be obtained?

    (A) 1 (B) 2 (C) 1/2 (D) 1/3 (E) 3

75. Suppose the following reaction occurs:

    1 mole zinc acetate → 1 mole zinc oxide + 3 moles water + $x$ moles X

    What is "$x$ moles X"?

    (A) 4 moles $CO_2$ (B) 4 moles CO (C) 2 moles CO (D) 2 moles $CO_2$
    (E) 4 moles of carbon

76. Suppose that one mole of iodide ions reacts with three moles of hypochlorite ions to yield three moles of chloride ions and one mole of X. What is X?

    (A) iodide ion (B) hypoiodite ion (C) iodate ion (D) periodate ion
    (E) hydrogen iodide

77. Covalent bonding is **least** important in which of the following?

    (A) $H_2$ (B) HF (C) $H_2O$ (D) $CH_4$ (E) CsCl

78. The order of **decreasing** (left to right) nitrogen-oxygen bond distances in $NO_2^-$, $NO_3^-$, $NO^+$, and $NO_2^+$ is:

(A) $NO_3^-, NO_2^-, NO_2^+, NO^+$ (B) $NO_2^+, NO_2^-, NO^+, NO_3^-$
(C) $NO_2^-, NO_2^+, NO^+, NO_3^-$ (D) $NO^+, NO_2^+, NO_3^-, NO_2^-$
(E) $NO_2^+, NO_2^-, NO_3^-, NO^+$

79. Aluminum and sulfur combine to form a binary compound. In forming the compound, 0.020 mole of aluminum would require a minimum of ______ mole of S.

(A) 0.020 (B) 0.015 (C) 0.030 (D) 0.013 (E) 0.010

80. The formula for a binary compound of aluminum with phosphorus is:

(A) AlP (B) $AlPO_4$ (C) $Al_5P_3$ (D) $Al_3P$ (E) $Al(PO_3)_3$

81. The element gadolinium (Gd, $Z = 64$) forms a sulfide $Gd_2S_3$. The corresponding fluoride of gadolinium would be expected to have the formula:

(A) $Gd_2F_3$ (B) $GdF_2$ (C) $GdF_3$ (D) $Gd_3F_2$ (E) none of these

82. The Lewis structure for the nitrite ion can be written:

(A) $:\ddot{\underset{..}{O}}:\ddot{\underset{..}{N}}:\ddot{\underset{..}{O}}:^-$ (B) $:\ddot{O}::\ddot{N}:\ddot{\underset{..}{O}}:^-$ (C) $:\ddot{O}::\ddot{N}::\ddot{O}:^-$ (D) $:\ddot{O}::\ddot{\underset{.}{N}}:\ddot{\underset{..}{O}}:^-$

(E) $:\ddot{\underset{..}{O}}:N::\ddot{\underset{..}{O}}:^-$

83. An element X whose atoms have the electron configuration $1s^2$, $2s^2$, $2p^6$, $3s^2$ would be expected to form which chloride?

(A) $X_2Cl$ (B) XCl (C) $XCl_2$ (D) $XCl_3$ (E) $XCl_4$

84. The total number of valence electrons that must be shown in the electronic (Lewis) formula for magnesium fluoride is:

(A) 16 (B) 8 (C) 14 (D) 20 (E) 12

85. In the fourth period atom Z forms $Z^{2-}$ with the electronic configuration $ns^2$, $np^6$. Atom Z is:

(A) Kr (B) Br (C) Se (D) As (E) Ni

86. Which of the following compounds is the most ionic?

(A) ICl (B) $BCl_3$ (C) $AsCl_3$ (D) $CCl_4$ (E) $CaCl_2$

87. Of the pairs of elements listed, which would form the most ionic bond?

(A) B, N (B) H, Ca (C) F, Cl (D) C, O (E) B, Cl

88. The maximum number of $N_2$ molecules obtainable from one formula unit of ammonium nitrite is:

(A) one (B) two (C) $3.0 \times 10^{23}$ (D) $6.0 \times 10^{23}$ (E) none of these

89. Suppose you set out to prepare the compound aluminum pyrophosphate. The formula expected would be:

(A) $AlP_2O_7$ (B) $Al_4(P_2O_7)_3$ (C) $Al(P_2O_7)_3$ (D) $Al_2(P_2O_7)_3$
(E) $Al_2P_2O_7$

90. Ignition of 1 mole of $MgNH_4PO_4$ yields 1 mole of ammonia, 1/2 mole of water, and 1/2 mole of compound X. Compound X is:

    (A) magnesium dihydrogenphosphate (B) magnesium phosphate
    (C) magnesium pyrophosphate (D) magnesium phosphide
    (E) magnesium hydrogenphosphate

91. Which molecule has **nonpolar bonds**?

    (A) $CO_2$ (B) CO (C) $H_2O$ (D) HF (E) none of these

92. In which of the following compounds is the bonding **ionic**?

    (A) HCl (B) $CH_4$ (C) $H_2O$ (D) $H_2O_2$ (E) NaH

93. A certain mouthwash contains zinc chloride. If this compound is the only source of zinc and of chlorine in the solution, the molar concentration of zinc ion should be ______ that of the chloride ion.

    (A) the same as (B) twice (C) three times (D) one-half
    (E) one-third

94. A sack of fertilizer contains a compound containing ammonium and nitrate ions. The formula of the compound is:

    (A) $NH_3NO_3$ (B) $NH_4NO_3$ (C) $NH_4NO_2$ (D) $NH_3NO_2$
    (E) $(NH_4)_2NO_3$

95. A reaction occurs as follows (unbalanced):

    lithium nitride + excess water → lithium hydroxide + ammonia

    How many moles of ammonia could be obtained from one mole of lithium nitride?

    (A) 1 (B) 2 (C) 3 (D) 1/2 (E) none of these

96. Upon heating 1 mole of compound X, 1 mole of $O_2$ and 1 mole of barium nitrite are formed. Compound X is:

    (A) $BaNO_3$ (B) $Ba(NO_2)_2$ (C) $Ba(NO_3)_2$ (D) $Ba(NO_3)_3$
    (E) none of these

97. Which of the following compounds is the most ionic?

    (A) RbCl (B) $H_2O$ (C) $AlCl_3$ (D) $CS_2$ (E) $N_2O_5$

98. The *halide ions* are derived from atoms in Group(s) ______ of the periodic table.

    (A) IA (B) IIA (C) VIA (D) VIIA (E) IB-VIIB

99. How many **unpaired** electrons are in a $S^{=}$ ion?

    (A) 0 (B) 1 (C) 2 (D) 6 (E) 8

100. Which of the following is a correct Lewis symbol for a silicon atom?

    (A) $\cdot\dot{\underset{\cdot}{\mathrm{Si}}}\cdot$ (B) $:\ddot{\underset{\cdot\cdot}{\mathrm{Si}}}:$ (C) $Si^{4+}$ (D) Si. (E) $:\ddot{\underset{\cdot}{\mathrm{Si}}}.$

101. Based on phosphorus, the maximum number of moles of $P_4O_{10}$ that could be obtained from one mole of aluminum phosphate is:

(A) one (B) one-half (C) one-fourth (D) one-third
(E) none of these

102. Which of the following contains a **triple** bond?

(A) $SO_3^{2-}$ (B) $SO_2$ (C) $CN^-$ (D) $ClF_3$ (E) $NO_2^+$

103. The sulfur atom is the central atom in each of the following species. In which does the sulfur have **1** lone (unshared) pair of electrons?

(A) $SO_3$ (B) $SO_4^{2-}$ (C) $SO_2$ (D) $H_2S$ (E) $SCl_2$

104. In which of the following molecules does the central atom violate the "octet rule"?

(A) $OF_2$ (B) $SF_4$ (C) $PF_3$ (D) ClF (E) $SiF_4$

---

## ANSWERS TO CHAPTER 5 PROBLEMS

| | | | | | |
|---|---|---|---|---|---|
| 1. D | 19. D | 37. A | 55. C | 73. B | 91. E |
| 2. C | 20. B | 38. E | 56. B | 74. A | 92. E |
| 3. A | 21. E | 39. E | 57. E | 75. E | 93. D |
| 4. B | 22. B | 40. C | 58. E | 76. C | 94. B |
| 5. D | 23. E | 41. B | 59. C | 77. E | 95. A |
| 6. B | 24. C | 42. A | 60. E | 78. A | 96. C |
| 7. D | 25. E | 43. D | 61. D | 79. C | 97. A |
| 8. A | 26. C | 44. B | 62. D | 80. A | 98. D |
| 9. E | 27. E | 45. E | 63. D | 81. C | 99. A |
| 10. C | 28. B | 46. B | 64. A | 82. B | 100. A |
| 11. C | 29. B | 47. C | 65. D | 83. C | 101. C |
| 12. B | 30. C | 48. D | 66. A | 84. A | 102. C |
| 13. B | 31. A | 49. B | 67. A | 85. C | 103. C |
| 14. C | 32. B | 50. C | 68. D | 86. E | 104. B |
| 15. E | 33. D | 51. A | 69. E | 87. B | |
| 16. D | 34. C | 52. B | 70. C | 88. A | |
| 17. B | 35. D | 53. A | 71. D | 89. B | |
| 18. C | 36. B | 54. A | 72. E | 90. C | |

# 6
# *Bonding and Shapes of Covalent Species*

The purely ionic bond is nondirectional in nature—the force of attraction exerted by a positive ion on a negative ion is equal in all directions. Thus, in ionic solids, the distance between centers of neighboring cations and anions is the sum of the radii ($r_{cation} + r_{anion}$), and the number of anions surrounding each cation depends on the relative sizes of the ions (the radius ratio, $r_{cation}/r_{anion}$).

On the other hand, covalent bonds are distinctly directional. Two ways to account for this are:

1. the *valence-bond-hybrid-atomic-orbital* approach rearranges the valence orbitals on the central atom to form an **equal** number of *hybridized* orbitals which geometrically point to, and overlap with, orbitals on the various bonded atoms. For the carbon in carbon tetrachloride, $CCl_4$,

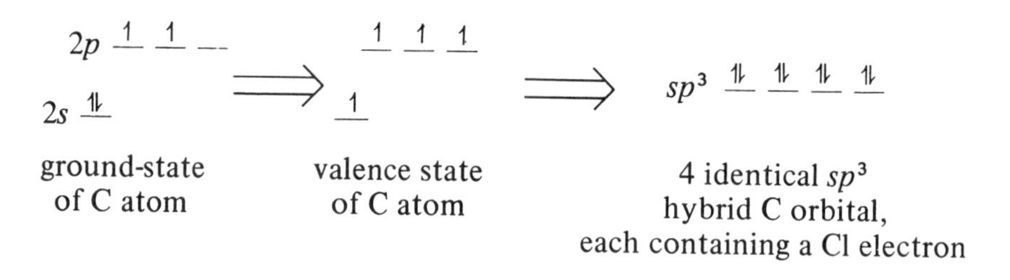


The result is that the *s* orbital has been mixed with the *p* carbon orbitals to give four new carbon orbitals that point to the corners of a

tetrahedron, resulting in the tetrahedrally shaped (dotted lines) molecule:

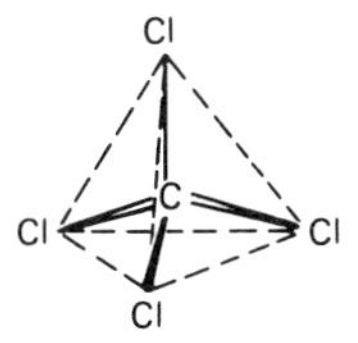


Each C–Cl bond is identical, having 1/4 *s* character and 3/4 *p* character. The most common hybridizations (where A = the central atom, X = a bonded atom, and E = a lone pair or a single electron) are:

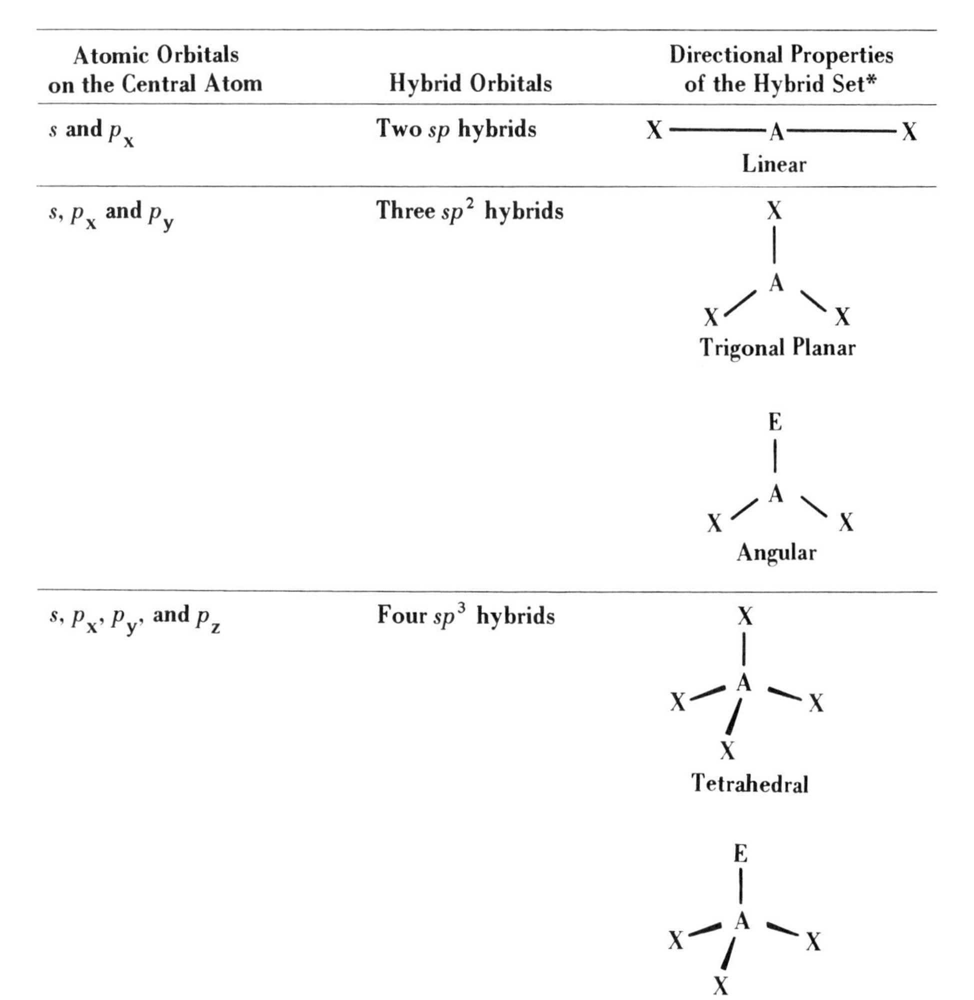

| Atomic Orbitals on the Central Atom | Hybrid Orbitals | Directional Properties of the Hybrid Set* |
|---|---|---|
| $s$ and $p_x$ | Two $sp$ hybrids | X——A——X<br>Linear |
| $s$, $p_x$ and $p_y$ | Three $sp^2$ hybrids | X, A, X, X<br>Trigonal Planar<br><br>E, A, X, X<br>Angular |
| $s$, $p_x$, $p_y$, and $p_z$ | Four $sp^3$ hybrids | X, X, A, X, X<br>Tetrahedral<br><br>E, X, A, X, X<br>Trigonal Pyramidal |

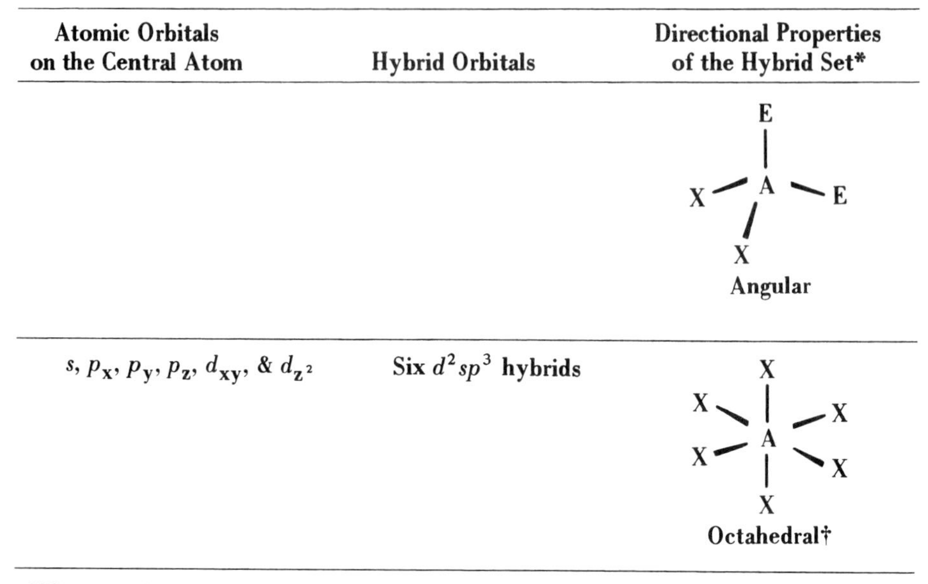

| Atomic Orbitals on the Central Atom | Hybrid Orbitals | Directional Properties of the Hybrid Set* |
|---|---|---|
| | | Angular |
| $s, p_x, p_y, p_z, d_{xy}, \& d_{z^2}$ | Six $d^2sp^3$ hybrids | Octahedral† |

*When specifying the shape of a species, only the positions of atoms (here A and X) are indicated. Lone pairs of electrons, though thought to be present in these positions, are not part of the shape of an atom or ion. †Additional geometries derived from this hybrid set can be obtained by substitution of one or more E's for X's.

2. the *Valence-Shell-Electron-Pair-Repulsion* (or *VSEPR*) theory considers the valence pairs about the central atom as localized into "domains" of negative charge. These domains consist of either a lone (unshared) pair of electrons or one or more bond (shared) pairs of electrons. Since each domain is negatively charged, they mutually repel one another. These domains spread themselves apart so that the repulsion between them is a minimum, leading to the most stable (lowest energy) arrangement. The shapes that result for the various $AX_mE_n$ combinations are identical to those predicted by the valence bond approach (paragraph (1)).

Another theory of the formation of covalent bonds is the *molecular orbital* theory which will not be discussed here. If your instructor takes this up, carefully read the section in your text.

As atoms approach one another close enough to begin to form a bond, orbitals on each atom overlap, forming new valence orbitals that can accommodate one electron pair. Two fundamental types of overlap are possible:

1. Head-on overlaps, leading to *sigma* shared pairs ($\sigma$-bond) in which the electron cloud lies along the line joining the two atoms, *e.g.*

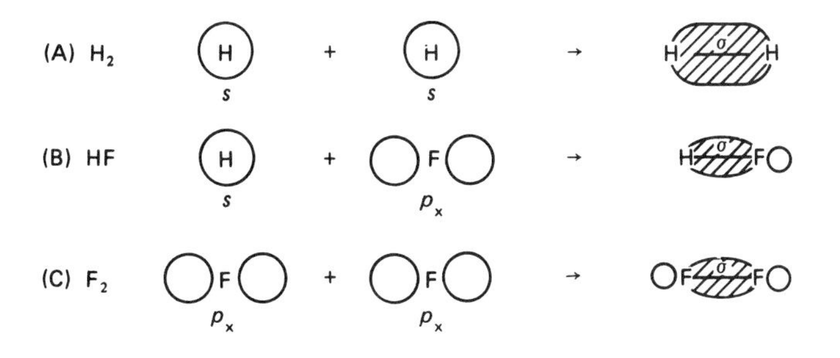


2. Side-wise overlaps due to overlap of $p$ orbitals (extending in the y and z directions) as the atoms are drawn together by $\sigma$-bond formation. Side-wise overlaps lead to charge distributions above and below (or, behind and in front) of the line joining the two atoms, the x axis:

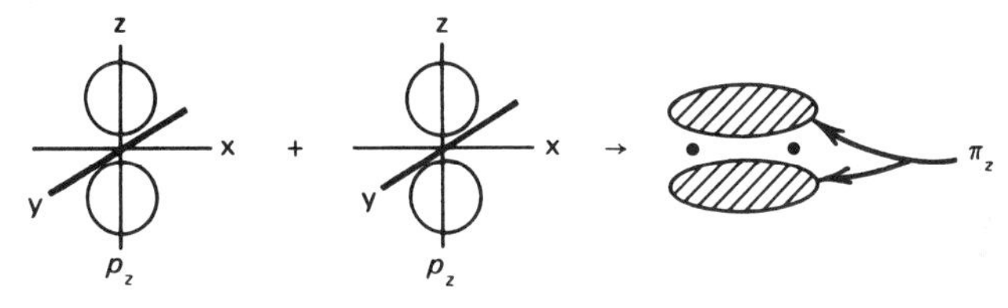


and

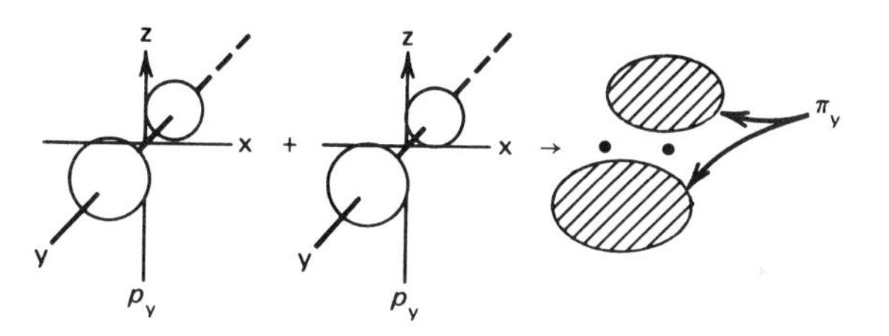


Single bonds result from head-on $\sigma$-overlaps. Multiple bonding results when sidewise $\pi$-overlaps occur in addition to a $\sigma$-overlap. The net result is that a single bond consists of only a $\sigma$-bond; a double bond is a $\sigma$- and a $\pi$-bond; a triple bond is a $\sigma$-bond and two $\pi$-bonds.

All of the bonds discussed so far are localized bonds, meaning the electron pair making up the bond is confined to just two atoms. There are numerous instances where the $\pi$-bond is de-localized, *i.e.* spread out over several atoms. This occurs when more than one equivalent Lewis structure can be written for a molecule or ion. Thus for $SO_3$, there are three equivalent (*resonance*) forms:

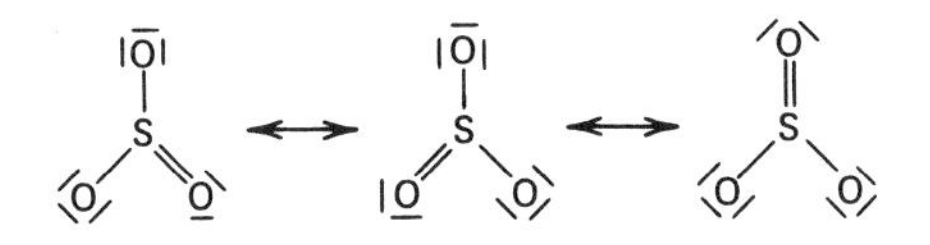

The framework of the molecule is made up of three sigma bonds ($sp^2$ hybridization), with the sidewise overlap of the $p_z$ orbitals on S and each O to form one de-localized $\pi$-bond. Since there are 3 equivalent linkages sharing one $\pi$-bond, the average linkage between S and O consists of 1 $\sigma$-bond and 1/3 of a $\pi$-bond, or 1-1/3 bonds. This is called the bond order and may be obtained readily by:

$$\text{Bond order} = \frac{\text{total number of electron pairs in equivalent linkages}}{\text{number of equivalent linkages}}$$

In a single bond, the bond order is 1; in a double bond, 2; in a triple bond, 3; in the above case of $SO_3$, 4/3 = 1-1/3. The larger the bond order, the shorter the bond distance and the stronger the bond.

An individual bond is polar if the atoms that are joined together differ in electronegativity. On the other hand, a **molecule** will be polar if the centers of positive and negative charge do not coincide, *i.e.* if the electron domains about the central atom are not equivalent or are located unsymmetrically. In simple cases, where only one type of atom is bound to the central atom and where there is an octet of electrons around the central atom, the presence of one or more lone pairs (E) will destroy the symmetry and the molecule will be *polar.* A polar molecule has a non-zero *dipole moment; non-polar molecules* have a zero dipole moment.

## Example Problem Involving Theories of Bonding and Shapes of Covalent Species

### 6-A. Bonding and Shape of a Molecule

The hybrid orbitals used by sulfur to accommodate the $\sigma$-bonds and the lone pairs in an $SO_2$ molecule are called __(1)__ hybrids. Using the A/X/E classification, $SO_2$ would be classified as __(2)__. This molecule has a __(3)__ shape and will have a dipole moment of __(4)__. The molecule contains __(5)__ $\sigma$-bonds and __(6)__ $\pi$-bonds. The respective blanks should be filled by the terms:

| | (1) | (2) | (3) | (4) | (5) | (6) |
|---|---|---|---|---|---|---|
| (A) | $sp^2$ | $AX_2E$ | Trigonal planar | >0 | 2 | 2 |
| (B) | $sp^2$ | $AX_2E$ | Angular | >0 | 2 | 1 |
| (C) | $sp$ | $AX_2$ | Linear | 0 | 2 | 0 |
| (D) | $sp^3$ | $AX_2E_2$ | Angular | >0 | 2 | 0 |
| (E) | $sp^2$ | $AX_2$ | Angular | 0 | 2 | 1 |

#### *Solution*

It is best to write the Lewis formula first, since that plus a knowledge of the terminology permits most of the questions to be answered. Thus, $SO_2$ is an

18-valence electron molecule (3 Group VIA atoms) and, distributing these to give octets of electrons around each atom, we find two equivalent structures:

$$:\ddot{O}::\ddot{S}:\ddot{\underset{..}{O}}: \quad \text{and} \quad :\ddot{\underset{..}{O}}:\ddot{S}::\ddot{O}:$$

Both Lewis formulas show two bond domains and one lone-pair domain about the sulfur–thus, the classification is $AX_2E$. These three domains, if equivalent, would be directed toward the corners of an equilateral triangle, $\overset{|}{S}$ /\ . By using any two of these three domains as bonding domains, this gives $\ddot{S}$ /\ O O. The shape, however, is determined by the **positions of atoms** (excluding any lone pairs); the shape of the $SO_2$ molecule is angular (rather than linear).

There are two $\sigma$ shared pairs and one unshared pair of electrons in the plane of the molecule, requiring three sulfur orbitals. Taken in order of increasing energy from the valence shell (3*s*, 3*p*, 3*d*), the three pairs require the 3*s* and two 3*p* orbitals on sulfur. Thus, the hybridization would be $sp^2$; since any two *p* orbitals define a plane and are not collinear, this molecule would (again) be angular by the valence-bond approach.

The $\pi$-pair of electrons belongs to the molecule as a whole, the electronic structure being half-way between the extremes shown by the two resonance forms (or the S—O bond order would be 1-1/2). Since the electronegativity of oxygen is greater than that of sulfur, the bonding electrons are drawn towards the oxygens. This leaves small fractional negative charges on the oxygens, a fractional positive charge on the sulfur,

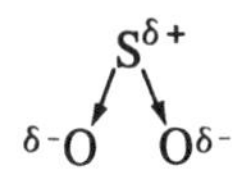


The center of positive charge is on the S atom; the center of negative charge is between the two oxygens, below the sulfur. The centers of positive and negative charge do not coincide and, hence, the molecule would be polar and have a non-zero dipole moment. **[B]**

## Exercises Involving Theories of Bonding and Shapes of Covalent Species

1. Which of the following are polar molecules?

$$PH_3 \quad OF_2 \quad HF \quad SO_3$$

(A) all except $SO_3$ (B) only HF (C) only HF and $OF_2$
(D) none of these (E) all of these

2. How many $\sigma$- and how many $\pi$-electron **pairs** are in a $C_6H_6$ molecule if the carbons are bonded in a hexagonal (six-membered) ring and one hydrogen is bonded to each carbon?

| | $\sigma$ | $\pi$ |
|---|---|---|
| (A) | 6 | 6 |
| (B) | 3 | 6 |
| (C) | 6 | 3 |
| (D) | 9 | 6 |
| (E) | 12 | 3 |

3. What would be the hybrid combination of sulfur orbitals used in the $SF_4$ molecule?

   (A) $sp$ (B) $sp^2$ (C) $sp^3$ (D) $sp^3d$ (E) $sp^3d^2$

4. If the triatomic molecule $XY_2$ is non-polar and has **each** Y atom bonded to the X atom by a $\sigma$- and a $\pi$-bond, what hybrid combination of X orbitals are used?

   (A) $sp$ (B) $sp^2$ (C) $sp^3$ (D) $sp^3d$ (E) $sp^3d^2$

5. The molecule $C_7H_{12}$ consists of a chain of bonded carbon atoms. How many $\sigma$- and how many $\pi$-bonds are present in the molecule?

| | $\sigma$ | $\pi$ |
|---|---|---|
| (A) | 19 | 1 |
| (B) | 18 | 2 |
| (C) | 7 | 3 |
| (D) | 12 | 7 |
| (E) | 19 | 3 |

6. What is the shape of the $NF_3$ molecule?

   (A) trigonal planar (B) tetrahedral (C) linear (D) angular
   (E) trigonal pyramidal

7. The molecule $AX_3$ is polar and obeys the octet rule; therefore

   (A) the central atom (A) has no lone pairs
   (B) the central atom (A) has one lone pair
   (C) the central atom (A) has two lone pairs
   (D) the central atom (A) has three lone pairs
   (E) the central atom (A) has four bond pairs

8. Which of the following molecules has a zero dipole moment?

   (A) HCl (B) $O_2$ (C) NO (D) ClF (E) HF

9. Which of the following species is planar?

   (A) $NH_3$ (B) $H_3O^+$ (C) $SO_3^{2-}$ (D) $PF_3$ (E) $NO_3^-$

10. The shape of the $PH_4^+$ ion is:

   (A) trigonal planar (B) linear (C) angular (D) trigonal pyramidal
   (E) tetrahedral

11. The shape of the phosphate ion is:

    (A) linear (B) angular (C) triangular (D) tetrahedral
    (E) none of the above

12. The shape of the carbonate ion is:

    (A) angular (B) tetrahedral (C) octahedral (D) trigonal pyramidal
    (E) trigonal planar

13. Which one of the following molecules has a dipole moment greater than zero?

    (A) $CCl_4$ (B) $CO_2$ (C) $N_2$ (D) $H_2S$ (E) all of these

14. Which of the following is linear?

    (A) $SO_2$ (B) $H_2O$ (C) $NO_2^-$ (D) HCN (E) none of these

15. Which of the following molecules would be tetrahedral?

    (A) $N_2O$ (B) $O_3$ (C) $SiH_4$ (D) $BCl_3$ (E) $SF_6$

16. Which of the following molecules would be trigonal planar?

    (A) $N_2O$ (B) $O_3$ (C) $SiH_4$ (D) $BCl_3$ (E) $NH_3$

17. What hybrid orbitals are used in bonding by S in $SO_3^{2-}$?

    (A) $sp^3d^2$ (B) $sp$ (C) $sp^2$ (D) $sp^3$ (E) none

18. How many $\pi$-bonds are there in $C_2Cl_2$ (Cl—C—C—Cl)?

    (A) 0 (B) 1 (C) 2 (D) 3 (E) 7

19. What hybrid orbitals are used in bonding by S in the sulfate ion?

    (A) $s$ (B) $sp^2$ (C) $sp$ (D) $d^2sp^3$ (E) none of these

20. Which of the following molecules is nonpolar, but contains polar bonds?

    (A) $H_2$ (B) $SO_2$ (C) $PCl_3$ (D) $SO_3$ (E) $H_2O$

21. A central atom (A) in a molecule is described as having $dsp^3$ hybrid orbitals. If each hybrid orbital is used to bond an atom X to the central atom, then the molecule must have the formula:

    (A) $AX_2$ (B) $AX_3$ (C) $AX_4$ (D) $AX_5$ (E) $AX_6$

22. The compound ammonium nitrate:

    (A) exhibits **only** ionic bonding (B) exhibits **only** covalent bonding
    (C) exhibits **both** ionic and covalent bonding
    (D) has the formula $NH_3NO_3$ (E) has the formula $NH_4NO_2$

23. In which of the following compounds is the bonding most ionic?

    (A) $AlCl_3$ (B) HCl (C) $SO_2$ (D) $CO_2$ (E) KCl

24. Which of the following molecules is a polar molecule?

(A) $BeCl_2$ (B) $BF_3$ (C) $CCl_4$ (D) $SF_6$ (E) $PCl_3$

25. The structure of $ClO_3^-$ is best described as being

    (A) tetrahedral (B) pyramidal (C) angular (D) trigonal bipyramidal (E) triangular planar

26. Which of the following is best represented by **three** resonance forms?

    (A) $N_2$ (B) $ClO_3^-$ (C) $NCl_3$ (D) $H_3O^+$ (E) $CO_3^{2-}$

27. Which of the following is a nonpolar molecule having one or more polar bonds?

    (A) $CO_2$ (B) $H_2$ (C) HCl (D) $SO_2$ (E) $PH_3$

28. In which of the following molecules is (are) the bond(s) the most polar?

    (A) $OF_2$ (B) $SF_4$ (C) $PF_3$ (D) ClF (E) $SiF_4$

29. According to modern bonding theory the carbon monoxide molecule, CO, contains ______ sigma ($\sigma$) and ______ pi ($\pi$) bond(s).

    (A) 3, 1 (B) 1, 2 (C) 2, 2 (D) 2, 1 (E) 3, 2

30. Which of the following would be expected to have the shortest nitrogen-oxygen bond length?

    (A) $NO_3^-$ (B) $NO_2^+$ (C) $NO_2^-$ (D) $NO^+$ (E) $HNO_3$, ($HONO_2$)

31. A $CN_2^{2-}$ ion has a structure that is best described as:

    (A) angular (B) linear (C) tetrahedral (D) pyramidal (E) planar

---

## ANSWERS TO CHAPTER 6 PROBLEMS

| | | | | | |
|---|---|---|---|---|---|
| 1. A | 7. B | 13. D | 19. E | 25. B | 31. B |
| 2. E | 8. B | 14. D | 20. D | 26. E | |
| 3. D | 9. E | 15. C | 21. D | 27. A | |
| 4. A | 10. E | 16. D | 22. C | 28. E | |
| 5. B | 11. D | 17. D | 23. E | 29. B | |
| 6. E | 12. E | 18. C | 24. E | 30. D | |

## DRILL ON BONDING AND SHAPES

***Student's Name*** ______________________

Draw the Lewis Formula for each of the given species and on the basis of the Lewis formula fill in the remaining blanks. You may be asked to turn in this sheet.

| Formula | Lewis Structure | Shape of Species | Hybrid Set of Orbitals Used by Central Atom | Polar Molecule Yes/No |
|---|---|---|---|---|
| $PCl_3$ | | | | |
| HCN | | | | |
| $ClO_2$ | | | | |
| $ClF_3$ | | | | |

# 7
# Solution Reactions—1

Reactions that occur in aqueous solution are an extremely important part of chemistry. Many of these can be classified into three different reaction types: proton transfer (*neutralization* or acid-base), electron transfer (*oxidation-reduction*, commonly shortened to *redox*), and precipitation (*metathesis*). To understand and deal with these quantitatively, it is necessary to write balanced equations. This, in turn, requires that we know the reactants and products and that we can write formulas for these. Equations where formulas of compounds are written for each reactant and each product are known as *total equations*. A total equation is particularly convenient when dealing with stoichiometric calculations. The total equation pays no heed to the actual chemical species in solution (*e.g.* an aqueous solution of HCl contains *no* HCl molecules; the solute species are $H_3O^+$ and $Cl^-$). A second type of chemical equation, the *net ionic equation*, is used to show only the actual molecules and ions that undergo chemical change. It is thus necessary to know something of the *electrolytic nature* of compounds in solution (the extent to which ions are present in aqueous solution).

## THE ELECTROLYTIC NATURE OF COMPOUNDS

In many instances the formula of a substance does represent the actual species that is present when this substance is dissolved. Frequently, however, the formula does not clearly indicate the species in the pure substance nor the species that are present in solution. Finally, in some cases the species in the pure substance reacts with the solvent to form different species that are in solution. To understand reactions in aqueous solution, one needs to know specifically the actual species that react.

*Non-Electrolytes, Strong Electrolytes and Weak Electrolytes.* The substance urea, $CO(NH_2)_2$, exists in the solid as $CO(NH_2)_2$ molecules, and in aqueous solution the only solute species are $CO(NH_2)_2$ molecules. The resulting solution is a non-conductor of electricity. The solute, urea, is categorized as a *non-electrolyte*. A compound which when dissolved in water exists solely as solute molecules is said to be a *non-electrolyte*.

In the case of sodium chloride there are *no* NaCl molecules in the solid nor in aqueous NaCl solution. Sodium chloride exists both in the solid and in solution solely as $Na^+$ and $Cl^-$; in solution, these ions are set free and act independently. The presence of these ions makes the solution a good conductor of electricity—a *strong electrolyte*. Hydrogen chloride exists as HCl molecules and is a gas at room temperature and pressure. In the presence of water it reacts completely to yield $H_3O^+$ and $Cl^-$ ions.*

Compounds that exist exclusively as 2 or more ions in solution are said to be *strong electrolytes*. A compound such as sodium perchlorate, $NaClO_4$, is also a strong electrolyte since in solution all of the sodium exists as $Na^+$ and all of the perchlorate exists as $ClO_4^-$ (the oxygen and chlorine, however, do not exist as individual ions, but are tightly bound together by covalent bonds). In general, the strong electrolytes consist of the ionic compounds plus a few covalent compounds of hydrogen that react *totally* with water to give $H_3O^+$ and an anion. These are the common, important acids, HCl, HBr, HI, $HClO_4$, $HNO_3$, and $H_2SO_4$. It should be carefully noted that in $H_2SO_4$ **only** one of the hydrogens reacts completely with water; thus in solution one would find $H_3O^+$ and $HSO_4^-$ ions as the major solute species.

There are many molecular substances which when dissolved in water slightly interact with the water so that the solution contains a few solute ions but mostly solute molecules. These solutions are poor conductors of electricity since there are relatively few ions present.

Thus, ammonia dissolves readily in water to yield a solution containing for the most part ammonia molecules; a few of the ammonia molecules interact with the water to yield ammonium ions and hydroxide ions. Such substances are known as *weak electrolytes*.

In summary, dissolved substances are divided into three categories:

(a) Non-electrolytes—most covalent compounds other than covalent compounds of hydrogen that are acids or bases (see discussion in the next section). These include sugars, alcohols, gases such as $H_2$, $O_2$, and $N_2$; even water itself is essentially a non-electrolyte.
(b) Weak electrolytes—ammonia and most acids
(c) Strong electrolytes—ionic compounds (salts†) and the six strong acids, HCl, HBr, HI, $HNO_3$, $HClO_4$, and $H_2SO_4$ (first hydrogen only).

---

*$H^+$ is commonly used as an abbreviation for $H_3O^+$. $H^+$, as a species, does not exist in aqueous solution. It is generally understood that when ions are written without subscripts, the *aqueous* ion is implied; thus $Cl^-$ refers to $Cl^-_{(aq)}$.

†As used in this book, the term "*salt*" will refer to those crystalline, ionic compounds that contain positive ions **other than $H^+$** and negative ions **other than $OH^-$, $H^-$ and $O^{2-}$**.

## DRILL ON ELECTROLYTES

***Student's Name*** ______________________

For each of the following compounds, (A) write the formula of the compound, (B) categorize it as to whether it would be a weak acid, strong acid, weak base, strong base, salt, or non-electrolyte in solution, and (C) indicate the principal solute species in an aqueous solution of the compound. In some cases in (B) more than one category may be needed. You may be asked to turn in this sheet.

| Compound | (A) Formula | (B) Nature | (C) Principal Solute Species |
|---|---|---|---|
| Sulfuric acid | | | |
| Colbalt (II) perchlorate | | | |
| Ammonia | | | |
| Nitrous acid | | | |
| Ammonium chloride | | | |
| | $KH_2PO_4$ | | |
| Methanol | | | |

*Total and Net Ionic Equations.* A *total equation* uses complete formulas of substances as reactants and complete formulas of substances that one might reasonably imagine as products of the reaction. Thus if a solution of potassium bisulfate is added to a solution of sodium hydroxide, water is formed and some mixture of sodium and/or potassium sulfate:

$$\text{TOTAL:} \qquad KHSO_{4(aq)} + NaOH_{(aq)} \rightarrow H_2O + KNaSO_{4(aq)}$$

The 'aq' written after formulas of some of the compounds indicates that potassium bisulfate and sodium hydroxide, in whatever form they may exist in solution, were used. This type of equation is particularly useful when making stoichiometric calculations.

The *net ionic* equation takes advantage of our knowledge of the electrolytic nature of the solutes and allows us to focus on the predominant reactive species in the solution. This is done by including in the equation **only** species that undergo chemical change. In the above case, one would predict that in solution potassium bisulfate exists as $K^+$ and $HSO_4^-$ ions, sodium hydroxide as $Na^+$ and $OH^-$, $KNaSO_4$ as $K^+$, $Na^+$ and $SO_4^{2-}$ ions, and that water is a non-electrolyte existing as $H_2O$ molecules. An expanded equation can now be written:

$$K^+ + HSO_4^- + Na^+ + OH^- \rightarrow H_2O + K^+ + Na^+ + SO_4^{2-}$$

It is readily noted that the sodium and potassium ions undergo no change and so these are cancelled from both sides of the equation. This leaves our reaction as merely,

$$\text{NET:} \qquad HSO_4^- + OH^- \rightarrow H_2O + SO_4^{2-}$$

This *net ionic* equation now shows only the actual species (ions and molecules) that are involved in the reaction. It also implies that the same reaction would occur with any soluble bisulfate compound and any soluble hydroxide compound.

In writing net ionic equations, the following guidelines should be observed:

(A) each reactant and product should be written so as to indicate the predominant chemical species representing that reactant or product in solution;
(B) entire formulas of the compound are written for substances **not** in solution, *e.g.* HCl gas bubbled into a reaction mixture or the formation of a precipitate of AgCl;
(C) any species that remains unchanged throughout the course of the reaction is cancelled from both sides of the equation.

## ACIDS, BASES AND ACID-BASE REACTIONS

A particularly useful definition of acids and bases for reactions in aqueous solution is that of Bronsted:

*acid*—any species that when added to water increases the hydronium ($H_3O^+$) ion concentration; ACIDS ARE PROTON ($H^+$) DONORS.
*base*—any species that when added to water increases the hydroxide ($OH^-$) ion concentration; BASES ARE PROTON ACCEPTORS.
*acid-base reaction*—the transfer of a proton from the best acid present to the best base present.

The student should remember the 6 common strong acids, hydrogen chloride, hydrogen bromide, hydrogen iodide, perchloric acid, sulfuric acid (first proton only) and nitric acid; the chances are very good that any other acid that you find in the laboratory will be a weak acid. In solution, highly charged metal cations (+3, +4 and even small cations with a +2 charge) are tightly bound to a number of water molecules by means of the oxygen electrons. The protons on the bound water molecules are repelled causing these species to be weak acids. The important strong bases to be remembered are the soluble hydroxides, oxides, and hydrides (those of the alkali metals and barium); other metal oxides and hydroxides are bases, but most of them are not particularly soluble in water. These slightly soluble bases will react with acids.

The products of an acid-base reaction are a new acid and a new base. In general the equation for an acid base reaction fits the form:

$$\text{ACID}_{(\text{reactant})} + \text{BASE}_{(\text{reactant})} \rightarrow \text{ACID}_{(\text{product})} + \text{BASE}_{(\text{product})}$$

If the reaction is to go to any appreciable extent, the product acid and product base must be weaker electrolytes (a weaker acid and a weaker base) than were the reactants. The reactant acid and the product base (formed by loss of a proton) are called a *conjugate* acid-base pair. The loss of a proton from the hydronium ion results in formation of a water molecule. $H_3O^+$ and $H_2O$ are a conjugate acid-base pair. Loss of a proton by water (hence, now acting as an acid) results in formation of $OH^-$. Water and hydroxide ion are another conjugate acid-base pair (differ by a proton).

Acids and bases have varying strengths. Relative strengths of acids and bases are frequently shown in an 'acid-base table' (*cf.* Table I). Acids are listed in the left-hand column in order of decreasing acid strength. Each respective conjugate base is placed across from the acid in the right-hand column. Base strength increases as one goes down the column. Thus, the stronger the acid, the weaker its conjugate base. The strong acids (species which in aqueous solution have completely reacted with $H_2O$ to form $H_3O^+$ and their conjugate base) occur above the hydronium ion in the table. In aqueous solution all of the strong acids are 'leveled' to the strength of the hydronium ion, the acid species actually in solution. The conjugate bases of the strong acids are weaker bases than in water itself; hence base properties of these anions are not observable in aqueous solution. The region including these "non-bases" is shaded on Table I. Likewise, the conjugate acids related to strong bases are weaker acids than water and in aqueous solution have no acid properties (also shaded in Table I).

The acid-base table helps predict products and the extent to which an acid-base reaction will proceed. If a solution of HF is added to a KOH solution, it should be noted that HF is a weak acid, meaning that the main solute species present will be HF molecules; but KOH is an ionic compound, meaning that the solute species will be $K^+$ and $OH^-$. The best acid will be HF and the best base $OH^-$. The $K^+$ is merely a 'spectator' and is not shown in the net ionic equation. The acid-base reaction then is the transfer of a proton from the acid HF (yielding $F^-$) to the base $OH^-$ (yielding a water molecule). The equation would be:

$$HF_{(aq)} + OH^- \rightarrow F^- + H_2O_{(\ell)}$$

Note that the product acid, $H_2O$, and the product base, $F^-$, are weaker than the reacting acid and base. This means that the reaction proceeds to a considerable extent. If we had used a KF solution as the base and a solution of $H_2S$ as the acid, transfer of a proton from the acid to the base would give us $HS^-$ and HF as products. $HS^-$ is a stronger base than is $F^-$ and HF is a stronger acid than $H_2S$; this reaction would *not* proceed to an appreciable extent.

## Example Problems Involving Aqueous Acid-Base Reactions

### 7-A. Equation Writing

Write the *total* and *net ionic* equation for the "neutralization" reaction occurring "when solutions of acetic acid and sodium hydroxide are mixed."

The *total* equation involves a 1:1 molar combination since the acid can furnish only one $H^+$ and the base only one $OH^-$ (per formula unit):

$$HC_2H_3O_{2(aq)} + NaOH_{(aq)} \rightarrow HOH + NaC_2H_3O_{2(aq)}$$

It is observed that only $Na^+$ is in the same chemical form on both the reactant and product "sides" of the equation—since it has undergone no chemical change, it may be canceled to give the net ionic equation (where all species are in aqueous medium):

$$HC_2H_3O_2 + OH^- \rightarrow C_2H_3O_2^- + H_2O$$

**TABLE I**
**RELATIVE STRENGTHS OF ACIDS AND BASES**

| Conjugate Acid | | Conjugate Base | |
|---|---|---|---|
| Name | Formula | Name | Formula |
| Perchloric acid$_{(\ell)}$ | $HClO_4$ | Perchlorate ion | $ClO_4^-$ |
| Hydrogen iodide$_{(g)}$ | $HI$ | Iodide ion | $I^-$ |
| Hydrogen bromide$_{(g)}$ | $HBr$ | Bromide ion | $Br^-$ |
| Hydrogen chloride$_{(g)}$ | $HCl$ | Chloride ion | $Cl^-$ |
| Sulfuric acid$_{(\ell)}$ | $H_2SO_4$ | Bisulfate ion* | $HSO_4^-$ |
| Nitric acid$_{(\ell)}$ | $HNO_3$ | Nitrate ion | $NO_3^-$ |
| Hydronium ion | $H_3^+O$ or ($H^+_{(aq)}$) | Water | $H_2O$ |
| Bisulfate ion* | $HSO_4^-$ | Sulfate ion | $SO_4^{2-}$ |
| Sulfurous acid | $H_2SO_3$ | Bisulfite ion* | $HSO_3^-$ |
| Phosphoric acid | $H_3PO_4$ | Dihydrogenphosphate ion | $H_2PO_4^-$ |
| Hexaquoiron(III) ion | $Fe(H_2O)_6^{3+}$ | | $Fe(H_2O)_5OH^{2+}$ |
| Hydrofluoric acid | $HF$ | Fluoride ion | $F^-$ |
| Nitrous acid | $HNO_2$ | Nitrite ion | $NO_2^-$ |
| Acetic acid | $HC_2H_3O_2$ | Acetate ion | $C_2H_3O_2^-$ |
| Hexaquoaluminum ion | $Al(H_2O)_6^{3+}$ | | $Al(H_2O)_5OH^{2+}$ |
| Carbonic acid | $H_2CO_3$ ($CO_{2(aq)}$) | Bicarbonate ion* | $HCO_3^-$ |
| Bisulfite ion* | $HSO_3^-$ | Sulfite ion | $SO_3^{2-}$ |
| Hydrogen sulfide | $H_2S$ | Hydrosulfide ion | $HS^-$ |
| Dihydrogenphosphate ion | $H_2PO_4^-$ | Monohydrogenphosphate ion | $HPO_4^{2-}$ |

Acid Strength (increasing upward) — Increasing

Decreasing

| Acid | | Conjugate base | |
|---|---|---|---|
| Phenolphthalein | HPhth | Phenolphthalein anion | $Phth^-$ |
| Ammonium ion | $NH_4^+$ | Ammonia | $NH_3$ |
| Hydrocyanic acid | HCN | Cyanide ion | $CN^-$ |
| Hexaquoiron(II) ion | $Fe(H_2O)_6^{2+}$ | | $Fe(H_2O)_5OH^+$ |
| Bicarbonate ion* | $HCO_3^-$ | Carbonate ion | $CO_3^{2-}$ |
| Hydrogen peroxide | $H_2O_2$ | Hydroperoxide ion | $HO_2^-$ |
| Monohydrogenphosphate ion | $HPO_4^{2-}$ | Phosphate ion | $PO_4^{3-}$ |
| Hydrosulfide ion | $HS^-$ | Sulfide ion | $S^{2-}$ |
| Water | $H_2O$ | Hydroxide ion | $OH^-$ |
| Methanol | $CH_3OH$ | Methoxide ion | $CH_3O^-$ |
| Ammonia | $NH_3$ | Amide ion | $NH_2^-$ |
| Hydroxide ion | $OH^-$ | Oxide ion | $O^{2-}$ |
| Hydrogen | $H_2$ | Hydride ion | $H^-$ |
| Amide | $NH_2^-$ | Imide ion | $NH^{2-}$ |
| Imide ion | $NH^{2-}$ | Nitride ion | $N^{3-}$ |
| Methane | $CH_4$ | Methide | $CH_3^-$ |

Base Strength

*Accepted systematic nomenclature of ions derived from diprotic acids upon removal of a single proton calls for attaching the word "hydrogen" directly to the name of the parent ion, *e.g.*, $HCO_3^-$ would be the hydrogencarbonate ion. Commonly, however, the prefix "bi" is used to denote this single hydrogen; hence $HCO_3^-$ more usually is called the bicarbonate ion.

## DRILL ON ELECTROLYTES & ACID-BASE REACTANTS

***Student's Name*** ______________________

Given in each case an acid and a base, write the formulas of the major solute species. In some cases there may be only one. Then write the *net ionic* equation for the reaction. You may be asked to turn in this sheet.

| Reactants | | Major Species, Acid | | Major Species, Base | | Net Ionic Equation |
|---|---|---|---|---|---|---|
| Hydrochloric Acid Solution | NaOH Solution | | | | | |
| Water | solid* Sodium Oxide | | | | | |
| Hydrogen chloride Gas | Water | | | | | |
| Hydrochloric Acid Solution | solid Sodium Hydride | | | | | |
| Hydrochloric Acid Solution | Ammonia Solution | | | | | |
| Hydrochloric Acid Solution | Gaseous Ammonia | | | | | |
| Acetic Acid Solution | Gaseous Ammonia | | | | | |
| ______________ | ______________ | | | | | |

***Remember to write the formula of the compound as the major species for solids, gases and weak electrolytes.**

It is observed that only $Na^+$ is in the same chemical form on both the reactant and product "sides" of the equation–since it has undergone no chemical change, it may be cancelled to give the net ionic equation (where all species are in aqueous medium):

$$HC_2H_3O_2 + OH^- \rightarrow C_2H_3O_2^- + H_2O$$

### 7-B. Equation for the Reaction of a Solid, Strong Base with Acid, Water

Write the equation for the reaction that takes place when solid sodium oxide is added to water.

#### *Solution*

Since $Na_2O$ is a **solid**, it is a major reactant and cannot be represented merely by $O^{2-}$ (which would represent an aqueous solution containing the oxide ion as a major component!). The best acid is water; the best base is the oxide ion in the solid. As the solid dissolves, the strong base $O^{2-}$ is leveled to the strongest base capable of existing in water, $OH^-$, thus,

$$Na_2O_{(s)} + H_2O_{(\ell)} \rightarrow 2\,Na^+ + 2\,OH^-$$

The reaction goes to completion. Note that the 2 $Na^+$ on the right must be included since they originate in the solid, but now are in solution.

### 14-D. An Acid and Base Reacting to Form a Common Conjugate

Write the equation for the reaction that occurs when equal volumes of 0.10 *M* $KH_2PO_4$ and 0.10 *M* $Na_3PO_4$ are mixed.

#### *Solution*

$K^+$ and $Na^+$ are merely spectator ions and should be omitted from the equation. Otherwise, the major species in each solution are $H_2O$, $H_2PO_4^-$, and $PO_4^{3-}$. The reaction is

$$H_2PO_4^- + PO_4^{3-} \rightarrow 2\,HPO_4^{2-}$$

since in terms of acid strength $H_2PO_4^- >> HPO_4^{2-} > H_2O$ and in terms of base strength $PO_4^{3-} >> HPO_4^{2-} > H_2O$. Thus, the "best acid"-"best base" combination is the one shown and, as indicated, the reaction will proceed to a large extent.

## Exercises on Writing Aqueous Acid-Base Equations

1. Write the TOTAL and the NET IONIC equation for the acid-base reaction that occurs, if any, when:
   (A) potassium hydroxide and nitric acid solutions are mixed.
   (B) sodium hydroxide and nitrous acid solutions are mixed.

(C) solutions of perchloric acid and ammonia are mixed.
(D) solutions of hydrofluoric acid and ammonia are mixed.
(E) solid iron (III) oxide is dissolved by excess hydrochloric acid
(F) solutions of $HC_2H_3O_2$ and $H_2SO_3$ are mixed.
(G) excess sodium hydroxide solution is added to sulfuric acid solution.
(H) excess sulfuric acid solution is added to a few drops of sodium hydroxide solution.
(I) solid magnesium hydroxide is added to excess sulfuric acid solution.
(J) a solution containing 1 mmole of phosphoric acid is added to a solution containing 2 mmoles of sodium hydroxide.
(K) pure liquid sulfuric acid is added to water.

## Exercises on Brønsted-Lowry Acids and Bases and Equation-Writing

For many of these exercises you will need to refer to the Acid-Base Table on pp. 162–163.

2. The conjugate acid of HF is:

   (A) HF (B) $H_2F^+$ (C) $F^-$ (D) $H^+$ (E) $H_3O^+$

3. The conjugate base of HF is:

   (A) HF (B) $H_2F^+$ (C) $F^-$ (D) $H_2O$ (E) $OH^-$

4. The conjugate base of $NH_3$ is:

   (A) $OH^-$ (B) $NH_2^-$ (C) $NH_4^+$ (D) $H_2O$ (E) none of these

5. All of the following are conjugate acid-base pairs EXCEPT:

   (A) $H_3O^+$, $OH^-$ (B) $NH_4^+$, $NH_3$ (C) $HCO_3^-$, $CO_3^{2-}$
   (D) $HC_2H_3O_2$, $C_2H_3O_2^-$ (E) HCl, $Cl^-$

6. Which of the following is the strongest acid in aqueous solution?

   (A) $CO_2$ (B) $H_2S$ (C) $H_2O_2$ (D) $H_2PO_4^-$ (E) HCN

7. A species that is a stronger base in aqueous solution than $NH_3$ is:

   (A) $CO_3^{2-}$ (B) $F^-$ (C) $NO_2^-$ (D) $HS^-$ (E) $HPO_4^{2-}$

8. In aqueous solution, which of the following species is a stronger acid than acetic acid?

   (A) $NH_4^+$ (B) $C_2H_3O_2^-$ (C) $H_2PO_4^-$ (D) $HPO_4^{2-}$ (E) $HSO_4^-$

9. The conjugate base of dihydrogenphosphate ion is:

   (A) $H_3PO_4$ (B) $H_2PO_4^-$ (C) $HPO_4^{2-}$ (D) $PO_4^{3-}$ (E) $OH^-$

10. If a solution containing one mole of ammonium dihydrogenphosphate is mixed with a solution containing one mole of ammonia, the final solution will contain:

(A) one mole $NH_4^+$ and one mole $HPO_4^{2-}$
(B) two moles $NH_4^+$ and one mole $H_2PO_4^-$
(C) one mole $NH_4^+$ and two moles $HPO_4^{2-}$
(D) two moles $NH_4^+$ and one mole $HPO_4^{2-}$
(E) two moles $NH_4^+$ and two moles $HPO_4^{2-}$

11. A solution of ammonium bisulfate is added to an **excess** of aqueous sodium hydroxide. The **net ionic** equation for this reaction would include (among others):

(A) **both** hydronium and sulfate ions
(B) **both** ammonia molecules and hydronium ions
(C) **both** $NH_4$ and $HSO_4$
(D) **both** hydroxide and bisulfite ions
(E) **both** ammonium and hydroxide ions

12. A hydrochloric acid solution containing two millimoles of hydronium ions is added to a solution containing one millimole of carbonate ions. The net result of this addition is the formation of:

(A) one mmole of bicarbonate ions
(B) two mmoles of bicarbonate ions
(C) one-half mmole of bicarbonate ions
(D) one mmole of "carbonic acid"
(E) two mmoles of "carbonic acid"

13. The equation for the reaction that takes place upon addition of an acetic acid solution to a sodium hydroxide solution is:

(A) $HC_2H_3O_{2\,(aq)} + H_2O_{(\ell)} \rightleftharpoons C_2H_3O^-_{2\,(aq)} + H_2O^+_{(aq)}$
(B) $HC_2H_3O_{2\,(aq)} + OH^-_{(aq)} \rightleftharpoons C_2H_3O^-_{2\,(aq)} + H_2O_{(\ell)}$
(C) $C_2H_3O^-_{2\,(aq)} + OH^-_{(aq)} \rightleftharpoons HC_2H_3O_{2\,(aq)} + O^{2-}_{(aq)}$
(D) $H_3O^+_{(aq)} + OH^-_{(aq)} \rightleftharpoons 2\,H_2O_{(\ell)}$
(E) $C_2H_3O^-_{2\,(aq)} + Na^+_{(aq)} \rightleftharpoons NaC_2H_3O_{2\,(s)}$

14. Hydrofluoric acid, when added to a sodium hydroxide solution, reacts in accord with the equation:

(A) $H_3O^+_{(aq)} + OH^-_{(aq)} \rightarrow 2\,H_2O_{(\ell)}$
(B) $HF_{(aq)} + OH^-_{(aq)} \rightarrow F^-_{(aq)} + H_2O_{(\ell)}$
(C) $F^-_{(aq)} + H_3O^+_{(aq)} \rightarrow HF_{(aq)} + H_2O_{(\ell)}$
(D) $F^-_{(aq)} + Na^+_{(aq)} \rightarrow NaF_{(s)}$
(E) $HF_{(aq)} + NaOH_{(s)} \rightarrow H_2O_{(\ell)} + NaF_{(s)}$

15. Given $HX_{(aq)} + Y^-_{(aq)} \rightleftharpoons HY_{(aq)} + X^-_{(aq)}$, which of the following is a conjugate acid-base pair involved in the indicated equilibrium?

(A) HX, $H_3O^+$ (B) HY, $X^-$ (C) HX, HY (D) $Y^-$, $X^-$ (E) HY, $Y^-$

16. If the reaction in Question 15 proceeds far to the right, it can be concluded that:

    (A) the extent of the reaction (left to right) is small
    (B) in terms of acid strength, HY >> HX
    (C) in terms of base strength, $Y^-$ >> $X^-$
    (D) $X^-$ is a better proton-acceptor than $Y^-$
    (E) none of the above is true

17. For each of the following, write the net ionic equation for the principal acid-base reaction. Indicate the extent of each reaction by using arrows of differing lengths.

    (A) A solution of barium hydroxide is added to hydrochloric acid.
    (B) Gaseous hydrogen chloride is bubbled into a solution of sodium cyanide.
    (C) Excess hydrofluoric acid is added to solid sodium oxide.
    (D) Hydrogen fluoride gas is added to water.
    (E) Solid calcium hydride is added to water.
    (F) Sulfuric acid solution is added to excess sodium hydroxide solution.
    (G) Solid calcium hydride is added to excess acetic acid solution.
    (H) A solution containing one mole of ammonia is added to a solution containing one mole of phosphoric acid.
    (I) Solid calcium hydride is added to excess ammonium chloride solution.
    (J) Ammonium nitrate solution is added to sodium fluoride solution.
    (K) A solution of potassium bicarbonate is added to acetic acid.
    (L) A solution of sodium dihydrogenphosphate is added to a solution of sulfurous acid.
    (M) Solid magnesium nitride is added to excess hydrochloric acid.
    (N) A solution of sodium bisulfite is mixed with a solution of potassium monohydrogenphosphate.
    (O) Hydrogen chloride is bubbled into excess ammonia solution.
    (P) A solution containing three moles of sodium acetate is added to a solution containing one mole of sulfuric acid.
    (Q) Hydrogen sulfide is bubbled through excess sodium hydroxide solution.
    (R) Carbon dioxide is added to excess potassium sulfide solution.
    (S) Solid sodium carbonate is added to excess nitric acid solution.
    (T) Hydrofluoric acid is added to sodium cyanide solution.
    (U) Hydrocyanic acid is added to sodium fluoride solution.
    (V) A solution containing one mole of sodium hydroxide is added to a solution containing one mole of iron(III) nitrite.

## LEWIS ACIDS AND BASES

An alternate concept of acids and bases is the *Lewis Electronic Theory*. According to this concept, an acid and a base are defined as:

**Lewis Base**—an electron pair donor. In principle, any species having an unshared pair of electrons, *e.g.*,

$$:\ddot{O}{-}H,\ :NH_3,\ CH_3{-}C(=O)\ddot{O}:^{-},\ :H^-$$

**Lewis Acid**—an electron pair acceptor. Any species that is electron-deficient, or contains an electron-deficient site, may function as a Lewis acid. Lewis acids may be *aprotic*—contain no hydrogen. Among the most prominent Lewis acids are metallic cations ($Al^{3+}$, $Ag^+$, $Zn^{2+}$, *etc.*), ions and compounds of elements from the 3rd or higher rows of the periodic table which may expand to more than an octet of electrons ($SbCl_3 + Cl^- \rightarrow SbCl_4^-$), and molecules that contain an atom that is multiply bonded to an electronegative atom (frequently oxygen: $H_2C{=}\ddot{O}:$, $O_2S{=}\ddot{O}:$, $:\ddot{O}{=}C{=}\ddot{O}:$).

The Lewis concept greatly expands the number of species that are considered acids. "Neutralization," in the Lewis sense, involves formation of a covalent bond between the donor atom of the base and the electron-deficient center of the acid:

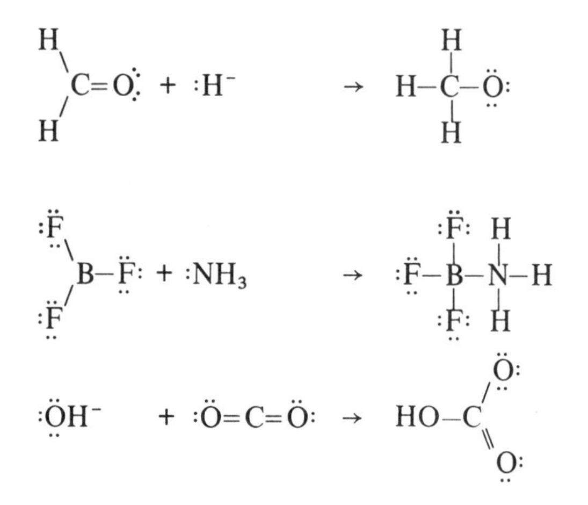

Many Lewis acid-base reactions are base displacement reactions, whereby one coordinated base is displaced by another. For example, the reaction,

$$:\ddot{\underset{..}{O}}\text{—}H^- + H_3O^+ \rightarrow H_2O + H_2O$$

can be viewed as the displacement of the base $H_2O$ by the stronger base, $OH^-$. Also, the formation of *complex ions* generally involves the displacement of coordinated water in aqueous solution by other bases (often called *ligands*), *e.g.*,

$$Ag(OH_2)_2^+ + 2\ NH_3 \rightarrow Ag(NH_3)_2^+ + 2\ H_2O$$

### Exercises on Lewis Acids and Bases

18. Which of the following could be a Lewis acid?

    (A) $Zn^{2+}$ (B) $BF_3$ (C) $H_3C\text{—}C(=O)\text{—}CH_3$ (H—C(H)(H)—C(=O)—C(H)(H)—H) (D) all of these

    (E) none of these

19. An example of a species that behaves as a Lewis acid due to its capability to increase the number of valence electrons on an atom beyond an octet is:

    (A) $BF_3$ (B) $BF_4^-$ (C) $NH_3$ (D) $PCl_3$ (E) $NO_3^-$

20. If the Lewis base, hydride ion, were to react with a $CO_2$ molecule, the donor-acceptor product would be:

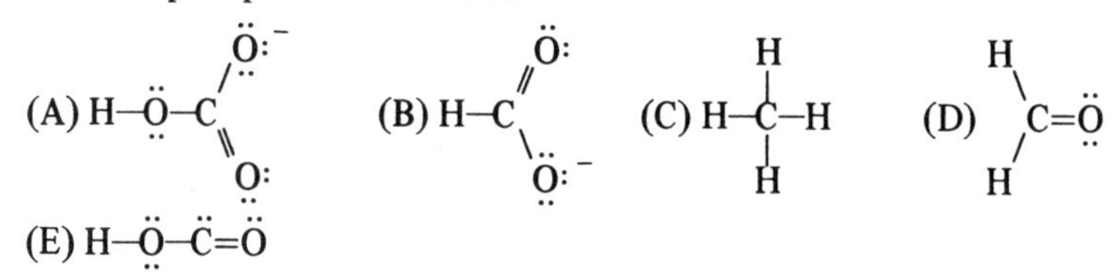

21. LiOH is an ionic hydroxide that serves to remove carbon dioxide from spacecraft by a Lewis acid-base reaction. The product of this reaction is:

    (A) lithium bicarbonate (B) lithium hydride (C) water
    (D) lithium oxide (E) hydrogen

22. A species that can act either as a Lewis acid or a Lewis base is:

    (A) $H^-$ (B) $NH_4^+$ (C) $SbF_3$ (D) $F^-$ (E) $Na^+$

## OXIDATION-REDUCTION

A large class of reactions involves electron-transfer between the reactants—these are called oxidation-reduction reactions. The processes of oxidation and of reduction, and the assignment of oxidation numbers has been mentioned in Chapter 5. It would be wise to review this before proceeding with this section. Oxidation and reduction processes *must* occur simultaneously and involve equal numbers of electrons. One reactant must include an atom whose oxidation number is increased—this reactant is said to be oxidized (lose electrons) and is called the *reducing agent*; the other reactant must include an atom whose oxidation number is decreased during the reaction—this reactant is said to be reduced (by gaining electrons) and is called the *oxidizing agent*. A convenient mnemonic for keeping the redox processes and the function of the reactants is "RALEAIO", meaning "*R*educing *A*gent *L*oses *E*lectrons *A*nd *I*s *O*xidized" in the reaction (the counter mnemonic is, of course, "OAGEAIR").

In terms of reaction-type recognition, a redox reaction is distinguished by oxidation number changes from Reactants→Products. In other reaction types (acid-base, precipitation) changes in oxidation numbers do not usually occur.

## OXIDATION-REDUCTION REACTIONS

These reactions involve electron transfer and may occur with pure substances or with species in solution. In the former case the *total equation* is written; in the latter case, the *net ionic equation* generally is used. The prediction of products, however, is not simple and requires considerable experience. Also, aqueous solution "*redox*" reactions are further complicated by the fact that the solvent ($H_2O$) and major species like $H^+$ (acidic solution) or $OH^-$ (basic solution) may take part in the reactions. Therefore, in the exercises and examples in this section, the major projects and the solution conditions (where applicable) will be given.

Commonly, one of two different approaches is used to balance an oxidation-reduction reaction:

a. The *change-in-oxidation-number* method, which may be generally used.

b. The *ion-electron* method, used only for net ionic equations.

These methods are discussed in some detail in most textbooks. The following examples illustrate both of these techniques.

## Example Problems Involving Balancing of Redox Equations

### Example 7-D.

Balance the following by the change-in-oxidation-number method.

$$\_\_\_K_2Cr_2O_7 + \_\_\_KI + \_\_\_H_2SO_4 \rightarrow \_\_\_K_2SO_4 + \_\_\_Cr_2(SO_4)_3 + \_\_\_I_2 + \_\_\_H_2O$$

This is a *total equation*—all reactants and products written as neutral formula units. Assignment of oxidation numbers (Ch. 5) reveals that all atoms except Cr and I remain constant in oxidation state. The change in oxidation number for Cr is from +6 to +3, corresponding to a gain of $3e^-$ **per Cr atom**; for I, the change is from -1 to 0, a loss of $1e^-$ **per I atom**. The basis for balancing an equation is that the electron gain must equal the electron loss. Since 1 $K_2Cr_2O_7$ gains $6e^-$, then 6 KI will be required to furnish the $6e^-$ to the oxidizing agent. This fixes the coefficients:

$$\underline{\ 1\ }\ K_2Cr_2O_7 + \underline{\ 6\ }\ KI + \_\_\_H_2SO_4 \rightarrow \underline{\ 4\ }\ K_2SO_4 + \underline{\ 1\ }\ Cr_2(SO_4)_3 + \underline{\ 3\ }\ I_2 + \_\_\_H_2O$$

Since the sulfate groups remain intact, the coefficient of $H_2SO_4$ must be 7, since there are 7 $SO_4$ groups on the right. Balancing hydrogen atoms gives 7 $H_2O$'s, with the final equation (coefficient **1** understood) being:

$$K_2Cr_2O_7 + 6\,KI + 7\,H_2SO_4 \rightarrow 4\,K_2SO_4 + Cr_2(SO_4)_3 + 3\,I_2 + 7\,H_2O$$

### Example 7-E.

By (a) the change-in-oxidation-number method and (b) the ion-electron method, develop a net ionic equation for the indicated transformation:

$$MnO_4^- + H_2O_2 \rightarrow Mn^{2+} + O_2 \qquad (\textit{acidic}\text{ solution}).$$

(a) "Acidic solution" indicates that the balanced equation may involve $H^+$ and/or $H_2O$ on **either** side of the equation. The oxidation number of Mn changes by 5 units per Mn atom, thus each $MnO_4^-$ gains $5e^-$. The oxidation number of oxygen changes from -1 in $H_2O_2$ to zero in $O_2$; thus each formula unit of $H_2O_2$ loses $2e^-$. Therefore, 2 $MnO_4^-$ are required for every 5 $H_2O_2$ to balance the electron gain and loss, and we can write:

$$2\,MnO_4^- + 5\,H_2O_2 \rightarrow 2\,Mn^{2+} + 5\,O_2 \qquad \text{(unbalanced)}$$

This leaves a net ionic charge of -2 on the left; +4 on the right. In acidic solution, the net ionic charge is balanced by adding the appropriate number of hydrogen ions to the side deficient in **positive** charge. In this case, adding 6 $H^+$ on the left and then balancing hydrogen atoms by placing 8 $H_2O$ on the right gives: •

$$2\,MnO_4^- + 6\,H^+ + 5\,H_2O_2 \rightarrow 2\,Mn^{2+} + 5\,O_2 + 8\,H_2O$$

The equation is now balanced—in terms of electron gain and loss; in terms of net ionic charge; and finally in terms of **all** atoms. It should be noted that the "charge balance" criterion preserves the gross electrical neutrality of the solution. For example, if $KMnO_4$ had been the source of $MnO_4^-$, and if $H_2SO_4$ the source of $H^+$, the *spectator* counterions required for neutrality on the left, would have been ($2K^+$, $3\ SO_4^{2-}$), exactly balancing the $2\ Mn^{2+}$ on the right.

(b) In the ion-electron method, the equation is divided into two 'skeletal half-reactions,' one for the reduction and one for the oxidation:

$$MnO_4^- \rightarrow Mn^{2+} \quad \text{(reduction)}$$

$$H_2O_2 \rightarrow O_2 \quad \text{(oxidation)}$$

In acidic solution, **each** of these is balanced by first adding 1 $H_2O$ for each oxygen deficiency, next balancing the hydrogens by adding $H^+$, and finally balancing the charge by adding an appropriate number of electrons, $e^-$. This gives:

$$MnO_4^- + 8\ H^+ + 5\ e^- \rightarrow Mn^{2+} + 4\ H_2O$$

$$H_2O_2 \rightarrow O_2 + 2\ H^+ + 2e^-$$

In order that the electron gain and loss may be balanced, the first half-reaction is multiplied by 2, the second by 5, and the half-reactions are summed to give:

$$2\ MnO_4^- + 16\ H^+ + 10e^- + 5\ H_2O_2 \rightarrow$$

$$2\ Mn^{2+} + 8\ H_2O + 5\ O_2 + 10\ H^+ + 10e^-$$

This gives, on algebraically collecting common terms,

$$2\ MnO_4^- + 6\ H^+ + 5\ H_2O_2 \rightarrow 2\ Mn^{2+} + 8\ H_2O + 5\ O_2$$

which is exactly the same result as obtained in (a).

## Example 7-F.

Develop by (a) the oxidation-number method and (b) the ion-electron method, a balanced equation for the oxidation of $Cr(OH)_4^-$ by $ClO^-$ in basic solution, yielding chromate and chloride ions.

The unbalanced, skeletal reaction can be indicated as:

$$Cr(OH)_4^- + ClO^- \rightarrow CrO_4^{2-} + Cl^- \quad (\textit{basic}\ \text{solution})$$

(a) The Cr changes by 3 oxidation-number units, each $Cr(OH)_4^-$ losing $3e^-$. The Cl changes by 2 units, each $ClO^-$ gaining $2e^-$. Consequently, for electron balance 2 $Cr(OH)_4^-$ will be required for every 3 $ClO^-$, giving

$$2\ Cr(OH)_4^- + 3\ ClO^- \rightarrow 2\ CrO_4^{2-} + 3\ Cl^- \quad \text{(unbalanced)}$$

"Basic solution" means that $OH^-$ and/or $H_2O$ may be required on **either** side of the equation. For net ionic charge balance, an appropriate number of $OH^-$ ions are added to the side deficient in **negative** charge. As the expression now stands, the net charge on the left is -5, on the right -7. Thus, adding 2 $OH^-$ on the left and balancing hydrogen with water gives as the final balanced equation:

$$2\,Cr(OH)_4^- + 3\,ClO^- + 2\,OH^- \rightarrow 2\,CrO_4^{2-} + 3\,Cl^- + 5\,H_2O$$

(Note that if each anion on the left had been supplied as the potassium compound, the reactants would include the 7 unshown $K^+$ ions, which are exactly the right number of positive charges to render the products, 2 $CrO_4^{2-}$ and 3 $Cl^-$, electrically neutral.)

(b) Separating into skeletal half reactions, we have:

$$Cr(OH)_4^- \rightarrow CrO_4^{2-} \quad \text{(oxidation)}$$

$$ClO^- \rightarrow Cl^- \quad \text{(reduction)}$$

In basic solution each of these is balanced by adding (a) ($2OH^-$ - $H_2O$) for each oxygen deficiency; (b) ($H_2O$ - $OH^-$) for each hydrogen deficiency; and (c) electrons, $e^-$, to achieve net charge balance. This gives for the oxidation,

$$Cr(OH)_4^- \rightarrow CrO_4^{2-} + 4(H_2O - OH^-) + 3e^-$$

or

$$Cr(OH)_4^- + 4\,OH^- \rightarrow CrO_4^{2-} + 4\,H_2O + 3\,e^-$$

and for the reduction,

$$ClO^- + 2\,e^- \rightarrow Cl^- + (2\,OH^- - H_2O)$$

or

$$ClO^- + 2\,e^- + H_2O \rightarrow Cl^- + 2\,OH^-$$

Multiplication of the oxidation half-reaction by 2, the reduction half-reaction by 3, and then summing gives:

$$2\,Cr(OH)_4^- + 8\,OH^- + 3\,ClO^- + 6\,e^- + 3\,H_2O \rightarrow$$

$$2\,CrO_4^{2-} + 8\,H_2O + 6\,e^- + 3\,Cl^- + 6\,OH^-$$

which upon collecting the common terms leads to the final equation:

$$2\,Cr(OH)_4^- + 3\,ClO^- + 2\,OH^- \rightarrow 2\,CrO_4^{2-} + 5\,H_2O + 3\,Cl^-$$

Again, this is the same result as in (a).

### Example 7-G.

Balance the equation for "the reaction of potassium dichromate with hydrogen iodide, forming as products potassium iodide, chromium(III) iodide, iodine, and water."

This calls for the total equation, and it is first necessary to write the formulas of the reactants and products:

$$K_2Cr_2O_7 + HI \rightarrow KI + CrI_3 + I_2 + H_2O \quad \text{(unbalanced)}$$

This reaction is somewhat tricky because HI must furnish both the $I^-$ that has been oxidized ($I_2$) as well as the $I^-$ that is unchanged in oxidation number (KI, $CrI_3$). There is a gain of $6e^-$ per formula unit of $K_2Cr_2O_7$ and a loss of $1e^-$ per formula unit of HI. Therefore, 6 HI will be required for **each $K_2Cr_2O_7$ to furnish the oxidized iodide**, forming 3 $I_2$ on the right; however, additional HI will be required to furnish the $I^-$ for both KI and $CrI_3$. Fortunately, both the coefficient of KI (2) and $CrI_3$ (2) have already been fixed by placing the coefficient **1** before $K_2Cr_2O_7$. Thus, 8 additional HI units will be required on the left to give the proper coefficient of HI, (8 + 6) = 14. The final equation will then be:

$$K_2Cr_2O_7 + 14\ HI \rightarrow 2\ KI + 2\ CrI_3 + 3\ I_2 + 7\ H_2O$$

(Note that in this case balancing could have been done by the "inspection method," *i.e.*, by placing the coefficient **1** before $K_2Cr_2O_7$, all other coefficients would have been fixed; however, the "inspection method" is only of value in balancing simple equations.)

## Exercises on Balancing Oxidation-Reduction Equations

23. Write balanced equations for the oxidation-reduction reaction that occurs when:

(A) a mixture of $KMnO_4$, $K_2C_2O_4$, and $H_2SO_4$ reacts to form $K_2SO_4$, $MnSO_4$, $H_2O$, and $CO_2$
(B) $MnO_4^-$ reacts with $H_2C_2O_4$ in acidic solution to produce $Mn^{2+}$ and $CO_2$
(C) $MnO_4^-$ reacts with $C_2O_4^{2-}$ in basic solution to produce $MnO_2$ and $CO_3^{2-}$
(D) $MnO_4^{2-}$ reacts with itself (*disproportionation* reaction) in acidic solution to produce $MnO_4^-$ and $MnO_2$
(E) a mixture of KOH, $KMnO_4$, and $MnO_2$, reacts to form $K_2MnO_4$ and $H_2O$
(F) elemental copper reacts with $HNO_3$ to produce $Cu(NO_3)_2$, NO, and $H_2O$
(G) elemental antimony reacts with nitrate ion in acidic solution to form $Sb_2O_5$, $NO_2$, and $H_2O$
(H) chromium(III) ion in acidic solution reacts with solid $NaBiO_3$ to produce dichromate ion, bismuth(III) ion, sodium ion, and water
(I) arsenic acid solution (acidic solution) reacts with metallic zinc to form arsine ($AsH_3$) and zinc ions
(J) iron(III) ion in aqueous solution reacts with iron to produce iron(II) ion
(K) $VO^{2+}{}_{(aq)}$ reacts with permanganate ion in acidic solution to form $V(OH)_4^+{}_{(aq)}$ and manganese(II) ion
(L) metallic vanadium reacts with chlorate ion in hot basic solution to produce chloride ion and vanadate ion. (Predict the formula of vanadate ion by analogy with a Group VA element)

(M) an acidic solution containing $H_2V_{10}O_{28}^{4-}$ reacts with $H_3PO_2$ to produce a solution containing vanadium(II) ions and phosphoric acid

(N) a mixture of chromium(II) iodide, potassium hydroxide, and $Cl_2$ reacts to form potassium chromate, potassium periodate, potassium chloride, and water

(O) white phosphorus, $P_4$, reacts in basic solution to form $H_2PO_2^-$ and $PH_3$

(P) hydrazoic acid, $HN_3$, reacts with nitrate ion in acidic solution to product nitrogen gas ($N_2$) and water

24. Which of the following conversions requires a reducing agent?

(A) $Fe^{2+} \rightarrow Fe^{3+}$ (B) $NH_3 \rightarrow NH_4^+$ (C) $H_2O_2 \rightarrow O_2$
(D) $Cr(OH)_4^- \rightarrow CrO_4^{2-}$ (E) $NO_2^- \rightarrow N_2O_2^{2-}$

25. The following change occurs in **acidic** solution:

$$H_2S + Cr_2O_7^{2-} \rightarrow S + Cr^{3+}$$

The properly balanced equation shows that for every mole of $Cr_2O_7^{2-}$ that reacts _______ moles of $H^+$ are consumed.

(A) 8 (B) 14 (C) 7 (D) 10 (E) 5

26. Consider the following unbalanced expression:

$$MnO_4^- + H^+ + Fe^{2+} \rightarrow Mn^{2+} + Fe^{3+} + H_2O$$

(A) $MnO_4^-$ is the reducing agent (B) The manganese has been oxidized
(C) $Fe^{2+}$ gains electrons (D) The iron has been reduced
(E) $Fe^{2+}$ is the reducing agent

27. In the basic solution $CrO_4^{2-}$ reacts with $HSnO_2^-$ to form $CrO_2^-$ and $HSnO_3^-$. The *change* in the oxidation number of chromium is:

(A) 0 (B) + 1 (C) − 1 (D) + 3 (E) − 3

28. In the balanced equation for the reaction given in Question 27, the coefficients for $CrO_4^{2-}$ and $HSnO_2^-$ are, respectively:
(A) 1, 3 (B) 1, 2 (C) 2, 1 (D) 3, 2 (E) 2, 3

29. In the balanced equation for the reaction described in Question 27, there will be:

(A) 2 $OH^-$ on the right (B) 2 $OH^-$ on the left (C) 3 $OH^-$ on the right
(D) 3 $OH^-$ on the left (E) no $OH^-$ on either side of the equation

30. The balanced equation for the reaction described in Question 27 will require:

(A) one water on the right (B) one water on the left
(C) two waters on the right (D) two waters on the left
(E) no water on either side of the equation

## DRILL ON OXIDATION-REDUCTION REACTIONS *Student's Name* ______

Balance each of the following equations and then identify the oxidizing agent and the reducing agent. You may be asked to turn in this sheet.

| REACTION | Oxidizing Agent | Reducing Agent |
|---|---|---|
| $MnO_{2\,(s)} + H_2SO_{3(aq)} \xrightarrow{Acid} S_2O_6^{2-} + Mn^{2+}$ | | |
| $Cr_2O_7^{2-} + Zn_{(s)} \xrightarrow{Neutral} Cr(OH)_{3(s)} + Zn(OH)_{2(s)}$ | | |
| $Fe_2O_{3(s)} + Cl_{2(g)} \xrightarrow{Base} FeO_4^{2-} + Cl^-$ | | |
| $FeO_4^{2-} \xrightarrow{Acid} Fe^{3+} + O_{2(g)}$ | | |
| | | |

31. The following change occurs in **alkaline** solution:

$$MnO_4^- + ClO_2^- \rightarrow MnO_2 + ClO_4^-$$

Write the balanced equation for the change and then add together **all** of the coefficients in the balanced equation. The sum of these coefficients is:

(A) 15 (B) 20 (C) 24 (D) 25 (E) 30

32. When zinc reacts with dilute nitric acid, zinc ion and ammonium ion are formed. The balanced equation indicates that, for every mole of zinc reacting, ______ mole(s) of $H^+$ will be consumed.

(A) 2.0 (B) 2.5 (C) 3.0 (D) 5.0 (E) 10.0

33. A student mixed solutions containing $MnO_4^-$ and $I^-$ under basic conditions and correctly concluded that $MnO_4^{2-}$ and $IO_3^-$ were formed. The balanced equation for this reaction would show that ______ mole(s) of $OH^-$ are required for **every one** mole of $I^-$.

(A) 1 (B) 2 (C) 4 (D) 6 (E) 8

34. The following reaction is used for the analytical determination of uranium in **acidic** solution:

$$UO^{2+} + Cr_2O_7^{2-} \rightarrow UO_2^{2+} + Cr^{3+} \quad \text{(unbalanced)}$$

For every millimole of $UO^{2+}$ in an unknown, ______ millimole(s) of $Cr_2O_7^{2-}$ would be required for the above reaction.

(A) 1/3 (B) 1/2 (C) 2 (D) 3 (E) 6

35. The following reaction occurs in **acidic** solution:

$$Sn + NO_3^- \rightarrow SnO_2 + NO_2 \quad \text{(unbalanced)}$$

The balanced equation indicates that for every mole of Sn that reacts, ______ mole(s) of $NO_2$ will be formed.

(A) 1 (B) 2 (C) 3 (D) 4 (E) 5

36. The following change occurs when $Sb_2O_3$ is treated with potassium permanganate in hot, basic solution:

$$Sb_2O_{3(s)} + MnO_4^- \rightarrow MnO_4^{2-} + Sb(OH)_6^-$$

For every mole of $Sb(OH)_6^-$ formed,

(A) 3 moles of $H_2O$ are also formed
(B) 3 moles of hydroxide ion are consumed
(C) 3 moles of permanganate are reduced
(D) 2 moles of $Sb_2O_3$ are oxidized
(E) 0.5 mole of $MnO_4^{2-}$ is also formed

37. The following redox process occurs in **basic** aqueous medium:

$$Bi(OH)_{3(s)} + HSnO_2^- \rightarrow Bi_{(s)} + Sn(OH)_6^{2-}$$

The properly balanced equation shows that for every **two moles** of $Bi(OH)_3$ reacting,

(A) 3 moles each of $OH^-$, $HSnO_2^-$, and $H_2O$ are consumed
(B) 2 moles of $OH^-$ are formed (C) 2 moles of $Sn(OH)_6^{2-}$ are formed
(D) 4 moles of $HSnO_2^-$ are oxidized (E) none of the above occurs

38. The redox reaction indicated below occurs in acidic solution:

$$MnO_4^- + Cl^- \rightarrow Mn^{2+} + Cl_2 \qquad \text{(unbalanced)}$$

Which of the following is **true** for this reaction?

(A) $Cl_2$ is the reducing agent. (B) $MnO_4^-$ undergoes oxidation
(C) $Cl^-$ is the reducing agent. (D) $Cl^-$ undergoes reduction
(E) The oxidation number of Mn increases by 3 units

39. Which of the following conversions would require a **reducing** agent?

(A) $H_3NOH^+ \rightarrow H_2N_2O_2$ (B) $H_3NOH^+ \rightarrow HNO_2$
(C) $H_2NOH^+ \rightarrow NH_4^+$ (D) $H_3NOH^+ \rightarrow H_2NOH$ (E) none of these

40. In acidic solution $ClO_3^-$ reacts with $I_2$ to form $IO_3^-$ and $Cl^-$. The oxidation numbers of chlorine in $ClO_3^-$ and in $Cl^-$ are, respectively:

(A) -1, -1 (B) -1, +1 (C) +3, -1 (D) +5, +1 (E) +5, -1

41. In the reaction described in Question 40, the oxidizing agent is:

(A) $ClO_3^-$ (B) $I_2$ (C) $Cl^-$ (D) $IO_3^-$ (E) $H_2O$

42. For the reaction described in Question 40, the coefficients of $ClO_3^-$ and $I_2$ in the balanced equation are, respectively:

(A) 5, 3 (B) 5, 6 (C) 6, 5 (D) 1, 1 (E) 1, 2

---

## REDOX CHEMISTRY OF METALS AND NONMETALS

*The Activity Series.* In general, metals tend to furnish electrons to form cations, i.e. act as reducing agents (RED), and non-metals tend to accept electrons to form anions, i.e. act as oxidizing agents (OX). The reduced form of a metal is the metal itself; the reduced form of a non-metal is an anion. These electron-rich species act as reducing agents of differing strengths. Table II lists reducing agents in order of decreasing strength and corresponding oxidizing agents in order of increasing strength and is called the "activity series" (also called the electromotive series). For the sake of comparison all ions in solution are at a concentration of 1 M. It is readily noted that the metals on the left and on center portion of the periodic table are better reducing agents than is $H_2$ and non-metals and ions of the noble metals ($Cu^{2+}$, $Ag^+$, and $Hg^{2+}$) are better oxidizing agents than $H^+$. This table (which includes neutral water for refer-

**TABLE II. The Activity Series**

| Reducing Agent | Oxidizing Agent |
|---|---|
| Li | $Li^+$ |
| K | $K^+$ |
| Ca | $Ca^{2+}$ |
| Na | $Na^+$ |
| Mg | $Mg^{2+}$ |
| Al | $Al^{3+}$ |
| Mn | $Mn^{2+}$ |
| Zn | $Zn^{2+}$ |
| Fe | $Fe^{2+}$ |
| $H_2$ | (neutral) $H_2O$ |
| Co | $Co^{2+}$ |
| Ni | $Ni^{2+}$ |
| Sn | $Sn^{2+}$ |
| Pb | $Pb^{2+}$ |
| $H_2$ | (1 M) $H^+$ |
| Cu | $Cu^{2+}$ |
| $I^-$ | $I_2$ |
| Ag | $Ag^+$ |
| $H_2O$ (neutral) | $O_2$ |
| Hg | $Hg^{2+}$ |
| $Br^-$ | $Br_2$ |
| $Cl^-$ | $Cl_2$ |
| $F^-$ | $F_2$ |

Left arrow (pointing up): Increasing strength as a Reducing Agent (Increasing tendency to be oxidized)

Right arrow (pointing down): Increasing strength as an Oxidizing Agent (Increasing tendency to be reduced)

ence) summarizes considerable aqueous redox chemistry. It shows that whenever a left-upper RED is mixed with a right-lower OX, transfer of electrons from RED to OX can occur:

$$RED_{(reactant)} + OX'_{(reactant)} \rightarrow RED'_{(product)} + OX_{(product)}$$

Using this table simple redox reactions would be predicted to occur:

TOTAL $Mg_{(s)} + 2HCl_{(aq)} \rightarrow H_{2(g)} + MgCl_{2(aq)}$

NET $Mg_{(s)} + 2H^+ \rightarrow H_{2(g)} + Mg^{2+}$

TOTAL $Zn_{(s)} + 2AgNO_{3(aq)} \rightarrow Zn(NO_3)_{2(aq)} + 2Ag_{(s)}$

NET $Zn_{(s)} + 2Ag^+ \rightarrow Zn^{2+} + 2Ag_{(s)}$

TOTAL $Cl_{2(g)} + 2KBr_{(aq)} \rightarrow Br_{2(aq)} + 2KCl_{(aq)}$

NET $Cl_{2(g)} + 2Br^- \rightarrow Br_{2(aq)} + 2Cl^-$

Conversely, the following will not react:

$$Mn + Mg^{2+}; \quad Ag + H^+; \quad Fe + Al^{3+}$$

The placement of neutral water as an oxidizing agent is interesting. Iron and all of the metals above it will reduce water to $H_2$; Li, K, Ca and Na react readily with cold water. Warm water or steam is required for Fe, Zn, Mn, Al, or Mg to undergo the same reaction. The metal containing product of these reactions is the metal hydroxide; with steam the product is more often the metal oxide since in general hydroxides are converted by heat to the metal oxide and water vapor:

$$2Al(OH)_{3(s)} \rightarrow Al_2O_{3(s)} + 3H_2O_{(g)}$$

**Exercises Involving the Activity Series**

43. What weight of chlorine gas is needed to react with one mole of bismuth(III) iodide?
(A)17.7 g (B) 35.5 g (C) 70.9 g (D) 106 g (E) 213 g

44. If 1.00 g of metallic magnesium is added to 100. mL of 1.00 M $CuSO_4$, how many grams of copper can be formed?
(A) 1.00 g (B) 2.61 g (C) 3.14 g (D) 4.71 g (E) 6.35 g

45. If the reaction in #44 (above) occurs without any change of solution volume, what concentration of copper(II) ions will be left in solution?
(A) 0.00300 M (B) 0.268 M (C) 0.589 M
(D) 0.613 M (E) 0.996 M

46. Which will displace the greatest mass of $Ag_{(s)}$ when added to excess silver nitrate solution?
(A) 1.00 g Al (B) 1.00 g Mg (C) 3.00 g Mn
(D) 4.00 g Zn (E) 5.00 g Pb

47. Which will displace the greatest number of moles of $H_{2(g)}$ when added to excess 1 M $HClO_4$?
(A) 2.00 g Al (B) 2.00 g Mg (C) 3.00 g Mn (D) 5.00 g Zn
(E) 10.0 g Sn

48. Which of the following is the best reducing agent?
(A) 1 *M* HCl (B) $Mn^{2+}$ (C) $Zn^{2+}$ (D) $Zn_{(s)}$ (E) $Ag_{(s)}$

49. Which of the following is the best oxidizing agent?
(A) $H_3O^+$ (B) $H^+$ (C) $H_{2(g)}$ (D) $Ag^+$ (E) $H_2O_{(\ell)}$

50. When mixed, which of the following pairs of reactants should give a spontaneous redox reaction?
(A) $Ni^{2+}$;$Cu^{2+}$ (B) Co;$Sn^{2+}$ (C) $H_2$;$Zn^{2+}$ (D) $H^+$;$Co^{2+}$
(E) $I_2$;$Br^-$

51. When mixed, **all** of the following pairs of reactants should result in a spontaneous oxidation-reduction EXCEPT:

    (A) Ag;$Cl_2$ (B) $Cl_2$;$Br^-$ (C) Fe;$H^+$ (D) $Ag^+$;$Hg^{2+}$
    (E) Mn;$Sn^{2+}$

52. From the table find a species that will convert $Cu^{2+}$ to Cu but **will not** convert $Fe^{2+}$ to Fe.

    (A) $Ag^+$ (B) $H^+$ (C) Pb (D) Zn (E) Ag

## ANSWERS TO CHAPTER 7 PROBLEMS

1. (A) $KOH + HNO_3 \rightarrow KNO_3 + H_2O$
   $OH^- + H_3O^+ \rightarrow 2\ H_2O$

   (B) $NaOH + HNO_2 \rightarrow NaNO_2 + H_2O$
   $OH^- + HNO_2 \rightarrow NO_2^- + H_2O$

   (C) $HClO_4 + NH_3 \rightarrow NH_4ClO_4$
   $H_3O^+ + NH_3 \rightarrow NH_4^+ + H_2O$

   (D) $HF + NH_3 \rightarrow NH_4F$
   $HF + NH_3 \rightarrow NH_4^+ + F^-$

   (E) $Fe_2O_3 + 6HCl \rightarrow 2FeCl_3 + 3H_2O$
   $Fe_2O_3 + 6H^+ \rightarrow 2Fe^{3+} + 3H_2O$

   (F) No reaction

   (G) $2NaOH^+ H_2SO_4 \rightarrow Na_2SO_4 + 2H_2O$
   $2OH^- + H_3O^+ + HSO_4^- \rightarrow SO_4^{2-} + 3H_2O$

   (H) $H_2SO_4 + NaOH \rightarrow NaHSO_4 + H_2O$
   $H_3O^+ + OH^- \rightarrow 2H_2O$

   (I) $Mg(OH)_2 + 2H_2SO_4 \rightarrow Mg(HSO_4)_2 + 2H_2O$
   $Mg(OH)_2 + 2H^+ \rightarrow Mg^{2+} + 2H_2O$

   (J) $H_3PO_4 + 2NaOH \rightarrow Na_2HPO_4 + 2H_2O$
   $H_3PO_4 + 2OH^- \rightarrow HPO_4^{2-} + 2H_2O$

   (K) $H_2SO_{4(\ell)} \rightarrow H_2SO_{4(aq)}$
   $H_2SO_{4(\ell)} + H_2O \rightarrow H_3O^+ + HSO_4^-$

2. B
3. C
4. B
5. A
6. A
7. A
8. E
9. C
10. D
11. E
12. D
13. B
14. B
15. E
16. C

17. Answers given only to selected equations.

    B. $HCl_{(g)} + CN^- \rightarrow HCN + Cl^-$
    F. $H_3O^+ + HSO_4^- \rightarrow 2\ OH^- \rightarrow SO_4^{2-} + 3\ H_2O$

I. $CaH_{2(s)} + 2\ NH_4^+ \rightarrow 2\ H_{2(g)} + 2\ NH_3 + Ca^{2+}$
L. $H_2PO_4^- + H_2SO_3 \rightarrow H_3PO_4 + HSO_3^-$
N. $HSO_3^- + HPO_4^{2-} \rightarrow SO_3^{2-} + H_2PO_4^-$
R. $CO_2 + 2\ S^{2-} + H_2O \rightarrow 2\ HS^- + CO_3^{2-}$

18. D 19. D 20. B 21. A 22. C

23. (A) $2\ KMnO_4 + 5\ K_2C_2O_4 + 8\ H_2SO_4 \rightarrow 6\ K_2SO_4 + 2\ MnSO_4 + 8\ H_2O + 10\ CO_2$

(B) $2\ MnO_4^- + 5\ H_2C_2O_4 + 6\ H^+_{(aq)} \rightarrow 2\ Mn^{2+} + 10\ CO_{2\ (g)} + 8\ H_2O$

(C) $2\ MnO_4^- + 3\ C_2O_4^{2-} + 4\ OH^- \rightarrow 2\ MnO_{2\ (s)} + 6\ CO_3^{2-} + 2\ H_2O$

(D) $3\ MnO_4^{2-} + 4\ H^+_{(aq)} \rightarrow 2\ MnO_4^- + MnO_{2\ (s)} + 2\ H_2O$

(E) $4\ KOH + 2\ KMnO_4 + MnO_2 \rightarrow 3\ K_2MnO_4 + 2\ H_2O$

(F) $3\ Cu + 8\ HNO_3 \rightarrow 3\ Cu(NO_3)_2 + 2\ NO + 4\ H_2O$

(G) $2\ Sb_{(s)} + 10\ NO_3^- + 10\ H^+_{(aq)} \rightarrow Sb_2O_{5\ (s)} + 10\ NO_{2\ (g)} + 5\ H_2O$

(H) $2\ Cr^{3+} + 3\ NaBiO_{3\ (s)} + 4\ H^+_{(aq)} \rightarrow Cr_2O_7^{2-} + 3\ Bi^{3+} + 3\ Na^+ + 2\ H_2O$

(I) $H_3AsO_{4\ (aq)} + Zn_{(s)} + 8\ H^+ \rightarrow AsH_{3\ (g)} + 4\ Zn^{2+} + 4\ H_2O$

(J) $2\ Fe^{3+} + Fe_{(s)} \rightarrow 3\ Fe^{2+}$

(K) $5\ VO^{2+} + MnO_4^- + 11\ H_2O \rightarrow 5\ V(OH)_4^+ + Mn^{2+} + 2\ H^+_{(aq)}$

(L) By analogy with $PO_4^{3-}$, take vanadate to be $VO_4^{3-}$,
$6\ V_{(s)} + 5\ ClO_3^- + 18\ OH^- \rightarrow 5\ Cl^- + 6\ VO_4^{3-} + 9\ H_2O$
or by analogy with nitrate, $NO_3^-$, take vanadate to be $VO_3^-$
$6\ V_{(s)} + 5\ ClO_3^- + 6\ OH^- \rightarrow 5\ Cl^- + 6\ VO_3^- + 3\ H_2O$
Note that the V to $ClO_3^-$ ratios are the same in either case.

(M) $2\ H_2V_{10}O_{28}^{4-} + 15\ H_3PO_2 + 48\ H^+_{(aq)} \rightarrow 20\ V^{2+} + 15\ H_3PO_4 + 26\ H_2O$

(N) $CrI_2 + 24\ KOH + 10\ Cl_{2\ (g)} \rightarrow K_2CrO_4 + 2\ KIO_4 + 20\ KCl + 12\ H_2O$

(O) $P_{4\ (s)} + 3\ OH^- + 3\ H_2O \rightarrow 3\ H_2PO_2^- + PH_{3\ (g)}$

(P) $5\ HN_3 + NO_3^- + H^+_{(aq)} \rightarrow 8\ N_{2(g)} + 3\ H_2O$

24. E
25. A
26. E
27. E
28. E
29. A
30. B
31. B
32. B
33. D
34. A
35. D
36. A
37. A
38. C
39. C
40. E
41. A
42. A
43. D
44. B
45. C
46. D
47. A
48. D
49. D
50. B
51. D
52. C

# 8
# *Solution Reactions—2*

The literal meaning of *metathesis* is "a process in which things are placed differently". In chemistry this refers to a reaction in which the positive part of one chemical compound combines with the negative part of second chemical compound; this now leaves the negative fragment of the first compound with the positive portion of the second compound. The positive fragment of a compound is usually written first; the negative fragment second, *e.g.* NaCl, $H_2SO_4$. Using this convention a simple metathesis is represented by:

$$AB + CD \rightarrow AD + CB$$

## SOLUBILITY RULES AND PRECIPITATION REACTIONS

The constituent parts of a compound determine its water solubility. Although there are many exceptions, Table I allows prediction of which compounds will be reasonably soluble in water; any compound containing fragments that are not mentioned should be deemed 'insoluble' (this really means the compound is only very slightly soluble; its constituent fragments cannot exist as independent ions in solution). Thus, in the case of $KNO_3$ one would predict that the compound is soluble in accord with both Rule 1 and Rule 2; $MgSO_4$ likewise would be soluble, Rule 4; CoS would be insoluble since no rule applies. There are numerous exceptions, *e.g.* in the case of AgF, no rule applies and we would predict that the compound is insoluble. We should not be extremely surprised to find that AgF is very soluble. The solubility rules are a good first guide and should be memorized. They will correctly predict the solubility for many inorganic compounds, including all that you are asked about in this course.

*Precipitation Reactions*. Any time the mixing of two solutions would result in the presence of two or more ions that the solubility rules predict to form an insoluble compound, then one should predict the formation of

TABLE I. SOLUBILITY RULES

| Compound Class | Solubility in Water |
|---|---|
| 1. All salts containing $(I\text{-}A)^+$ or $NH_4^+$ ions and all common Acids | Soluble |
| 2. All salts containing $NO_3^-$, $ClO_4^-$, $ClO_3^-$, or $C_2H_3O_2^-$ ions | Soluble |
| 3. All salts containing $Cl^-$, $Br^-$, or $I^-$ ions | Soluble, EXCEPT in combination with $Pb^{2+}$, $Ag^+$, $Hg_2^{2+}$, and $Tl^+$ |
| 4. All salts containing $SO_4^{2-}$ | Soluble, EXCEPT in combination with $Pb^{2+}$, $Ca^{2+}$, $Sr^{2+}$, and $Ba^{2+}$ |
| 5. All others (*e.g.*, $OH^-$, $CO_3^{2-}$, $S^{2-}$, $PO_4^{3-}$) | INsoluble, unless exempted above |

this compound as a precipitate. Thus, if a solution of cadmium nitrate is mixed with a solution of sodium hydroxide, our knowledge of the electrolytic nature of the solutes would give as potential reactants:

$$Cd^{2+} + NO_3^- + Na^+ + OH^-$$

Taking the cation from the first solution and the anion from the second solution, the solubility rules predict that $Cd(OH)_2$ is insoluble—Rule 5; similarly, taking the anion from the first solution and the cation from the second, the solubility rules predict that $NaNO_3$ would be soluble—Rules 1 and 2. Therefore, we need not include the $Na^+$ and $NO_3^-$ in the net ionic equation. Our reaction would be:

$$\text{NET} \quad Cd^{2+} + 2OH^- \rightarrow Cd(OH)_{2(s)}$$

Note that we did need to use 2 hydroxides to balance the 2 positive charges on the cadmium ion. To write the total equation, we would have to write the complete formulas for the reactants, cadmium nitrate and sodium hydroxide. Since the sodium and nitrate groups are included with the reactants, they would have to show up as the product, $NaNO_3$:

$$\text{TOTAL} \quad Cd(NO_3)_{2(aq)} + 2NaOH_{(aq)} \rightarrow Cd(OH)_{2(s)} + 2NaNO_{3(aq)}$$

We can now readily note the metathetical behavior of the reaction. The cadmium ions started out with the nitrate ions and ended up in a precipitate with the hydroxide ions; the sodium ions started out in solution with the hydroxide ions and ended up still in solution but now with nitrate ions. Colorless sodium nitrate crystals could be obtained by filtering off the precipitate and then evaporating water from the filtrate. Sometimes inexperienced students write the two sodium nitrates incorrectly as $Na_2(NO_3)_2$. It should be remembered that ionic compounds (salts) are written as empirical formulas. It should be noted that the driving force causing this type of metathesis to occur is the removal of ions from solution by incorporating them into the solid particles of the precipitate (similar to the driving force in acid-base reactions being the formation of weakly-ionized products from more highly-ionized reactants).

In some cases we can have two reactions occur simultaneously. Thus, if a solution of silver sulfate is mixed with a solution of barium chloride, the silver ions would precipitate with the chloride ions (Rule 3) and the barium ions would precipitate with the sulfate ions (Rule 4). This is shown by the equation:

$$\text{NET} \quad 2Ag^+ + SO_4^{2-} + Ba^{2+} + 2Cl^- \rightarrow 2AgCl_{(s)} + BaSO_{4(s)}$$

Actually two precipitation reactions have occurred and the precipitate formed would be a mixture of the two products.

An example of two simultaneous reactions where one is a precipitation and one an acid-base reaction occurs when a barium hydroxide solution is mixed with hydrofluoric acid. The HF donates a proton to the $OH^-$ forming $F^-$ and water. Barium fluroide is insoluble and precipitates. The equation for the reactions is:

$$\text{NET} \quad Ba^{2+} + 2OH^- + 2HF \rightarrow BaF_{2(s)} + 2H_2O$$

## 8-A. Equation Writing

Write equations for the "reaction expected when solutions of lead nitrate and hydrochloric acid are mixed."

Precipitation of $PbCl_2$ is expected by Solubility Rule 3. For the total equation, neutrality of the precipitate requires that the $Pb(NO_3)_2$ : HCl stoichiometry be 1:2. The total equation then is:

$$Pb(NO_3)_{2\,(aq)} + 2\,HCl_{(aq)} \rightarrow PbCl_{2\,(s)} + 2\,HNO_{3\,(aq)}$$

Expanding, we obtain

$$(Pb^{2+}, 2\,NO_3^-) + (2\,H^+, 2\,Cl^-) \rightarrow PbCl_{2\,(s)} + (2H^+, 2\,NO_3^-)$$

$$\text{or} \qquad\qquad \text{or}$$

$$(2\,H_3O^+, 2\,Cl^-) \qquad\qquad (2\,H_3O^+, 2\,NO_3^-)$$

Cancelling those species undergoing no change ($NO_3^-$ and $H^+$, or $H_3O^+$) gives the new ionic equation:

$$Pb^{2+} + 2\,Cl^- \rightarrow PbCl_{2\,(s)}$$

$$\text{or}$$

$$PbCl_2\downarrow$$

## 8-B. Equation Writing

Write equations for what happens "when lead hydroxide is dropped into sulfuric acid solution."

Note that $Pb(OH)_2$ is added as the pure substance, not as a solution (it is "insoluble" anyway, Solubility Rule 5). This is an acid-base reaction that also involves conversion of one solid into another (Solubility Rule 4). The total equation is:

$$Pb(OH)_{2\,(s)} + H_2SO_{4\,(aq)} \rightarrow PbSO_{4\,(s)} + 2\ HOH$$

Expanding, we obtain

$$Pb(OH)_{2\,(s)} + (H^+, HSO_4^-) \rightarrow PbSO_{4\,(s)} + 2\ H_2O$$

However, no species can be deleted from the above expression since in no instance does the same species appear on both the right and left sides of the equation. Thus the net equation is:

$$Pb(OH)_{2\,(s)} + H^+ + HSO_4^- \rightarrow PbSO_{4\,(s)} + 2\ H_2O$$

[$H^+$, meaning $H^+_{(aq)}$, could just as properly been written $H_3O^+$–in which case the coefficient for $H_2O$ would have been 3.]

## Exercises on Aqueous Acid-Base and Precipitation Reactions

1. Write total and net ionic equations for the acid-base and/or precipitation reaction that occurs (if any) when:

   (A) potassium hydroxide and nitric acid solutions are mixed
   (B) solutions of iron(III) chloride and sodium hydroxide are mixed
   (C) iron(III) hydroxide is dissolved by excess sulfuric acid solution
   (D) aqueous copper(II) sulfate is mixed with aqueous sodium sulfide
   (E) aluminum hydroxide is dissolved in perchloric acid solution
   (F) a solution of calcium chloride is mixed with a solution of sodium carbonate
   (G) copper(II) chloride solution is added to a solution of hydrogen sulfide
   (H) a solution of aluminum chloride is added to a solution of silver nitrate
   (I) solid, white cadmium hydroxide is added to hydrosulfuric acid, and a yellow solid is formed
   (J) solutions of silver perchlorate and sodium chromate are mixed
   (K) magnesium hydroxide is dissolved by a large excess of sulfuric acid solution
   (L) excess barium hydroxide solution is added to a solution of phosphoric acid
   (M) solutions of mercury(I) nitrate and calcium chloride are mixed
   (N) solutions of magnesium sulfate and barium hydroxide are mixed
   (O) solutions of zinc nitrate and ammonium sulfate are mixed

2. The net ionic equation for the reaction of **solid** barium hydroxide with **aqueous** HCl is:

   (A) $Ba(OH)_{2\,(s)} + 2\ H^+ + 2\ Cl^- \rightarrow BaCl_{2\,(s)} + 2\ H_2O$

## DRILL ON PRECIPITATE FORMATION

***Student's Name*** ______________________

Write formulas of any compounds that precipitate (possibly more than one) when each of the following pairs of solutions is mixed. If no precipitate forms, write "NP". You may be asked to turn in this sheet.

| Pairs of 0.2 M Solutions | Formula of Precipitate | Pairs of 0.2 M Solutions | Formula of Precipitate |
|---|---|---|---|
| Ammonium sulfate and Lead(II) acetate | | Silver perchlorate and Iron(III) chloride | |
| Sodium acetate and Zinc sulfate | | Magnesium chloride and Barium hydroxide | |
| Silver nitrate and Ammonium chromate | | Cadmium acetate and Sodium sulfide | |
| Zinc nitrate and Sodium carbonate | | Mercury(II) nitrate and Cadmium chloride | |
| ______ and ______ | | ______ and ______ | |

(B) $Ba(OH)_{2\,(s)} + 2\,HCl \rightarrow BaCl_{2\,(s)} + 2\,H_2O$
(C) $Ba^{2+} + 2\,OH^- + 2\,H^+ \rightarrow Ba^{2+} + 2\,H_2O$
(D) $Ba(OH)_{2\,(s)} + 2\,H^+ \rightarrow Ba^{2+} + 2\,H_2O$
(E) $OH^- + H^+ \rightarrow H_2O$

3. An aqueous solution of magnesium sulfate is added to an aqueous solution of each of the following: barium chloride, potassium hydroxide, lead nitrate, and sodium carbonate. In which case(s) would a precipitate be formed?

   (A) barium chloride and sodium carbonate
   (B) barium chloride, lead nitrate, and sodium carbonate
   (C) barium chloride and lead nitrate
   (D) only lead nitrate
   (E) a precipitate forms with each of the four reagents

4. An aqueous solution of ammonium sulfate is added to an aqueous solution of the reagents listed in Question 3. In which case(s) would a precipitate be formed?

   (A) barium chloride and sodium carbonate
   (B) barium chloride, lead nitrate, and sodium carbonate
   (C) barium chloride and lead nitrate
   (D) only lead nitrate
   (E) a precipitate forms with each of the four reagents

5. A precipitate would be expected when an aqueous solution of ammonium sulfate is added to an aqueous solution of:

   (A) zinc chloride (B) ammonium bromide (C) potassium carbonate
   (D) magnesium perchlorate (E) none of these

6. The net ionic equation for the neutralization reaction occurring when $HC_2H_3O_2$ and sodium hydroxide solutions are mixed is:

   (A) $H^+ + OH^- \rightarrow H_2O$
   (B) $HC_2H_3O_2 + NaOH \rightarrow NaC_2H_3O_2 + H_2O$
   (C) $H_3O^+ + OH^- \rightarrow 2\,H_2O$
   (D) $H^+ + NaOH \rightarrow H_2O + Na^+$
   (E) $HC_2H_3O_2 + OH^- \rightarrow H_2O + C_2H_3O_2^-$

7. If two drops of 0.10 *M* $Pb(NO_3)_2$ are added to a mL of 0.25 *M* $K_2SO_4$, a white precipitate forms. Which of the following formulas should appear in the net ionic equation for this reaction?

   (A) $PbSO_4$ (B) $Pb(NO_3)_2$ (C) $Pb(SO_4)_2$ (D) $K_2SO_4$
   (E) none of these

8. Consider the following unbalanced expression:

$$MnO_4^- + H^+ + Fe^{2+} \rightarrow Mn^{2+} + Fe^{3+} + H_2O$$

   (A) $MnO_4^-$ is the reducing agent (B) The manganese has been oxidized

(C) $Fe^{2+}$ gains electrons (D) The iron has been reduced
(E) $Fe^{2+}$ is the reducing agent

9. If you wished to make a **solution** containing $Pb^{2+}$ ions starting from $Pb(OH)_2$, which of the following reagents could be used?

   (A) sulfuric acid solution (B) hydrochloric acid
   (C) carbonic acid solution (D) hydrosulfuric acid (E) nitric acid

---

## TITRATIONS, ANALYSIS, AND SOLUTION STOICHIOMETRY

A very common laboratory task is to determine the amount of a substance in a sample. The most important method of doing this is to dissolve a weighed amount of the sample and then add a fine stream of a solution containing a reagent (the titrant, T) that rapidly reacts with the substance (the *analyte*, A). The addition is stopped as closely as possible to the point at which the reactants are present in their stoichiometric ratios. This is frequently done by adding an *indicator*, a substance that changes color when the first minute excess of titrant is added. From a knowledge of (a) the concentration and volume of the titrant, (b) the volume of the analyte solution, and (c) the balanced chemical equation, the number of millimoles (or % or concentration) of the analyte can be calculated.

Any of the three solution reaction types studied (acid-base, redox, and precipitation) may be used as the basis for a titration.

**Acid-Base Titrations.** Acids may be titrated with bases or bases with acids. The titrant is almost always either a strong acid or a strong base.

**Precipitation Titrations.** In principle, any ion can be used as a titrant that will form a very insoluble precipitate with an oppositely charged analyte ion (See Solubility Rules, p. 188). Some common precipitation titrations involve $Ag^+$ and $Cl^-$, $Br^-$, $I^-$, $SCN^-$, or $CrO_4^{2-}$; $Pb^{2+}$ and $OH^-$, $CrO_4^{2-}$ or $SO_4^{2-}$; and $Ba^{2+}$ and $SO_4^{2-}$ or $CrO_4^{2-}$.

**Redox Titrations.** Usually oxidizing agents are used as titrants; four of the most important of these are listed in Table II along with some common analytes and the changes each of these undergo.

The remainder of the exercises in this chapter will require you to think of actual species in solution, to balance equations, and to recall (from Ch. 2) molarity, reaction stoichiometry, and percentage yields of reactions.

### Example Problems Involving Titrations and Analysis

#### 8-C. Mass of Impure Reactant Needed to React with a Solute

The following reaction occurs quantitatively in aqueous solution:

$$Al^{3+} + 3\,OH^- \rightarrow Al(OH)_3\downarrow$$

**TABLE II. Redox Titrations**

| Titrant (Oxidizing Agent) | Analyte (Reducing Agent) |
|---|---|
| $MnO_4^- \rightarrow Mn^{2+}$ (>1 M $H^+$) | $Fe^{2+} \rightarrow Fe^{3+}$ |
| | $Sn^{2+} \rightarrow Sn^{4+}$ |
| or | $VO^{2+} \rightarrow VO_2^+$ |
| | $Ti^{3+} \rightarrow TiO^{2+}$ |
| $Ce^{4+} \rightarrow Ce^{3+}$ (>1 M $H^+$) | $U^{4+} \rightarrow UO_2^{2+}$ |
| | $H_2O_2 \rightarrow O_2$ |
| or | $H_2C_2O_4 \rightarrow CO_2$ |
| | $HNO_2 \rightarrow NO_3^-$ |
| $Cr_2O_7^{2-} \rightarrow 2Cr^{3+}$ (>1 M $H^+$) | $H_2SO_3 \rightarrow HSO_4^-$ |
| | $H_3AsO_3 \rightarrow H_3AsO_4$ |
| | $I^- \rightarrow I_3^-$ |
| $I_2 \rightarrow 2I^-$ (neutral to slightly acid) | $2\ S_2O_3^{2-} \rightarrow S_4O_6^{2-}$ |
| | $H_2AsO_3^- \rightarrow HAsO_4^{2-}$ |
| | $H_2SbO_3^- \rightarrow Sb(OH)_6^-$ |
| | $H_2S \rightarrow S$ |

Given 50.0 mL of 0.500 *M* aluminum sulfate, how many grams of 85.0% pure KOH (15.0% inert material) need be dissolved to insure precipitation of the aluminum?

(A) 4.95 g (B) 9.90 g (C) 7.15 g (D) 3.58 g. (E) 3.30 g

### *Solution*

mL $Al_2(SO_4)_3$ → mmoles $Al_2(SO_4)_3$ → mmoles $Al^{3+}$ → mmoles $OH^-$ → mmoles KOH → mg KOH → mg impure KOH

$$[50.0\ \cancel{\text{mL Al}_2(\text{SO}_4)_3}]\left(0.500\ \frac{\text{mmoles}\ \cancel{\text{Al}_2(\text{SO}_4)_3}}{\cancel{\text{mL Al}_2(\text{SO}_4)_3}}\right)\left(\frac{2\ \cancel{\text{mmoles Al}^{3+}}}{\cancel{\text{mmole Al}_2(\text{SO}_4)_3}}\right)\times$$

$$\left(\frac{3\ \cancel{\text{mmole OH}^-}}{1\ \cancel{\text{mmole Al}^{3+}}}\right)\left(\frac{1\ \cancel{\text{mmole KOH}}}{1\ \cancel{\text{mmole OH}^-}}\right)\left(\frac{56.1\ \cancel{\text{mg KOH}}}{1\ \cancel{\text{mmole KOH}}}\right)\left(\frac{100\ \text{mg imp. KOH}}{85\ \cancel{\text{mg pure KOH}}}\right) =$$

### *Answer*

| | | |
|---|---|---|
| Calculator | 9.9000····· | |
| Significant figures required | 3 | |
| Answer reported | 9.90 g | [B] |

## 8-D. Solution Resulting from Mixing Acid and Base Solutions

Equal volumes of 0.0500 *M* $Ba(OH)_2$ (base) and 0.200 *M* HCl (acid) are mixed. The reaction that quantitatively ensues yields water. All other substances remain dissolved in solution as ions. If we assume that the final volume is the sum of the volumes of the acid and the base mixed, what are the identities and concentrations of the ions that remain in solution?

(A) 0.200 $M$ $Cl^-$, 0.0500 $M$ $Ba^{2+}$, 0.100 $M$ $H^+$
(B) 0.100 $M$ $Cl^-$, 0.0250 $M$ $Ba^{2+}$, 0.0750 $M$ $H^+$
(C) 0.100 $M$ $Cl^-$, 0.0250 $M$ $Ba^{2+}$, 0.0500 $M$ $H^+$
(D) 0.100 $M$ $Cl^-$, 0.0250 $M$ $Ba^{2+}$, (a neutral solution)
(E) 0.200 $M$ $Cl^-$, 0.0500 $M$ $Ba^{2+}$, (a neutral solution)

### *Solution*

The net reaction is:

$$H^+_{(from\ acid)} + OH^-_{(from\ base)} = H_2O$$

$V$ mL of 0.0500 $M$ $Ba(OH)_2$ contains $0.100V$ mmoles of $OH^-$
$V$ mL of 0.200 $M$ HCl contains $0.200V$ mmoles of $H^+$

### *Answer*

Therefore, in $2V$ mL of solution there would be:

$0.100V$ mmoles excess $H^+$, or $[H^+] = 0.0500\ M$
$0.0500V$ mmoles $Ba^{2+}$ or $[Ba^{2+}] = 0.0250\ M$
$0.200V$ mmoles $Cl^-$ or $[Cl^-] = 0.100\ M$ **[C]**

## 8-E. Volume of Oxidant Solution Required

How many milliliters of 0.0200 $M$ $KMnO_4$ would be needed to oxidize 25.0 mL of 0.110 $M$ $FeSO_4$ in acidic solution yielding $Fe^{3+}$ and $Mn^{2+}$ as products?

(A) 11.0 (B) 20.0 (C) 25.0 (D) 27.5 (E) none of these

### *Solution*

(1) Write and balance the equation
(2) mL $FeSO_4$ → mmoles $FeSO_4$ → mmoles $Fe^{2+}$ → mmoles $MnO_4^-$ → mmoles $KMnO_4$ → mL $KMnO_4$

$$5\ Fe^{2+} + MnO_4^- + 8\ H^+ \rightarrow 5\ Fe^{3+} + Mn^{2+} + 4\ H_2O$$

$$[25.0\ \cancel{mL\ FeSO_4}]\left(0.110\frac{\cancel{mmole\ FeSO_4}}{\cancel{mL\ FeSO_4}}\right)\left(\frac{1\ mmole\ Fe^{2+}}{\cancel{1\ mmole\ FeSO_4}}\right) =$$

2.75 mmole $Fe^{2+}$

$$[2.75\ \cancel{mmole\ Fe^{2+}}]\left(\frac{\cancel{1\ mmole\ MnO_4^-}}{\cancel{5\ mmole\ Fe^{2+}}}\right)\left(\frac{1\ mL\ KMnO_4}{0.0200\ \cancel{mmole\ KMnO_4}}\right) =$$

### *Answer*

| | |
|---|---|
| Calculator | 27.5 |
| Significant figures required | 3 |
| Answer reported | 27.5 mL 0.0200 $M$ $KMnO_4$ **[D]** |

### Example 8-F. % of an Analyte from an Acid-Base Titration

An unknown mixture consisted of solid NaCl and solid $H_2C_2O_4 \cdot 2H_2O$. The unknown was analyzed by titrating the oxalic acid dihydrate with sodium hydroxide solution, the oxalic acid being converted to oxalate ion in the reaction. A 2.500-g sample of the unknown was dissolved in water. It required 20.50 mL of 1.547 M NaOH to react with the oxalic acid in the unknown. What is the % $H_2C_2O_4 \cdot 2H_2O$?

(A) 20.00% (B) 40.00% (C) 60.00% (D) 80.00% (E) 95.17%

#### *Solution*

(1) Write the balanced equation for the reaction that occurs. The sodium chloride does not react and should be ignored.

$$H_2C_2O_4 \cdot 2H_2O + 2NaOH \rightarrow Na_2C_2O_4 + 4H_2O$$

(2) mL NaOH $\rightarrow$ mmol NaOH $\rightarrow$ mmol $H_2C_2O_4 \cdot 2H_2O$ $\rightarrow$ mg $H_2C_2O_4 \cdot 2H_2O$
Then mg $H_2C_2O_4 \cdot 2H_2O$ $\rightarrow$ % $H_2C_2O_4 \cdot 2H_2O$

$$(20.50 \text{ mL NaOH}) \left(\frac{1.547 \text{ mmol NaOH}}{\text{mL NaOH}}\right) \left(\frac{1 \text{ mmol } H_2C_2O_4 \cdot 2H_2O}{2 \text{ mmol NaOH}}\right) \times$$

$$\left(\frac{126.1 \text{ mg } H_2C_2O_4 \cdot 2H_2O}{\text{mmol } H_2C_2O_4 \cdot 2H_2O}\right) = 2000. \text{ mg } H_2C_2O_4 \cdot 2H_2O$$

Then,

$$\% \ H_2C_2O_4 \cdot 2H_2O = \frac{2000. \text{ mg } H_2C_2O_4 \cdot 2H_2O}{2500. \text{ mg unknown}} \times 100 =$$

80.00% $H_2C_2O_4 \cdot 2H_2O$

#### *Answer*

| | | |
|---|---|---|
| Calculator | 80 | |
| Significant Figures Required | 4 | |
| Answer Reported | 80.00% $H_2C_2O_4 \cdot 2H_2O$ | [D] |

## Exercises on Solution Stoichiometry

10. A solution is prepared by dissolving 4.20 g of NaF (FW = 42.0) and 12.0 g of $Na_2SO_4$ in water and finally making the volume up to 100.0 mL. What is the concentration of sodium ion ($Na^+$) in the solution?

    (A) 0.100 *M* (B) 1.00 *M* (C) 2.00 *M* (D) 3.00 *M* (E) 2.69 *M*

## DRILL ON REDOX STOICHIOMETRY

***Student's Name*** ____________________

A number of possible redox titrations have been suggested in Table II. How many mL of 0.08000 M titrant (T) would be needed to react with a 25.00 mL sample that is 0.1000 M in the analyte (A)?

| (A) 25.00 mL of 0.100 M | (T) Titrant | mL of Titrant | mmoles of Titrant |
|---|---|---|---|
| $VO^{2+}$ | $MnO_4^-$ | | |
| $H_2O_2$ | $Ce^{4+}$ | | |
| $Sn^{2+}$ | $I_2$ | | |
| $I_2$ | $S_2O_3^{2-}$ | | |
| $I^-$ | Enter your choice | | |
| | | | |

11. If 40.0 mL of 0.125 *M* acid is completely neutralized by 50.0 mL of a 0.200 *M* base, how many moles of the base would be required to completely neutralize 1.00 mole of the acid?

    (A) 1.00 (B) 2.00 (C) 3.00 (D) 1.50 (E) 0.500

12.*When 0.10 *M* solutions of potassium oxalate and lanthanum nitrate are mixed, a precipitate forms. The solutions are to be mixed so that the final volume will be 100 mL (assume additive volumes) and so that the maximum amount of precipitate will be formed. What proportions should be used?

    (A) 40 mL $La(NO_3)_3$ solution; 60 mL $K_2C_2O_4$ solution
    (B) 50 mL $La(NO_3)_3$ solution; 50 mL $K_2C_2O_4$ solution
    (C) 30 mL $La(NO_3)_3$ solution; 70 mL $K_2C_2O_4$ solution
    (D) 67 mL $La(NO_3)_3$ solution; 33 mL $K_2C_2O_4$ solution
    (E) 33 mL $La(NO_3)_3$ solution; 67 mL $K_2C_2O_4$ solution

13. A solution containing 1.000 mmole of $Na_2B_4O_7$ was titrated with 0.05761 *M* $H_2SO_4$ in the presence of methyl red–bromcresol green mixed indicator. 17.36 mL of the acid were required to reach the endpoint. If the equation is of the form:

$$\_\_\_\ Na_2B_4O_{7\,(aq)} + \_\_\_\ H_2SO_{4\,(aq)} + \_\_\_\ H_2O_{(\ell)} \rightarrow \_\_\_\ H_3BO_{3\,(aq)} + \_\_\_\ Na_2SO_{4\,(aq)},$$

    the ratio of the coefficients, $H_2SO_4 : Na_2B_4O_7$, must be:

    (A) 2:1 (B) 3:2 (C) 1:1 (D) 1:2 (E) 3:5

14. If the solution in Question 13 has a volume of ~93 mL at the endpoint of the titration, the molarity of the $H_3BO_3$ in the titration flask at the endpoint would be:

    (A) 0.0108 *M* (B) 0.058 *M* (C) 0.11 *M* (D) 0.043 *M* (E) 0.230 *M*

15. Two solutions are prepared as follows:

    Solution A: 221 mg of $Pb(NO_3)_2$ in 52.8 mL of water
    Solution B: 175 mg of KBr in 47.2 mL of water

    The solutions are mixed, precipitating $PbBr_2$. The precipitate was collected, dried, and found to weigh 200. mg. What was the % yield of $PbBr_2$ in the experiment?

    (A) 81.6% (B) 74.1% (C) 37.0% (D) 50.5% (E) 73.0%

16. Sufficient 0.100 *M* KOH is added to 25.0 mL of 0.0250 *M* $H_3PO_4$ to convert the phosphoric acid into $HPO_4^{2-}$. How many milliliters of the KOH solution are required?

    (A) 6.25 (B) 20.0 (C) 10.0 (D) 7.75 (E) 12.5

17. When aqueous $KMnO_4$ and acidified $FeSO_4$ are mixed, $Mn^{2+}$ and $Fe^{3+}$ are formed. Suppose 50.0 mL of 0.0200 *M* $FeSO_4$ and 50.0 mL of 0.0200 *M*

$KMnO_4$ are mixed. The final solution, after reaction, would have which of the following concentrations:

(A) $8.00 \times 10^{-3}$ *M* $MnO_4^-$ (B) $5.00 \times 10^{-3}$ *M* $Fe^{2+}$
(C) $5.00 \times 10^{-3}$ *M* $Fe^{3+}$ (D) $8.00 \times 10^{-3}$ *M* $Mn^{2+}$
(E) $1.00 \times 10^{-3}$ *M* $K^+$

18. When solutions of $Cu^{2+}$ and $I^-$ are mixed, iodine and $CuI_{(s)}$ are formed. How many mmoles of $I_2$ would be formed by dissolving 2.0 mmoles of $CuSO_4 \cdot 5H_2O$, then adding 25 mL of 0.200 *M* KI?

(A) 2.0 (B) 1.0 (C) 5.0 (D) 2.5 (E) 4.0

19. What is the maximum amount of $Al(OH)_3$ (FW = 78.0) that can be prepared by mixing 20.0 mL of 2.00 *M* $NH_3$ with 40.0 mL of 0.500 *M* $Al^{3+}$ solution. The reaction is

$$Al^{3+} + 3\,NH_3 + 3\,H_2O \rightarrow Al(OH)_{3\,(s)} + 3\,NH_4^+$$

(A) 1.56 g (B) 1.56 mg (C) 3.12 g (D) 3.12 mg (E) 1.04 g

20. A 50.00-mL sample of 0.2500 *M* $H_2SO_4$ was diluted to 75 mL and titrated with a NaOH solution of unknown concentration. If 40.00 mL of the NaOH solution is required to neutralize the acid, what is the molarity of the base?

(A) 0.4688 *M* (B) 0.9376 *M* (C) 0.3244 *M* (D) 0.6250 *M*
(E) $4.167 \times 10^{-3}$ *M*

21.*When acidic solutions of $K_2Cr_2O_7$ and KI are mixed, $I_2$ and $Cr^{3+}$ are formed quantitatively. If equal volumes of acidified 0.100 *M* $K_2Cr_2O_7$ and 0.100 *M* KI are mixed, what concentration of $Cr^{3+}$ should result? (Assume that the volumes are additive.)

(A) 0.10 *M* (B) 0.017 *M* (C) 0.033 *M* (D) 0.050 *M* (E) none of these

22. The following reaction occurs upon heating the reactants strongly in an iron crucible:

$$MnO_2 + KClO_3 + KOH \rightarrow K_2MnO_4 + KCl + H_2O \quad \text{(unbalanced)}$$

The directions call for 0.30 mole of $MnO_2$ along with the theoretical (calculated) quantities of $KClO_3$ and KOH. How many **moles** of $KClO_3$ and KOH should be used?

(A) 0.10 mole of $KClO_3$ and 0.017 mole of KOH
(B) 0.10 mole of $KClO_3$ and 0.60 mole of KOH
(C) 0.30 mole of $KClO_3$ and 0.30 mole of KOH
(D) 0.90 mole of $KClO_3$ and 0.50 mole of KOH
(E) 0.30 mole of $KClO_3$ and 0.10 mole of KOH

23. What is the molar concentration (molarity) of NaOH in a solution prepared by mixing 50.0 mL of 5.00 *M* NaOH with 150.0 mL of 2.00 *M* NaOH? (Assume additive volumes)

(A) 11.0 *M* (B) 3.67 *M* (C) 3.50 *M* (D) 2.75 *M* (E) 0.364 *M*

24. Which of the following conversions would be a reduction?

    (A) $H_3NOH^+ \rightarrow H_2N_2O_2$ (B) $H_3NOH^+ \rightarrow HNO_2$
    (C) $H_3NOH^+ \rightarrow NH_4^+$ (D) $H_3NOH^+ \rightarrow H_2NOH$
    (E) none of these

25. In which of the following changes would it be said that electrons are **lost** by the nitrogen atom?

    (A) $NF_3 \rightarrow NH_3$ (B) $HNO_2 \rightarrow N_2O_3$ (C) $KN_3 \rightarrow KNH_2$
    (D) $BaN_2O_2 \rightarrow Ba(NO_2)_2$ (E) $H_2NOH \rightarrow N_2H_4$

26. In which of the conversions in Q-25 (above) would it be said that the substance on the left is oxidized?

    (A) the process in 25-A (B) the process in 25-B
    (C) the process in 25-C (D) the process in 25-D
    (E) the process in 25-E

27.*A basic copper carbonate of formula $[Cu(OH)_2 \cdot xCuCO_3]$ reacts quantitatively with hydrochloric acid, producing $CuCl_{2\,(aq)}$, $CO_2$, and $H_2O$. If 2.000 g of the basic carbonate requires 34.82 mL of 1.000 *M* HCl, what is the value of *x*? (FW of $Cu(OH)_2$ = 97.56; FW of $CuCO_3$ = 123.55)

    (A) 1/2 (B) 1 (C) 2 (D) 3 (E) 5/3

28. For the reaction *(Pharaoh's serpents)*:

$$2\,KSCN_{(aq)} + Hg(NO_3)_{2\,(aq)} \rightarrow Hg(SCN)_{2\,(s)} + 2\,KNO_{3\,(aq)}$$

What would be the theoretical yield of $Hg(SCN)_2$ if 30.0 mL of 0.750 *M* KSCN is mixed with 20.0 mL of 0.700 *M* $Hg(NO_3)_2$?

    FW's: KSCN, 97.18; $Hg(SCN)_2$, 316.75; $Hg(NO_3)_2$, 324.60;
    $KNO_3$, 101.10

    (A) 4.43 g (B) 3.56 g (C) 0.225 g (D) 7.13 g (E) 14.3 g

29.*Aqueous $XSO_4$ reacts with aqueous NaOH to form quantitatively a precipitate of $X_3(OH)_4SO_4$ and a solution of $Na_2SO_4$. 1.0 *M* $XSO_4$ and 1.0 *M* NaOH are to be mixed so as to give a total number of 50. mL of solution (assume additive volumes). What volumes of these two solutions should be mixed in order to obtain the **maximum** weight of precipitate?

    (A) 29 mL NaOH + 21 mL $XSO_4$ (B) 21 mL NaOH + 19 mL $XSO_4$
    (C) 15 mL NaOH + 35 mL $XSO_4$ (D) 20 mL NaOH + 30 mL $XSO_4$
    (E) 11 mL NaOH + 39 mL $XSO_4$

30. Antimony trichloride is quantitatively *hydrolyzed* by water, yielding a white precipitate and a solution of HCl. In a particular experiment a 9.653-g sample of $SbCl_3$ (FW = 228.1) was treated with excess water (exactly 100 mL) and allowed to react. The precipitate was removed by filtration and 20.00 mL of the 100-mL solution was titrated with 0.5123 *M* NaOH; 41.30 mL of the latter was required to neutralize the HCl sample

taken. Which of the following balanced equations best represents the hydrolysis reaction? [See *Journal of Chemical Education*, 52, 189 (1975).]

(A) $2\ SbCl_{3(s)} + 3\ H_2O_{(\ell)} \rightarrow Sb_2O_{3(s)} + 6\ HCl_{(aq)}$
(B) $SbCl_{3(s)} + H_2O_{(\ell)} \rightarrow SbOCl_{(s)} + 2\ HCl_{(aq)}$
(C) $4\ SbCl_{3(s)} + 5\ H_2O_{(\ell)} \rightarrow Sb_4O_5Cl_{2(s)} + 10\ HCl_{(aq)}$
(D) $2\ SbCl_{3(s)} + H_2O_{(\ell)} \rightarrow Sb_2OCl_{4(s)} + 2\ HCl_{(aq)}$
(E) $SbCl_{3(s)} + 3\ H_2O_{(\ell)} \rightarrow Sb(OH)_{3(s)} + 3\ HCl_{(aq)}$

31. When steel wool is placed in a solution of $CuSO_4$, the steel becomes coated with copper and the characteristic blue color of the solution fades as the following reaction proceeds:

$$Fe_{(s)} + Cu^{2+} \rightarrow Cu_{(s)} + Fe^{2+}$$

For this reaction, it can be said that:

(A) Fe is the oxidizing agent (B) Fe is reduced (C) $Cu^{2+}$ loses electrons
(D) $Cu^{2+}$ is an oxidizing agent (E) Fe gains electrons

32. 10. mL of 0.25 *M* $H_3PO_4$ and 18 mL of 0.25 *M* NaOH are mixed and then diluted to a volume of 100 mL. After the acid-base reaction occurs, what phosphorus species are in solution and what are their molar concentrations? (Note that in the neutralization reaction the second proton in $H_3PO_4$ is not removed until the first one has been neutralized, *etc.*)

(A) 0.025 *M* $H_2PO_4^-$ (B) 0.005 *M* $H_2PO_4^-$; 0.020 *M* $HPO_4^{2-}$
(C) 0.007 *M* $H_2PO_4^-$; 0.045 *M* $HPO_4^{2-}$
(D) 0.015 *M* $H_2PO_4^-$; 0.010 *M* $HPO_4^{2-}$
(E) 0.001 *M* $H_2PO_4^-$; 0.0024 *M* $PO_4^{3-}$

33. The formula of magnesium chlorate is:

(A) $Mg(ClO_3)_2$ (B) $MgClO_4$ (C) $Mg(ClO_2)_2$ (D) $Mg_2(ClO_3)_3$
(E) $Mg_3(ClO_3)_3$

34. Calculate the total number of moles of ions (cations **plus** anions) in one liter of a 1.0 *M* solution of sodium phosphate.

(A) 1.0 mole (B) 2.0 moles (C) 4.0 moles (D) 6.0 moles
(E) 8.0 moles

35. Which one of the following reactions is an oxidation-reduction reaction?

(A) $HC_2H_3O_2 + OH^- \rightarrow H_2O + C_2H_3O_2^-$
(B) $Pb^{+2} + 2\ Br^- \rightarrow PbBr_{2(s)}$
(C) $HSO_4^- \rightarrow H^+ + SO_4^{2-}$
(D) $AgNO_3 + NaCl \rightarrow AgCl + NaNO_3$
(E) $H_2 + Cl_2 \rightarrow 2\ HCl$

36. Which of the following statements best describes the result of mixing solutions of barium hydroxide and iron(III) sulfate?

(A) no chemical reaction occurs

(B) $Fe(OH)_3$ will precipitate
(C) $Ba_2SO_4$ will precipitate
(D) both $Fe(OH)_3$ and $BaSO_4$ will precipitate
(E) this is a typical oxidation-reduction reaction

37. You are given $1.00 \times 10^{-3}$ mole of $[Ni(NH_3)_6]Cl_2$ to analyze for chloride by precipitation of AgCl. The sample is dissolved in nitric acid and AgCl precipitated by addition of 0.20 *M* $AgNO_3$. What is the **minimum** volume of silver nitrate solution that need be added to precipitate all of the chloride?

    (A) 232 mL (B) 50.0 mL (C) 0.00400 mL (D) 10.0 mL
    (E) 25.0 mL

38. The following oxidation-reduction reaction occurs in **acidic** solution:

$$Cr_2O_7^{2-} + I_2 \rightarrow IO_3^- + Cr^{3+}$$

The coefficients of $I_2$ and $H^+$ in the balanced equation are, respectively:

    (A) 5 $I_2$, 16 $H^+$ (B) 3 $I_2$, 16 $H^+$ (C) 3 $I_2$, 34 $H^+$ (D) 5 $I_2$, 34 $H^+$
    (E) 6 $I_2$, 20 $H^+$

39. The reducing agent in the reaction in Question 38 is:

    (A) $Cr_2O_7^{2-}$ (B) $I_2$ (C) $H^+$ (D) $IO_3^-$ (E) $Cr^{3+}$

40. The oxidation number of iodine in the compound $Ca(IO_3)_2$ is:

    (A) +7 (B) +6 (C) -1 (D) -2 (E) +5

41.*A 2.00-g sample of $X_6(OH)_8(CO_3)_2$ was ignited, eliminating $H_2O$ and $CO_2$ and leaving 1.787 g of XO as residue. What is element X?

    (A) Sn (B) Cu (C) Pb (D) Ni (E) Ag

42.*A sample of $PbCO_3 \cdot xPb(OH)_2$ was treated with **excess** sulfuric acid, yielding $PbSO_{4\,(s)}$, $H_2O_{(\ell)}$, and $CO_{2\,(g)}$. It was observed that 0.0726 g of $CO_2$ was liberated **per gram** of $PbSO_4$ formed. What is the value of "*x*"?

    (A) 1 (B) 2 (C) 3 (D) 4 (E) 5

43. 0.1500 g of a sample contains 19.48% sodium and 25.98% potassium by weight. These metals are present as the chlorides, and no other sources of chlorine are present. How many milliliters of 0.04965 *M* $Hg(NO_3)_2$ would be needed to titrate the chloride in the sample? The titration reaction forms $HgCl_2$ molecules. See J. Chem. Ed. **52**, 714 (1975).

    (A) 23.24 mL (B) 16.75 mL (C) 12.02 mL (D) 19.48 mL
    (E) 22.83 mL

44. The following reaction occurs in **basic** solution:

$$SbO_2^- + ClO_2 \rightarrow ClO_2^- + Sb(OH)_6^-$$

The balanced equation indicates that for every mole of $ClO_2$ consumed, ______ mole(s) of $SbO_2^-$ will be required.

(A) 1/2 (B) 1 (C) 1-1/2 (D) 2 (E) 2-1/2

45. Consider the reaction given in Question 44. Starting with 100. mL of 1.00 *M* $SbO_2^-$ (as $KSbO_2$), what would be the maximum weight of $KSb(OH)_6$ (FW = 263) that could be formed?

    (A) 26.3 g (B) 5.26 g (C) 52.6 g (D) 13.2 g (E) 132 g

46. 2.00 g of a mixture of potassium nitrate and potassium chloride is dissolved in 25.0 mL of water. 100. mL of a 0.500 *M* solution of $AgNO_3$ is added. If 0.958 g of precipitate is formed, what percentage (by weight) of the original mixture was potassium nitrate?

    (A) 25% (B) 45% (C) 65% (D) 75% (E) 85%

47. A student in Laboratory Section 195 reported that a white precipitate was formed when solutions of $La(NO_3)_3$ and $K_2C_2O_4$ were mixed, and that a red precipitate was formed upon mixing solutions of $AgNO_3$ and $K_2CrO_4$. The most likely formulas of the white and red precipitates, respectively, are:

    (A) $LaC_2O_4$ and $Ag(CrO_4)_2$ (B) $La(C_2O_4)_2$ and $AgCrO_4$
    (C) $La_2C_2O_4$ and $Ag(CrO_4)_2$ (D) $La_2(C_2O_4)_3$ and $Ag_2CrO_4$
    (E) $La(C_2O_4)_3$ and $Ag_2CrO_4$

---

## "Chemical Equivalents" and Normality

A *chemical equivalent* (or *equivalent weight, EW*) of a substance is that weight of substance that in a **specified** reaction accepts or supplies one mole of unit charges, *i.e.*, the quantity of charge on $6.02 \times 10^{23}$ electrons. Commonly, in acid-base reactions this charge is represented by 1 mole of $H^+$, or 1 mole of $OH^-$; in redox reactions by 1 mole of $e^-$. Thus, the equivalent weight of a substance **may vary, depending on its behavior in a particular chemical reaction,** as shown in the following table. The number of equivalents of a substance will always be given by the product of some small integer ($n$) times the number of moles of the substance, where the integer represents the number of protons donated or accepted or the number of electrons donated or accepted, per formula unit of the substance.

Solution concentration can be expressed in terms of equivalent weights of solute per liter of solution; this is said to be the *normality, N,* of the solution:†

---

†At times, people are careless about specifying the reaction for which a reagent, labeled as to normality, is to be used. If the reagent is an acid or a base, then the labeled normality would be taken as meaning the use of the maximum number of $H^+$ or $OH^-$ groups available from the reagent. For instance, a bottle labeled merely "2.0 *N* $H_2SO_4$" would be likely to contain 98 g of $H_2SO_4$/liter.

$$N = \frac{\text{equivalents}}{\text{L of solution}} = \frac{n(\text{moles of solute})}{\text{L of solution}} = nM$$

The convenience of normality is that **one** equivalent (one EW) of substance A will exactly react with **one** equivalent (one EW) of substance B, so that:

$$(\text{L soln A})(N_A) = (\text{L soln B})(N_B) \text{ or } (\text{mL}_A)(N_A) = (\text{mL}_B)(N_B)$$

It also follows that:

$$(\text{mL}_A)(N_A) = \frac{\text{mg of A}}{\text{EW of A}} = \frac{\text{mg of B}}{\text{EW of B}}$$

Thus, for the acid-base reaction,

$$3\ Ba(OH)_{2\,(aq)} + 2\ H_3PO_{4\,(aq)} \rightarrow Ba_3(PO_4)_{2\,(s)} + 6\ H_2O$$

3 moles require ↑ 2 moles ↑

$\times \frac{2 \text{ equiv.}}{\text{mole}}$ ↓ $\times \frac{3 \text{ equiv.}}{\text{mole}}$ ↓

6 equiv. require 6 equiv.; therefore, 1 : 1 by equivalents (EW's)

and for the redox reaction,

$$2\ KMnO_{4\,(aq)} + 10\ KI_{(aq)} + 8\ H_2SO_{4\,(aq)} \rightarrow 5\ I_{2\,(s)} + 6\ K_2SO_{4\,(aq)} + 2\ MnSO_{4\,(aq)} + 8\ H_2O$$

2 moles require 10 moles

$\times \frac{5 \text{ equiv.}}{\text{mole}}$ ↓ $\times \frac{1 \text{ equiv.}}{\text{mole}}$ ↓

10 equiv. require 10 equiv; Again, 1 : 1 by equivalents (EW's)

## Normality Problems

48. If potassium dichromate is to be used as an oxidizing agent as in case No. 5 in the Table of p. 208, the number of grams of compound needed to prepare 500. mL of a 0.100 $N$ solution would be:

    (A) 14.7 g (B) 2.45 g (C) 0.294 g (D) 8.33 g (E) 1.47 g

49. 48.50 mL of 0.1125 $N$ $KMnO_4$ is required for the following titration:

    $$VO^{2+} + MnO_4^- \rightarrow VO_2^+ + Mn^{2+} \quad \text{(unbalanced)}$$

    How many milligrams of **vanadium** were present in the solution?

    (A) 5.456 (B) 55.58 (C) 277.9 (D) 101.9 (E) 152.8

50. Suppose that barium hydroxide is used in a neutralization reaction that

| No. | Solute | Reaction | Behavior | Product | EW = | Footnote |
|---|---|---|---|---|---|---|
| 1. | $H_2SO_4$ | Acid-base | Acid | $HSO_4^-$ | FW/1 | (a) |
| 2. | $H_2SO_4$ | Acid-base | Acid | $SO_4^{2-}$ | FW/2 | (b) |
| 3. | NaOH | Acid-base | Base | $Na^+$ | FW/1 | (c) |
| 4. | $Al(OH)_3$ | Acid-base | Base | $Al^{3+}$ | FW/3 | (d) |
| 5. | $K_2Cr_2O_7$ | Redox | Oxidizing agent | $Cr^{3+}$ | FW/6 | (e) |
| 6. | $H_2SO_3$ | Redox | Reducing agent | $HSO_4^-$ | FW/2 | (f) |
| 7. | $H_2SO_4$ | Redox | Oxidizing agent | $SO_2$ | FW/2 | (g) |

(a) Since 1 mole of $H_2SO_4$ has furnished 1 mole of positive charge (1 mole $H^+$)
(b) Since 1 mole of $H_2SO_4$ has furnished 2 moles of positive charges (2 moles $H^+$)
(c) Since 1 mole of NaOH has furnished 1 mole of negative charge (1 mole $OH^-$)
(d) Since 1 mole of $Al(OH)_3$ furnishes 3 moles of negative charge (3 moles $OH^-$)
(e) Since $Cr_2O_7^{2-} + 14\,H^+ + 6\,e^- \rightarrow 2\,Cr^{3+} + 7\,H_2O$, *i.e.*, 1 mole of $K_2Cr_2O_7$ has accepted 6 moles of negative charge (6 moles of $e^-$).
(f) Since $H_2SO_3 + H_2O \rightarrow HSO_4^- + 3\,H^+ + 2\,e^-$, *i.e.*, 1 mole of $H_2SO_3$ has lost 2 moles of negative charge (2 moles of $e^-$).
(g) Since $H_2SO_4 + 2\,H^+ + 2\,e^- \rightarrow SO_2 + 2\,H_2O$, *i.e.*, 1 mole of $H_2SO_4$ has gained 2 moles of negative charges (2 moles $e^-$).

---

utilizes all of the hydroxide ions. A solution that is labeled 0.0100 *N* would contain:

(A) 0.0100 mole of barium hydroxide per liter
(B) 1.71 g of barium hydroxide per liter
(C) 0.857 g of barium hydroxide per liter
(D) 0.0200 equivalents of barium hydroxide per liter
(E) 0.342 g of barium hydroxide per liter

51. If the reaction of 1.00 millimole of $H_nX$ requires 20.00 mL of 0.2000 N NaOH for

$$H_nX + n\,OH^- \rightarrow X^{n-} + n\,H_2O,$$

the equivalent weight of $H_nX$ is equal to its formula weight divided by:

(A) 1 (B) 2 (C) 3 (D) 4 (E) 5

52. If a 1.02-g sample of an acid, $H_nX$, requires 39.1 mL of 0.1279 *N* base for neutralization, what is the equivalent weight of the acid?

(A) 28 (B) 56 (C) 102 (D) 158 (E) 204

53. If the solution described in Question 48 is to be used as previously indicated to titrate a sample containing 50.0 mg of iron (II) to iron (III), the number of milliliters required would be:

(A) 8.95 (B) 13.4 (C) 26.8 (D) 10.0 (E) none of these

54. If 20.00 mL of 0.1000 *M* phosphoric acid is titrated with 0.1250 *N* sodium hydroxide to the point where it has just been converted to $HPO_4^{2-}$ (phenolphthalein endpoint), then the volume of base required is:

    (A) 20.00 mL (B) 12.50 mL (C) 25.00 mL (D) 32.00 mL
    (E) 37.50 mL

55. If a solution containing 2.50 millimoles of $X^n$ (where *n* may be either a positive or negative integer) requires 40.0 mL of 0.125 *N* $KMnO_4$ solution for oxidation to $XO_2^+$, then the charge on the original $X^n$ ion is:

    (A) -1 (B) +1 (C) +2 (D) +3 (E) +4

56. The solid compound "potassium tetroxalate" has the formula $KH_3(C_2O_4)_2$ with FW = 218.16. It can be used to determine accurately the concentration of either sodium hydroxide solutions (all three protons are neutralized) or potassium permanganate solutions (oxalate oxidized to $CO_2$ in acid solution; $Mn^{2+}$ also formed) by titration. If one liter of solution contains 4.3632 g of this substance, the normality of the solution would be _____ N when titrated with NaOH and _____ N when titrated with $KMnO_4$. Fill in the blanks, respectively.

    (A) 0.02000 N; 0.03000 N
    (B) 0.06000 N; 0.04000 N
    (C) 0.06000 N; 0.08000 N
    (D) 0.03000 N; 0.02000 N
    (E) 0.02000 N; 0.02000 N

---

## ANSWERS TO CHAPTER 8 PROBLEMS

1. The first equation is the net ionic equation; the second one, the total equation.

    (A) $OH^- + H_3O^+ \rightarrow 2\,H_2O$
    $KOH_{(aq)} + HNO_{3\,(aq)} \rightarrow KNO_{3\,(aq)} + H_2O$

    (B) $Fe^{3+} + 3\,OH^- \rightarrow Fe(OH)_{3\,(s)}$
    $FeCl_{3\,(aq)} + 3\,NaOH_{(aq)} \rightarrow Fe(OH)_{3\,(s)} + 3\,NaCl_{(aq)}$

    (C) $Fe(OH)_{3\,(s)} + 3\,H_3O^+ \rightarrow Fe^{3+} + 6\,H_2O$
    $Fe(OH)_{3\,(s)} + 3\,H_2SO_{4\,(aq)} \rightarrow Fe(HSO_4)_{3\,(aq)} + 3\,H_2O$

    (D) $Cu^{2+} + S^{2-} \rightarrow CuS_{(s)}$
    $CuSO_{4\,(aq)} + Na_2S_{(aq)} \rightarrow CuS_{(s)} + Na_2SO_{4\,(aq)}$

    (E) $Al(OH)_{3\,(s)} + 3\,H_3O^+ \rightarrow Al^{3+} + 6\,H_2O$
    $Al(OH)_{3\,(s)} + 3\,HClO_{4\,(aq)} \rightarrow Al(ClO_4)_{3\,(aq)} + 3\,H_2O$

(F) $Ca^{2+} + CO_3^{2-} \rightarrow CaCO_{3\,(s)}$
$CaCl_{2\,(aq)} + Na_2CO_{3\,(aq)} \rightarrow CaCO_{3\,(s)} + 2\ NaCl_{(aq)}$

(G) $Cu^{2+} + H_2S_{(aq)} + 2\ H_2O \rightarrow CuS_{(s)} + 2\ H_3O^+$
$CuCl_{2\,(aq)} + H_2S_{(aq)} \rightarrow CuS_{(s)} + 2\ HCl_{(aq)}$

(H) $Cl^- + Ag^+ \rightarrow AgCl_{(s)}$
$AlCl_{3\,(aq)} + 3\ AgNO_{3\,(aq)} \rightarrow 3\ AgCl_{(s)} + Al(NO_3)_{3\,(aq)}$

(I) $Cd(OH)_{2\,(s)} + H_2S_{(aq)} \rightarrow CdS_{(s)} + 3\ H_2O$
The net ionic and total equations are the same!

(J) $2\ Ag^+ + CrO_4^{2-} \rightarrow Ag_2CrO_{4\,(s)}$
$2\ AgClO_{4\,(aq)} + Na_2CrO_{4\,(aq)} \rightarrow Ag_2CrO_{4\,(s)} + 2\ NaClO_{4\,(aq)}$

(K) $Mg(OH)_{2\,(s)} + 2\ H_3O^+ \rightarrow Mg^{2+} + 4\ H_2O$
$Mg(OH)_{2\,(s)} + 2\ H_2SO_{4\,(aq)} \rightarrow Mg(HSO_4)_{2\,(aq)} + 2\ H_2O$

(L) $3\ Ba^{2+} + 6\ OH^- + 2\ H_3PO_4 \rightarrow Ba_3(PO_4)_{2\,(s)} + 6\ H_2O$
$3\ Ba(OH)_{2\,(aq)} + 2\ H_3PO_{4\,(aq)} \rightarrow Ba_3(PO_4)_{2\,(s)} + 6\ H_2O$

(M) $Hg_2^{2+} + 2\ Cl^- \rightarrow Hg_2Cl_{2\,(s)}$
$Hg_2(NO_3)_{2\,(aq)} + CaCl_{2\,(aq)} \rightarrow Hg_2Cl_{2\,(s)} + Ca(NO_3)_{2\,(aq)}$

(N) $Mg^{2+} + SO_4^{2-} + Ba^{2+} + 2\ OH^- \rightarrow Mg(OH)_{2\,(s)} + BaSO_{4\,(s)}$
$MgSO_{4\,(aq)} + Ba(OH)_{2\,(aq)} \rightarrow Mg(OH)_{2\,(s)} + BaSO_{4\,(s)}$

(O) No Reaction

2. D
3. E
4. C
5. E
6. E
7. A
8. E
9. E
10. E
11. B
12. A
13. C
14. D
15. A
16. E
17. A
18. B
19. A
20. D
21. B
22. B
23. D
24. C
25. D
26. D
27. C
28. B
29. A
30. C
31. D
32. B
33. A
34. C
35. E
36. D
37. D
38. C
39. B
40. E
41. C
42. A
43. E
44. A
45. A
46. D
47. D
48. B
49. C
50. C
51. D
52. E
53. A
54. D
55. D
56. C

# 9
# Gases

The gaseous state is a *dispersed* state of matter, meaning that the gas molecules, on the average, are separated by distances that are many times greater than the actual molecular diameters. The result of this is that the volume occupied by the gas ($V$) is primarily dependent on pressure ($P$), temperature ($T$), and number of moles ($n$, or molecules). Since this is the case, it is convenient to describe real gases in terms of an *ideal gas,* a hypothetical system of widely separated point masses (molecules) that exert no influence on each other. The *kinetic theory* of gases describes this model and, in addition, states that the **average** kinetic energy is proportional to the *absolute* temperature in *kelvins* (K, where K = °C + 273.16):

$$\text{Average kinetic energy} = 1/2m\bar{s}^2 = \text{Constant}\cdot T$$

$m$ is the molecular mass and is proportional to $MW$; $\bar{s}$ is the average molecular speed.

Thus, for an ideal gas the expression relating the variables,

$$PV = nRT$$

is called the *ideal gas law.* $R$ is the proportionality constant, the *ideal gas constant,* the numerical value of which depends upon the units of $P$ and $V$ ($R$ = 0.08205 L-atm/mole-K = 82.05 cm³-atm/mole-K = 1.9869 cal/mole-K = 62.4 L-torr/mole-K, *etc.*). It should be noted that this equation reduces to *Boyle's law* (at constant $T$ and $n$), to *Charles' law* (at constant $P$ and $n$), to *Avogadro's law* (at constant $T$ and $P$), and to *Dalton's law* of partial pressures (for nonreacting gas mixtures). Note that 1 atm is equivalent to 760. *millimeters of Hg* (also called 760 *torr*).

Dalton's law emphasizes that the gas molecules act independently of their chemical properties and, hence, for a mixture of several components:

$$P_{total}\cdot V = n_{total}\cdot RT$$

where

$$P_{total} = p_i + p_j + p_k + \cdots \quad \text{and} \quad n_{total} = n_i + n_j + n_k + \cdots$$

$p_i$ is the *partial pressure* of an individual gas, *i.e.*, the pressure that gas "i" would have if it were present in the container alone. It is convenient to remember Dalton's law in the form:

$$p_i = X_i P_{total}$$

where $X_i$, the *mole fraction* of component "i," is a concentration unit,

$$X_i = \frac{n_i}{n_i + n_j + n_k + \cdots} = \frac{n_i}{n_{total}}$$

A common application of Dalton's law is the calculation of the pressure of a gas that has been collected over water. The total pressure is given by $P_{total} = p_{gas} + p_{H_2O}$. $p_{gas}$ is found by subtracting $p_{H_2O}$, the vapor pressure of water at the experimental temperature, from the measured gas pressure.

Gas law problems fall into three categories: (1) those involving **changes** from a set of initial conditions ($P_1$, $V_1$, $T_1$, $n_1$) to a final set of conditions ($P_2$, $V_2$, $T_2$, $n_2$) and, hence,

$$\frac{P_1 V_1}{n_1 T_1} = \frac{P_2 V_2}{n_2 T_2}$$

(2) those involving **static systems** in which data are given for all but one of the variables, which may be calculated by direct application of the ideal gas law, $PV = nRT$ [Note that various algebraic rearrangements of the equation lead to values of the molecular weight ($n$ = g/MW), the molar volume ($V/n$), which at STP is equal to 22.4 liters, and gas density (g/V)] ; and (3) those involving **molecular transport** (*e.g.*, rate of diffusion) that are based on Graham's law (proportionality of kinetic energy to temperature):

$$\frac{\text{Rate of diffusion}_A}{\text{Rate of diffusion}_B} = \frac{\bar{s}_A}{\bar{s}_B} = \sqrt{\frac{\text{density}_B}{\text{density}_A}} = \sqrt{\frac{\text{MW}_B}{\text{MW}_A}}$$

---

## Example Problems Involving Units and Type 1 Gas Law Problems

### 9-A. Pressure Units and the Use of a Water Manometer

A gas sample was collected in a 500-mL bulb at 22°C and attached to a manometer containing water as shown in the figure. The difference in the water column heights in the legs of the manometer was 55 cm. The atmospheric pressure was 764 mm of Hg. What is the pressure, in atmospheres, of the gas in the bulb? (The density of Hg is 13.6 g/cc; of water, 1.00 g/cc.)

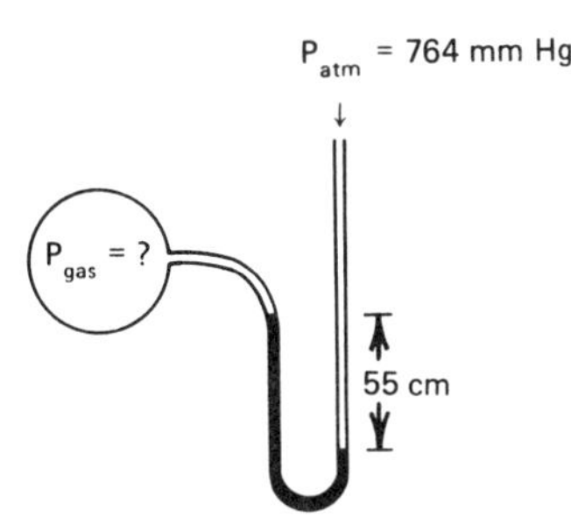


(A) 1.01 atm (B) 0.953 atm (C) 1.06 atm (D) 1.00 atm (E) 0.932 atm

### *Solution*

First, the diagram shows that the pressure exerted by the confined gas sample in the bulb is **less than** the atmospheric pressure. This comes from the fact that the atmospheric pressure supports the weight of a column of water 55 cm high against the opposing pressure of the confined gas. Therefore,

$$P_{gas} = P_{atm} - \text{(a correction).}$$

The "correction term" in this equation **must** be in the same units as $P_{atm}$, namely millimeters of mercury. The column of water then is expressed as 550 **mm of water**; however, since Hg is much more dense than water,

$P_{gas}$ = 764 mm Hg - (the mercury equivalent of a water column 550 mm high)

A column height of one barometric fluid may be converted into the equivalent column height of a second barometric fluid by the formula,

$$h_{(\text{fluid A})} \times d_{(\text{fluid A})} = h_{(\text{fluid B})} \times d_{(\text{fluid B})}$$

where h is the column height and d is the fluid density. Here A = Hg and B = water, so that

$$h_{Hg} = h_{H_2O}\left(\frac{1.0 \text{ g/cc}}{13.6 \text{ g/cc}}\right) = 40 \text{ mm Hg}$$

Thus, 764 - 40 = 724 mm Hg (or 724 torr); converting to units of atmospheres,

$$724 \text{ mm Hg}\left(\frac{1 \text{ atm}}{760 \text{ mm Hg}}\right) = 0.953 \text{ atm} \qquad [E]$$

## 9-B. Pressure-Temperature Relationship at Constant Volume

A 0.200-g sample of carbon dioxide exerts a pressure of 844 torr in a sealed glass tube at 25.0°C. It is known that the tube can withstand a maximum pressure of 2.24 atm without bursting. To what maximum temperature (°C) may the tube be safely heated without exceeding this pressure?

(A) 601°C (B) 573°C (C) 50.4°C (D) 212°C (E) 328°C

*Solution*

$$PV = nRT\ (n,\ V \text{ constant}), \text{ then } \frac{P_2}{P_1} = \frac{T_2}{T_1} \text{ or } T_2 = \frac{T_1 P_2}{P_1}$$

(**Caution!** $T$ in kelvins; $P$'s in consistent units)

$$\frac{(2.24\ \text{atm})(760\ \text{torr/atm})(25.0 + 273\ \text{K})}{844\ \text{torr}} =$$

*Answer*

| | | |
|---|---|---|
| Calculator | 601.0843602 K; 327.9243602°C | |
| Significant figures required | 3 | |
| Answer reported | 328°C | [E] |

## 9-C. Volume of "Wet" Gas to Volume of "Dry" Gas—Boyle's Law

A gas sample of oxygen collected over water at 30.0°C exerts a total pressure of 764 torr. Its volume is 750. mL. What volume would the **dried** oxygen occupy at the same conditions of pressure and temperature? ($p_{H_2O}$ = 32 torr at 303 K)

(A) 719 mL (B) 819 mL (C) 783 mL (D) 742 mL (E) 717 mL

*Solution*

Since $n$ and $T$ are constant, $P \propto 1/V$. Initially $p_{O_2}$ = 732 torr; after drying, $p_{O_2} = P_{total}$ = 764 torr. Thus, the pressure has been increased by 764/732 and, according to Boyle's law, the volume is reduced by the factor 732/764:

$$(750\ \text{ml})\left(\frac{732\ \text{torr}}{764\ \text{torr}}\right) =$$

*Answer*

| | | |
|---|---|---|
| Calculator | 718.5863874 | |
| Significant figures required | 3 | |
| Answer reported | 719 mL | [A] |

## Exercises on Units and Type 1 Gas Law Problems

1. Mercury has a density of 13.6 g/cc. A pressure that will support a column of Hg 15 mm high will support a column of water to a height of:

   (A) 760 mm (B) 20 mm (C) 1.1 mm (D) 11 cm (E) 20 cm

2. The temperature in kelvins corresponding to 212°F is:

   (A) 373 K (B) 485 K (C) 32 K (D) 173 K (E) 100 K

3. A barometric pressure of 30.2 inches of mercury is reported by the weather bureau. This corresponds to a pressure of:

   (A) 1.01 atm (B) 754 torr (C) 74.2 cm of mercury
   (D) 56.4 mm of water (E) 2.72 ft of mercury

4. The volume of a sample of gas measured at 30.0°C and 760 torr is 10.0 L. What final temperature would be required to reduce the volume to 9.0 L at constant pressure?

   (A) 36°C (B) 27°C (C) 24°C (D) 0°C (E) -30°C

5. If a given sample of gas, collected over water at 20°C and 758 torr pressure, occupies a volume of 150 mL, what volume will the same sample of dry gas occupy at STP? (The vapor pressure of water at 20°C is 18 torr.)

   (A) 136 mL (B) 161 mL (C) 140 mL (D) 166 mL (E) 157 mL

6. A sample of He occupies 600. mL at 27°C and 570. torr. The volume is reduced to 450. mL and the sample cooled until the pressure is 380. torr. What is final temperature in °C?

   (A) 0°C (B) 27°C (C) -123°C (D) -195°C (E) -80°C

---

## Example Problem Involving Type 2 Gas Law Calculations

### 9-D. Gas Density from Pressure-Temperature Data

What is the density of fluorine gas (g/L) in a sample that exerts a pressure of 95 torr at 0°C?

(A) 1.6 g/L (B) 0.21 g/L (C) 0.16 g/L (D) 1.1 g/L (E) 0.11 g/L

*Solution*

$$PV = nRT = \frac{\text{g}}{\text{MW}}RT;\ d = \frac{\text{g}}{V} = \frac{(\text{MW})P}{RT}$$

[Remember that the fluorine molecule is $F_2$ (FW = 38.0).]

$$\frac{(38.0\ \text{g}/\cancel{\text{mole}})\left(\dfrac{95\ \cancel{\text{torr}}}{760\ \cancel{\text{torr}}/\cancel{\text{atm}}}\right)}{(0.08205\ \text{L-}\cancel{\text{atm}}/\cancel{\text{mole}}\text{-}\cancel{\text{K}})(273\ \cancel{\text{K}})} =$$

| | |
|---|---|
| Calculator | 0.2120568848 |
| Significant figures required | 2 |
| Answer reported | 0.21 g/L [B] |

## Exercises on Type 2 Gas Law Calculations

7. How many moles are in a gas sample occupying 0.500 L at 170 torr and 25°C?

   (A) 0.00457 (B) 0.00500 (C) 2.18 (D) 3.48 (E) 3.85

8. Calculate the pressure exerted by 1.00 g of carbon dioxide gas at 125°C in a 1.00-liter vessel.

   (A) 32.6 atm (B) 1.25 atm (C) 0.743 atm (D) 0.510 atm
   (E) 22.4 atm

9. The number of molecules in 1.00 liter of oxygen gas at 56°C and 821 mm of Hg pressure is:

   (A) $4.00 \times 10^{-2}$ (B) $1.83 \times 10^{25}$ (C) 32 (D) $2.24 \times 10^{23}$
   (E) $2.41 \times 10^{22}$

10. The ideal gas law predicts that the molar volume (volume of one mole) of gas equals:

    (A) $gRT/PV$ (B) $(MW)P/RT$ (C) $1/2m\bar{s}^2$ (D) $RT/P$
    (E) 22.4 at any temperature and pressure

11. An unknown liquid is vaporized at 100°C to fill a 25.0-mL container at a pressure of 750 torr. If the vapor weighs 0.0564 g, what is the molecular weight of the liquid?

    (A) 85.0 (B) 18.8 (C) 22.2 (D) 70.0 (E) 60.0

12. The density of neon will be **greatest** at:

    (A) STP (B) 0°C and 2 atm (C) 273°C and 1 atm
    (D) 273°C and 2 atm
    (E) The density of neon is the same in all of the above cases

---

## Example Problem Involving Type 3 Gas Law Calculations

### 9-E. Graham's Law—Diffusion Rate and Molecular Weight

An unknown gas, X, has a rate of diffusion measured to be 1.14 times that of carbon dioxide at the same conditions of $T$ and $P$. Which of the following gases might X be?

(A) $O_2$ (B) $C_2H_6$ (C) CO (D) $PH_3$ (E) $NO_2$

*Solution*

From kinetic theory $1/2m_x\bar{s}^2_x = 1/2m_{CO_2}\bar{s}^2_{CO_2}$ ; hence $\frac{s_x}{s_{CO_2}} = \sqrt{\frac{MW_{CO_2}}{MW_x}}$

## DRILL ON THE IDEAL GAS LAW

***Student's Name*** ____________

For each of the gas samples, complete the table. Be sure to include units. You may be asked to turn in this sheet.

| Sample No. | Pressure P | Volume V | Number of moles n | Temperature T | Mass g | MW | Density g/L | Volume Sample would have at STP |
|---|---|---|---|---|---|---|---|---|
| 1 | 1.00 atm | | | 25.0°C | 2.00 g | 44.0 | | 1.02 L |
| 2 | | 3.00 L | 0.0500 | 400.0 K | 7.30 g | | | |
| 3 | | 1.00 L | | 0.00°C | 6.44 g | 92.0 | 6.44 g/L | |
| 4 | 608 torr | 7.00 L | 0.223 | | | 64.0 | | 5.00 L |
| 5 | 380. torr | 15.0 L | | 298 K | | 4.00 | | |
| 6 | | | | | | | | |

(Graham's law), since the actual **molecular** masses are proportional to the molecular weights (MW). Thus,

$$1.14 = \sqrt{\frac{44.01}{MW_x}}$$

If we square both sides and solve,

$$MW_X = 44.01/(1.14)^2$$

*Answer*

| | |
|---|---|
| Calculator | 33.86426593 |
| Significant figures required | 3 |
| Answer reported | 33.9, $PH_3$ (agreeing within 0.3%) [D] |

## Exercises on Type 3 Gas Law Calculations

13. The rate of diffusion of helium gas is about:

    (A) 16 times that of sulfur dioxide (B) 1/16 that of sulfur dioxide
    (C) 64 times that of sulfur dioxide (D) 4 times that of sulfur dioxide
    (E) 1/4 that of sulfur dioxide

14. If the rate of diffusion of methane gas ($CH_4$, MW = 16) is 2.0 times the rate of diffusion of gas X, what is the molecular weight of X?

    (A) 32 (B) 16 (C) 4.0 (D) 64 (E) 8.0

15. Which one of the following gases would effuse most rapidly through a small opening in a container?

    (A) chlorine (B) nitrogen (C) fluorine (D) oxygen
    (E) carbon dioxide

16. It takes 36 sec for 0.01 moles of He to effuse through a pinhole in an effusion apparatus. It takes 72 sec for 0.01 moles of another gas to effuse through the same system. What is the molecular weight of the unknown gas? (Hint: time and speed are inversely proportional)

    (A) 8.0 (B) 16 (C) 1.0 (D) 2.0 (E) none of these

17. In which of the following gas samples is the average molecular speed **least**? (MW's: He, 4.0; Ne, 20; $N_2$, 28; $F_2$, 38)

    (A) helium at 0.5 atm and 160 K (B) neon at 1.0 atm and 600 K
    (C) nitrogen at 0.5 atm and 560 K (D) fluorine at 1.0 atm and 570 K
    (E) neon at 0.5 atm and 320 K

## DRILL ON KINETIC MOLECULAR THEORY

***Student's Name*** ______________________

Compare gas samples A and B in each case and complete the table. In the columns labeled 'greatest average speed' and 'greatest average kinetic energy' enter 'A', 'B', or 'Same'. You may be asked to turn in this sheet.

| A | B | MW of B | Greatest Average Speed | Greatest Average K.E. | $\frac{\text{Average Speed B}}{\text{Average Speed A}}$ |
|---|---|---|---|---|---|
| $SO_3$ at 25°C | $SO_3$ at 100.°C | | | | |
| $SO_3$ at 25°C | $SF_6$ at 25°C | | | | |
| $O_2$ at 25°C | ?? at 25°C | | | | 2.83 |
| $SO_3$ at 25°C | $SF_6$ at 300.°C | | | | |
| | | | | | |

## Example Problems Involving Gas Laws and Reaction Stoichiometry

### 9-F. Weight and Gas Volume Relationship in a Chemical Reaction

A sample of solid $Pb(NO_3)_2$ is heated, yielding oxygen, $NO_2$ gas, and solid PbO. A 293-mL gas sample, measured at 200.°C and 1.00 atm, is collected. What was the mass of the $Pb(NO_3)_2$ sample?

(A) 1.75 g (B) 2.50 g (C) 1.00 g (D) 1.25 g (E) 5.00 g

*Solution*

(1) Write the balanced equation

(2) $\frac{PV}{RT} \rightarrow$ moles gas $\rightarrow$ moles $Pb(NO_2)_3 \rightarrow$ g $Pb(NO_3)_2$

$$2\,Pb(NO_3)_{2\,(s)} \rightarrow 2\,PbO_{(s)} + O_{2\,(g)} + 4\,NO_{2\,(g)}$$

$$\left(\frac{(1.00\text{ atm})(0.293\text{ L})}{(0.08205\text{ L-atm/mole-K})(473\text{ K})}\right) = \text{moles gas (total)};$$

$$\left(\frac{0.293}{(0.08205)(473)}\text{ moles gas}\right)\left(\frac{2\text{ moles }Pb(NO_3)_2}{5\text{ moles gas}}\right)\left(\frac{331\text{ g }Pb(NO_3)_2}{1\text{ mole }Pb(NO_3)_2}\right) =$$

*Answer*

| | | |
|---|---|---|
| Calculator | 0.9995761363 | |
| Significant figures required | 3 | |
| Answer reported | 1.00 g $Pb(NO_3)_2$ | [C] |

### 9-G. Combining Gas Volumes—Avogadro's Hypothesis

Given 6.0 L of $CH_3OH_{(g)}$ at 200.°C and 1.0 atm. How many liters of $O_2$, measured at the same temperature and pressure, would be needed for combustion of the $CH_3OH$ to $CO_2$ and water?

(A) 9.0 L (B) 6.0 L (C) 12 L (D) 22 L (E) 15 L

*Solution*

Since $T$ and $P$ are constant, $V \propto n$. Thus volumes can be read directly from the equation as would be moles.

$$2\,CH_3OH_{(g)} + 3\,O_{2\,(g)} \rightarrow 2\,CO_{2\,(g)} + 4\,H_2O_{(g)}$$

$$(6.0\,\cancel{\text{L }CH_3OH})\left(\frac{3\text{ L }O_2}{2\,\cancel{\text{L }CH_3OH}}\right) = 9.0\text{ L }O_2$$

Note that the stoichiometric link here is "L $O_2$/L$CH_3OH$." This can be done with any balanced equation **involving gases** (assuming ideality) **when all gases involved are measured at the same temperature and pressure.**

*Answer* Answer reported 9.0L$O_2$ (2 significant figures) [A]

## Exercises on Gas Laws and Reaction Stoichiometry

18. What volume of gaseous $PH_3$ (at STP) could be formed by the reaction of 100. g of calcium phosphide with excess water?

$$Ca_3P_{2\,(s)} + 6\,H_2O_{(\ell)} \rightarrow 3\,Ca(OH)_{2\,(s)} + 2\,PH_{3\,(g)}$$

(A) 6.15 L (B) 12.3 L (C) 24.6 L (D) 81.5 L (E) 40.8 L

19. What volume of HCl gas is produced by the reaction of 2.4 L of hydrogen with 1.5 L of chlorine? (Volumes are at the same *T* and *P*.)

(A) 2.4 L (B) 1.5 L (C) 0.75 L (D) 3.0 L (E) 4.8 L

20. The volume of $O_2$ at STP necessary to react completely with 0.200 mole of $SO_2$ to form $SO_3$ is:

(A) 22.4 L (B) 2.24 L (C) 4.48 L (D) 11.2 L (E) 7.47 L

21. What volume of oxygen at 300 K and 760 torr reacts with 4.00 L of NO gas at 400 K and 380 torr to form nitrogen dioxide gas?

(A) 2.00 L (B) 8.00 L (C) 5.33 L (D) 4.00 L (E) 0.750 L

22. How many liters of nitrogen would be collected at 600°C and 1000 torr from the decomposition of 50.0 g of ammonium dichromate to $Cr_2O_3$, $N_2$, and $H_2O$?

(A) 10.8 L (B) 7.4 L (C) 98 mL (D) 142 mL (E) none of these

23. If 0.0920 g of a metal (M) replaces $2.00 \times 10^{-3}$ mole of $H_2$ from an acid and forms $M^+$ ions in the reaction, what is the atomic weight of the metal?

(A) 6.94 (B) 39.1 (C) 23.0 (D) 108 (E) 85.5

---

## GENERAL EXERCISES ON CHAPTER 9—GAS LAWS

24. What volume would the gas sample in Question 7 occupy if the temperature was maintained constant and the pressure changed to 100 mm of Hg?

(A) 0.850 L (B) 0.670 L (C) 0.500 L (D) 0.330 L (E) 0.284 L

25. What would be the pressure if the gas sample in Question 7 were placed in a 65.0-mL container at 25°C?

(A) 1300 mm of Hg (B) 670 mm of Hg (C) 495 mm of Hg
(D) 235 mm of Hg (E) 220 mm of Hg

26. How many liters of $NH_3$ gas at STP would be needed to prepare 25 mL of 2.5 *M* "ammonium hydroxide" (solution of $NH_3$ in water)?

## DRILL ON GAS MIXTURES

***Student's Name*** ________________

Consider each of the following mixtures of methane ($CH_4$) and helium (He) gas in a fixed volume and maintained at 25.0°C. Fill in the blanks to complete the table. You may be asked to turn in this sheet.

| Sample No | $P_{total}$ atm | $g_{CH_4}$ | $g_{He}$ | Mole Fraction | | Partial Pressure | | g He that should be added to adjust $P_{total}$ to 2.00 atm |
|---|---|---|---|---|---|---|---|---|
| | | | | $X_{CH_4}$ | $X_{He}$ | $P_{CH_4}$ atm | $P_{He}$ atm | |
| 1 | 1.00 | 4.00 g | 4.00 g | | | | | |
| 2 | 1.00 | 4.00 | | 0.750 | | | | |
| 3 | | 4.00 | | 0.750 | | 0.800 | | |
| 4 | 1.00 | | | | | 0.800 | | 4.00 g |
| 5 | | | | | | | | |

(A) 1.4 L (B) 56 L (C) 0.56 L (D) 0.060 L (E) 0.062 L

27. If 0.385 g of a gas occupies 200. mL at -73°C and 750 torr, what is the molecular weight of the gas?

    (A) 48.0 (B) 20.8 (C) 32.0 (D) 21.4 (E) 44.0

28. A gas sample weighing 0.5280 g is collected in a flask having a volume of 126 mL. At 75°C the pressure of the gas is 754 torr. The molecular weight of the gas is closest to which one of the following?

    (A) 100 (B) 120 (C) 140 (D) 160 (E) 180

29. What volume of $CO_2$ gas at 477°C and 750 torr can be produced by the combustion of 32.5 g of $C_2H_2$? (**MW's**: $CO_2$, 44.0; $C_2H_2$, 26.0)

    $$C_2H_2 + O_2 \rightarrow CO_2 + H_2O \quad \text{(unbalanced)}$$

    (A) 312 liters (B) 156 liters (C) 78.0 liters (D) 112 liters
    (E) 624 liters

30.*Based on the kinetic theory of gases, the rate of diffusion of nitrogen at **25**°C would be ______ times that of carbon dioxide at **75**°C.

    (A) 1.16 (B) 0.877 (C) 1.73 (D) 1.41 (E) 0.97

31.*A 180.-mg sample of an alloy of iron and metal X is treated with dilute sulfuric acid, liberating hydrogen and yielding $Fe^{2+}$ and $X^{3+}$ ions in solution. It is known that the alloy contains 20.0% iron, by weight. The alloy yields 50.9 mL of hydrogen collected over water at 22°C (vapor pressure of water at this temperature is 20 torr) and a total pressure of 750. torr. What is element X?

    (A) Al (B) La (C) Sm (D) Ga (E) Gd

32. One method for the commercial preparation of chlorine gas is the electrolysis of molten sodium chloride: $2NaCl \rightarrow 2Na + Cl_2$. How many liters of chlorine, measured at 27.0°C and 1.00 atm, can be produced from 702 g of NaCl?

    (A) 148 L (B) 0.00164 L (C) 134 L (D) 44.8 L (E) 89.6 L

33. A given sample of gas is at 270. K and 360. torr. If this gas sample is heated at constant volume to 480. K and 640. torr, the average speed of the molecules would be increased by a factor of:

    (A) 1.78 (B) 1.33 (C) 1.44 (D) 3.16 (E) 2.25

34. If a given gas occupies 5.00 liters at 62.5 torr and 27°C, what volume would the same mass of gas occupy at 375 torr and 177°C?

    (A) 45.0 L (B) 22.5 L (C) 5.00 L (D) 2.50 L (E) 1.25 L

35. The molar volume of an ideal gas:

    (A) is decreased by increasing the temperature
    (B) has a constant value of 22.4 L and is unaffected by changes in temperature and pressure

(C) is decreased by decreasing the pressure of the gas
(D) is lower for a gas of low molecular weight (at STP) as compared to a gas of higher molecular weight (at STP)
(E) is decreased by decreasing the kinetic energy of the gas molecules

36. An ideal-gas mixture contains 4 millimoles of $H_2$ for every mmole of Ne. The partial pressure of Ne is:

(A) one-fourth of the total pressure
(B) three-fourths of the total pressure (C) one atmosphere
(D) one-fifth of the total pressure (E) four-fifths of the total pressure

37. A mixture contains 2.00 moles of $O_2$ and 4.00 moles of $N_2$ gases. If the total pressure of the mixture is 3.00 atm, what is the partial pressure of oxygen in the mixture?

(A) 2.00 atm (B) 6.00 atm (C) 1.00 atm (D) 1.50 atm (E) 12.0 atm

38. To produce one volume of ammonia from the elements would require what volume of hydrogen measured at the same temperature and pressure?

(A) 3 (B) 2 (C) 3/2 (D) 1 (E) 1/2

39. Consider the figure at the right for the collection of hydrogen gas over water at 25°C (vapor pressure, 24 torr). If atmospheric pressure is 757 torr and if mercury is 13.6 times as dense as water, the partial pressure of hydrogen is:

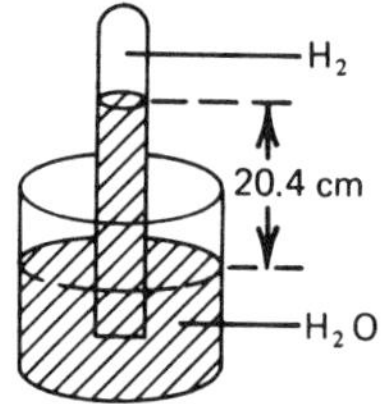


(A) 748 torr (B) 732 torr (C) 718 torr (D) 713 torr
(E) none of these

40. What is the density of $N_2$ gas at 227°C and 5.00 atm pressure?

(A) 2.93 g/L (B) 0.293 g/mL (C) 2.30 g/L (D) 3.41 g/L
(E) 1.25 g/L

41.*Solid ammonium chloride is deposited instantly when ammonia (MW = 17.0) and hydrogen chloride (MW = 36.5) gases come into contact. Suppose that these gases were simultaneously released from opposite ends of a one-meter diffusion tube as shown below:

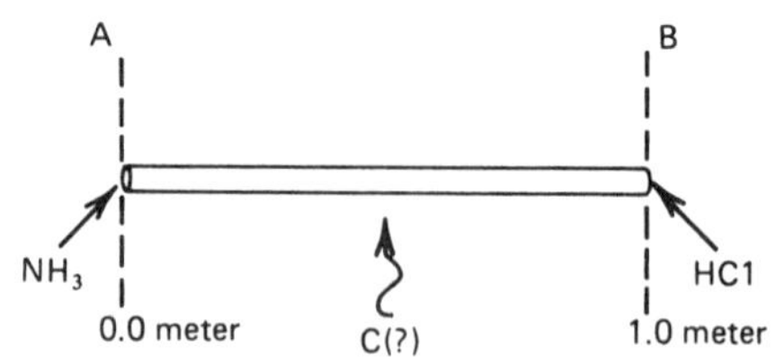

At approximately what point (C) should the first deposit of solid ammonium chloride be observed?

(A) at ~ 45 cm (B) at ~ 50 cm (C) at ~ 59 cm (D) at ~ 68 cm
(E) at ~ 90 cm

42. If 1.00 g of dry oxygen occupies 0.700 L at STP, what volume would it occupy if collected over water at 24°C at a total pressure of 726 torr? (The vapor pressure of water at 24°C is 22.4 torr.)

(A) 26.2 L (B) 0.822 L (C) 0.700 L (D) 0.846 L (E) 22.4 L

43. The volume occupied by 1.00 g of propane gas ($C_3H_8$) at STP is:

(A) 0.508 L (B) 1.01 L (C) 0.988 L (D) 22.4 L (E) 1.12 L

44. A certain gas has a density of 1.96 g/L at STP. Which of the following gases might it be?

(A) $O_2$ (B) $SO_2$ (C) $CO_2$ (D) $AsH_3$ (E) $N_2$

45. A mixture of $N_2$ and $O_2$ gases in a 200-mL container exerts a total pressure of 720 torr at 35°C. If there are 0.0020 moles of $N_2$ gas, what is the mole fraction of $N_2$ gas?

(A) 0.73 (B) 0.50 (C) 0.27 (D) 0.10 (E) 0.90

46. In a gaseous mixture at 2.0 atm pressure, 15% of the molecules are oxygen, 35% are nitrogen, 45% are carbon monoxide and the remainder are carbon dioxide. What is the partial pressure of nitrogen?

(A) 0.35 atm (B) 0.70 atm (C) 2.0 atm (D) 1.3 atm (E) 1.0 atm

47. At 27.2°C and 1.00 atm, the molecules of which of the following gases have the largest average kinetic energy?

(A) $O_2$ (B) $O_3$ (C) Ar (D) $CO_2$
(E) All have the same average kinetic energy

48. The molecules of which of the following gases have the greatest **average** molecular speed at 300 K?

(A) Argon (B) Nitrogen (C) Fluorine (D) Neon
(E) All have the same average molecular speed at 300 K.

49.*A hydrogen-oxygen mixture is passed over excess hot CuO, which causes **all** of the hydrogen to be removed by the reaction

$$CuO_{(s)} + H_{2\,(g)} \rightarrow Cu_{(s)} + H_2O_{(g)}$$

Some oxygen is removed by reaction with copper:

$$Cu_{(s)} + 1/2\ O_{2\,(g)} \rightarrow CuO_{(s)}$$

If 100. mL of the original mixture, measured at 750 torr and 298 K, yields 85.0 mL of **dry** oxygen, measured under the same conditions, what was the partial pressure of $H_2$ (in torr) in the original mixture?

(A) 75.0 (B) 113 (C) 190 (D) 637 (E) 675

50. How many moles of oxygen are required for the complete combustion of one gallon of octane ($C_8H_{18}$)? (1.00 gal weighs 2700 g)

    (A) 592 (B) 296 (C) 201 (D) 25 (E) 12.5

51. If air is 21 mole percent oxygen, what volume of air at 25°C and 1.0 atm is required to burn one gallon of octane (see Question 50)?

    (A) $2.6 \times 10^3$ L (B) $7.2 \times 10^3$ L (C) $3.2 \times 10^4$ L (D) $3.4 \times 10^4$ L (E) $2.9 \times 10^5$ L

52. A 100.0-mg sample of pure metal M was treated with excess dilute sulfuric acid, forming $M^{3+}$ and 25.0 mL of hydrogen (collected over water at 22.0°C and 715 torr). The vapor pressure of water at 22.0°C is 20 torr. What is the chemical symbol for the metal M?

    (A) Al (B) La (C) Ga (D) Dy (E) Tb

53. A real gas most closely approaches the behavior of an ideal gas under conditions of:

    (A) high $P$ and low $T$ (B) low $P$ and high $T$ (C) low $P$ and $T$ (D) high $P$ and $T$ (E) STP

54. The number of impacts of $N_2$ molecules against a container wall per unit time is:

    (A) increased by cooling the gas
    (B) increased by decreasing the pressure so that individual molecules will not get in the path of molecules headed toward the wall
    (C) the same when helium is added to the container
    (D) decreased when helium is added to the container
    (E) decreased when the average kinetic energy of the molecules is increased

55. The rate of diffusion of gas X is found to be about 1.3 times that of $SF_{6\,(g)}$ (MW = 146). The MW of X is about:

    (A) 190 (B) 110 (C) 86 (D) 55 (E) 32

56. Which of the following samples consists of the **largest** number of molecules?

    (A) 1.00 liter of hydrogen at STP (B) 1.00 liter of neon at STP
    (C) 1.00 liter of hydrogen at 20.0°C and 760 torr
    (D) 1.00 liter of hydrogen at 0.0°C and 800 torr
    (E) All of the above have the same number of gas molecules

57. The molar volume of gaseous neon is **greatest** at:

    (A) STP (B) 0°C and 2.0 atm (C) 273°C and 2.0 atm
    (D) 127°C and 2.0 atm (E) 127°C and 1.0 atm

58. A rigid 5.00-L cylinder contains 0.100 mole of argon gas at 25.0°C and 372 torr. The cylinder is heated to 34.0°C and 2.80 g of $N_2$ is pumped in. What is the partial pressure of Ar in the final gas mixture?

    (A) 744 torr (B) 383 torr (C) 372 torr (D) 316 torr (E) 190 torr

59. A 1.00-g sample of $KMnO_4$ (FW = 158) was completely decomposed by heating, yielding 99.2 mL of oxygen collected over water at 22°C and a total pressure of 755 torr. At 22°C, the vapor pressure of water is 20 torr. Which of the following equations best fits these data for the decomposition reaction?

    (A) $2\ KMnO_{4\,(s)} \rightarrow K_2MnO_{4\,(s)} + MnO_{2\,(s)} + O_{2\,(g)}$
    (B) $2\ KMnO_{4\,(s)} \rightarrow K_2MnO_{3\,(s)} + MnO_{(s)} + 2\ O_{2\,(g)}$
    (C) $6\ KMnO_{4\,(s)} \rightarrow 3\ K_2MnO_{4\,(s)} + Mn_3O_{4\,(s)} + 4\ O_{2\,(g)}$
    (D) $8\ KMnO_{4\,(s)} \rightarrow 4\ K_2MnO_{4\,(s)} + 2\ Mn_2O_{3\,(s)} + 5\ O_{2\,(g)}$
    (E) $9\ KMnO_{4\,(s)} \rightarrow 3\ K_3MnO_{4\,(s)} + 2\ Mn_3O_{4\,(s)} + 8\ O_{2\,(g)}$

60. According to the kinetic theory of gases the average speed of the molecules of a given gas is proportional to the:

    (A) absolute temperature (B) square root of the absolute temperature
    (C) square of the absolute temperature (D) volume of the container
    (E) reciprocal of the absolute temperature ($1/T$)

61. For a substance that remains a gas under the conditions listed, deviation from the ideal gas law would be most pronounced at:

    (A) 100°C and 2.0 atm (B) 0°C and 2.0 atm (C) -100°C and 2.0 atm
    (D) -100°C and 4.0 atm (E) 100°C and 4.0 atm

62. Arsine, $AsH_3$, is a very poisonous gas that decomposes when heated to form elemental arsenic and hydrogen:

$$2\ AsH_{3\,(g)} \xrightarrow{\text{heat}} 2\ As_{(s)} + 3\ H_{2\,(g)}$$

    An 0.0128-mole sample of $AsH_3$ exerts a pressure of 186 torr in a rigid 1.00-liter flask at -40°C. The flask was heated to 250°C for a short time and then returned to -40°C; the pressure was now measured to be 250 torr. Approximately what percent of the $AsH_3$ had decomposed?

    (A) ~17% (B) ~26% (C) ~34% (D) ~69% (E) ~99%

63. Given a gas sample containing $n$ moles at absolute temperature $T$, volume $V$, and pressure $P$. You now remove $1/2\ n$ moles of gas from the container, but wish to maintain $P$ and $V$ constant. You must:

    (A) increase the absolute temperature to $2T$
    (B) reduce the absolute temperature to $1/2\ T$
    (C) maintain the absolute temperature at $T$
    (D) have more data (E) none of the above

64. A compound containing only boron and hydrogen ($B_xH_y$) reacted completely with excess water, yielding 30.9 mg of $H_3BO_3$ and 30.26 mL of

dry hydrogen measured at 22°C and 730 torr. What is the empirical formula of the compound?

(A) $B_2H_6$ (B) $B_{10}H_{14}$ (C) $B_4H_{10}$ (D) $B_5H_9$ (E) $B_6H_{10}$

65. At 0.0°C and 200 torr, a pure gas sample has a density of 1.174 g/L. The sample could consist of which of the following gases?

(A) $N_2O$ (B) $SF_6$ (C) CO (D) Kr (E) $C_2F_4$

66. When a mixture of the solids Al, $NaNO_3$, and NaOH are dumped into water, ammonia and $NaAl(OH)_4$ are formed. Given excess water and 1.00 mole **each** of the solids, how many liters of ammonia, measured at STP, could be formed?

(A) 11.2 (B) 8.40 (C) 5.60 (D) 22.4 (E) 13.4

---

## ANSWERS TO CHAPTER 9 PROBLEMS

| | | | | | |
|---|---|---|---|---|---|
| 1. E | 12. B | 23. C | 34. E | 45. C | 56. D |
| 2. A | 13. D | 24. A | 35. E | 46. B | 57. E |
| 3. A | 14. D | 25. A | 36. D | 47. E | 58. B |
| 4. D | 15. B | 26. A | 37. C | 48. D | 59. D |
| 5. A | 16. B | 27. C | 38. C | 49. A | 60. B |
| 6. C | 17. D | 28. B | 39. C | 50. B | 61. D |
| 7. A | 18. C | 29. B | 40. D | 51. D | 62. D |
| 8. C | 19. D | 30. A | 41. C | 52. E | 63. A |
| 9. E | 20. B | 31. E | 42. B | 53. B | 64. D |
| 10. D | 21. E | 32. A | 43. A | 54. C | 65. E |
| 11. D | 22. A | 33. B | 44. C | 55. C | 66. B |

# 10
# *Liquids, Solids, and Changes of State*

Matter frequently is classified into three states: *solids, liquids,* and *gases.* A gas is a diffuse phase (>99% empty space) because of the existence of only very weak forces between neighboring molecules and relatively high average kinetic energy of the molecules. Increasing the forces of attraction between neighboring particles (*i.e.*, changing the nature of the substance) or decreasing the average kinetic energy possessed by the particles (*i.e.*, decreasing the temperature) allows formation of a liquid or a solid, a *condensed* phase (25 to 50% empty space). In the condensed phases the particles (atoms, molecules, or ions) are very close to one another, which means that the proximity of one particle influences the orientation of other neighboring particles (*ordering*). The properties of solids and liquids are therefore strongly dependent upon the nature of the individual particles.

## LIQUIDS AND INTERMOLECULAR FORCES

Most common liquids are molecular assemblies. The kinetic model of a liquid depicts a random, disordered, and closely packed array of mobile molecules. The molecules are capable of motion over, under, and around neighboring molecules. Most commonly the forces of attraction in liquids are *intermolecular* forces between neutral molecules. These forces (sometimes termed *secondary forces*) are weak when compared to the forces involved in *intramolecular* covalent bonds and include *dispersion forces, dipole-dipole forces,* and *"hydrogen bonding" forces.* The magnitudes of properties such as boiling point, heat of vaporization, critical temperature, vapor pressure, surface tension, and viscosity of liquids are manifestations of the strengths of these intermolecular forces.

## Exercises on Liquids and Intermolecular Forces

1. The normal boiling point for a liquid:

   (A) is 100°C
   (B) is the boiling point at STP
   (C) is the boiling point at one atmosphere pressure
   (D) varies with pressure
   (E) is the temperature where the vapor pressure of the liquid equals the external pressure

2. The vapor pressure of a liquid in a closed container:

   (A) depends upon the amount of liquid
   (B) depends upon the surface area
   (C) depends upon the temperature and the nature of the liquid
   (D) depends upon the shape of the container
   (E) depends upon none of these

3. Assuming ideal behavior, water vapor at 100°C and 1.00 atm has a molar volume about ______ times that of liquid water under the same conditions ($d_{100°C}$ = 0.9584 g/cc).

   (A) 20 (B) $1.2 \times 10^2$ (C) $1.6 \times 10^3$ (D) $2.5 \times 10^3$ (E) $1.8 \times 10^4$

4. Which of the following indicates very weak intermolecular forces of attraction in a liquid?

   (A) a very high boiling point (B) a very high vapor pressure
   (C) a very high critical temperature (D) a very high heat of vaporization
   (E) none of these

5. Liquid bromine and the interhalogen ICl have almost identical molecular weights, yet the latter boils at a temperature 38°C higher than does bromine. The best explanation is that:

   (A) the dispersion forces in ICl are much stronger than in $Br_2$
   (B) ICl is an ionic compound; $Br_2$ is molecular
   (C) the bond in the ICl molecule is stronger than that in $Br_2$
   (D) the bond length in the $Br_2$ molecule is greater than that in ICl
   (E) ICl is a polar molecule; $Br_2$ is nonpolar

6. Which of the following would have the highest vapor pressure at 25°C?

   (A) a nonpolar molecular solid having a formula weight of 100
   (B) a polar molecular solid having a formula weight of 100
   (C) a solid made up of a network of covalent bonds (FW = 100)
   (D) an ionic solid (FW = 100)
   (E) all of the above would have essentially the same vapor pressure

## SOLIDS

The primary differences between liquids and solids are (a) the high degree of order in *crystalline* solids and (b) the relative *immobility* of the solid state units (essentially only vibrational motion about a fixed center known as a *lattice point*). The structure of a crystalline solid may be described in terms of a three-dimensional *space lattice,* an array of points representing the centers of gravity of the units (atoms, molecules, or ions). A small repeating portion of the space lattice is a *unit cell.* Solids may be classified as being *molecular* (composed of molecules), as *ionic* (an alternating, three-dimensional array of positive and negative ions), as *network* (one-, two-, and three-dimensional arrays of atoms interconnected by covalent bonds), or as *metallic* (a three-dimensional array of metallic cations immersed in an orderly way in a common "sea" of mobile electrons–the "free-electron" theory). Properties such as hardness, melting point, heat of sublimation, and conductivity (in both the molten and the solid state) may be correlated with this classification or with the magnitude of the intermolecular, interionic, or interatomic forces of attraction operative in the solid. For example, the melting temperatures of solids range, on the one extreme, from very low for molecular solids (where weak intermolecular forces exist) to very high, on the other extreme, for network solids (where strong covalent bonds exist). Melting points of metals vary widely dependent to some extent upon the amount of covalent bonding superimposed upon the metallic bonding. Ionic solids have intermediate-to-high melting points with variations closely following Coulomb's law:

$$f = \frac{q_+ \times q_-}{(r_+ + r_-)^2}$$

where $f$ is the force of attraction between a cation of charge $q_+$ and an anion of charge $q_-$ ; $r_+$ and $r_-$ are, respectively, the radii of the cation and anion (the sum being the distance between the centers of the tangent ionic spheres).

Although 32 crystal systems are known, we shall consider only the three cubic systems: the *primitive* (*simple cubic*); the *body-centered cubic* (*bcc*); and the *face-centered cubic* (*fcc*), or "*cubic closest packing*" systems. The unit cell for each of these arrangements may be described in terms of the unit cell population ($P_{uc}$), with each corner lattice point contributing the equivalent of 1/8 of a formula unit to a given unit cell; each edge point, 1/4; each face-center position, 1/2; and each body-center position 1 formula unit to the unit cell.

The geometry of a cube requires that the ratio $E:FD:BD = 1:\sqrt{2}:\sqrt{3}$ (by the Pythagorean theorem) where *E, FD,* and *BD* refer, respectively, to the length of the edge, the face diagonal, and the body diagonal of the cube. See the figure below:

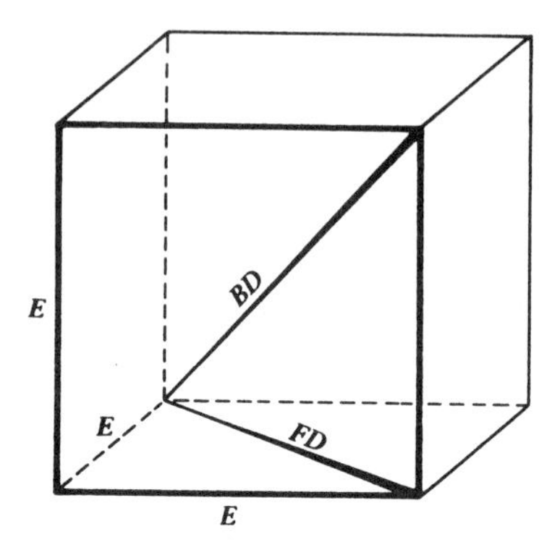


By the Pythagorean theorem,

$$(FD)^2 = E^2 + E^2 = 2E^2 \qquad E = \frac{FD}{\sqrt{2}}$$

$$(BD)^2 = E^2 + (FD)^2$$

$$\text{or } (BD)^2 = E^2 + E^2 + E^2 = 3E^2$$

$$E = \frac{BD}{\sqrt{3}}$$

Therefore,

$$E = E$$

$$FD = E\sqrt{2}$$

$$BD = E\sqrt{3}$$

and

$$E{:}FD{:}BD :: 1{:}\sqrt{2}{:}\sqrt{3}$$

These distances may be expressed in terms of the radii ($r$) of equivalent spheres that just touch (are tangent) in each of these arrangements. The following table summarizes these relationships:

| Lattice Type | $P_{uc}$ | Contact of Equivalent Spheres | Distance (as a function of $r$) |
|---|---|---|---|
| Primitive | 1 | Along $E$ | $E = 2r$ |
| *BCC* | 2 | Along $BD$ | $BD = 4r$ |
| *FCC* | 4 | Along $FD$ | $FD = 4r$ |

From the above relationships, it is easily shown that the density ($d$) of a cubic solid may be calculated from the contents and dimension of the unit cell by:

$$d = \frac{P_{uc} \times FW}{E^3 \times N_0} = \text{g/cm}^3$$

where FW is the formula weight of the unit making up the cell and $N_0$ is Avogadro's number. Also, it should be emphasized that for solids composed of **different units** (*e.g.*, ionic compounds), $P_{uc}$ must be consistent with the stoichiometry of the compound.

## Example Problems Involving Solids

### 10-A. Atomic Radius from Crystal Data

Vanadium crystallizes in a body-centered cubic system. If the edge of the unit cell is 3.09 Å, what is the radius of a vanadium atom?

(A) 1.09 Å (B) 1.34 Å (C) 1.54 Å (D) 7.14 Å (E) 6.18 Å

*Solution*

The "body-centered cubic structure" described for crystalline vanadium means that the V atoms, regarded as spheres, are packed in a cubic arrangement with vanadium atoms centered on corners and at the body-center of a unit cube called the unit cell. This structure does not allow vanadium atoms to touch along the edge or face-diagonal of the cube, but it does allow the atoms to be in contact along the body-diagonal. Showing only two opposite corner atoms and the body-center atom (in perspective) in the following figure:

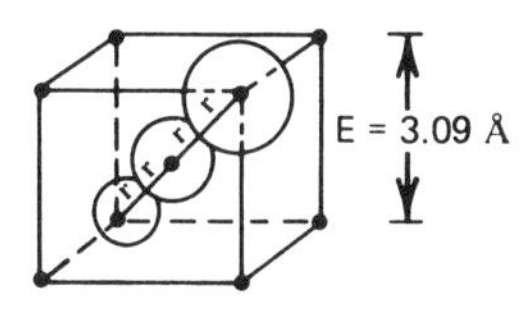


we see that the body-diagonal can be expressed as 4r, where r is the radius of the atom. By the Pythagorean theorem,

$$\frac{\text{E}}{\text{BD}} = \frac{1}{\sqrt{3}} = \frac{3.09}{4\text{r}}$$

Thus, $4\text{r} = 3.09\sqrt{3}$ and r = 1.34 Å, which is the radius of the vanadium atom. (Note that the edgelength is greater than 2 × 1.34 Å; as noted earlier the atoms do not touch along an edge, but have been "squeezed apart" to accommodate the body-center atom.) 1.34 Å = $1.34 \times 10^{-8}$ cm = 0.134 nm = 134 pm—the nanometer and picometer are two linear units that are coming into frequent use to express atomic dimensions. **[B]**

## 10-B. Density of a Solid from Crystal Data

Based on the data from Q. 8-A, what is the density of metallic vanadium?

(A) 2.96 g/cm³ (B) 4.01 g/cm³ (C) 5.74 g/cm³ (D) 6.35 g/cm³
(E) 7.94 g/cm³

### *Solution*

The volume of the cubic unit cell of vanadium = $E^3 = (3.09 \times 10^{-8}\ \text{cm})^3$, since 1 Å ≡ $10^{-8}$ cm. Each of the eight corner atoms contributes 1/8 of a vanadium atom to the **interior** of the cube, and the body-centered vanadium atom lies entirely within this volume. Thus, there is the **equivalent of** 8(1/8) + 1 = 2 vanadium atoms within the boundaries of the unit cell, *i.e.* the population of the unit cell is said to be 2 for this crystal structure. Since density (d) is defined as mass/volume, we have:

$$d = \frac{(2\ \text{atoms V})\left(\dfrac{1\ \text{mole V}}{6.02 \times 10^{23}\ \text{atoms V}}\right)\left(\dfrac{50.94\ \text{g V}}{1\ \text{mole V}}\right)}{(3.09 \times 10^{-8})^3\ \text{cm}^3} = 5.74\ \text{g/cm}^3 \quad [C]$$

## Exercises on Solids

7. Which of the following types of solids generally have the lowest melting points?

   (A) those composed of small symmetrical molecules
   (B) those composed of small positive and small negative ions
   (C) those composed of polar molecules
   (D) those in which atoms are interconnected with covalent bonds to form a network
   (E) those that consist of positive ions and mobile electrons

8. Polonium exists in a simple cubic crystal lattice. The number of polonium atoms per unit cell is:

   (A) 1 (B) 2 (C) 4 (D) 6 (E) 0

9. What is the volume of a face-centered cubic unit cell made up of atoms having a radius of $1.00 \times 10^{-8}$ cm?

   (A) $8.00 \times 10^{-24}$ cm³ (B) $1.23 \times 10^{-23}$ cm³ (C) $2.26 \times 10^{-23}$ cm³
   (D) $1.00 \times 10^{-24}$ cm³ (E) $1.53 \times 10^{-24}$ cm³

10. Solid carbon dioxide crystallizes in a cubic lattice with four molecules of $CO_2$ per unit cell. It could be classified as:

    (A) a simple cubic (B) a primitive cubic (C) a body-centered cubic
    (D) a face-centered cubic (E) any of the above

# DRILL ON CRYSTAL STRUCTURE CALCULATIONS

***Student's Name*** ______________________

Chromium, iron, and silver crystallize in cubic lattices. Your instructor will assign you one (or you may be allowed to select one) of these. From the given data for that metal, calculate the density from its unit cell edge length for three possible unit cells (simple cubic, body-centered cubic, and face-centered cubic). For each of the different unit cells, calculate the radius the metal atom would have. Compare the calculated densities with the given measured density and decided which unit cell is the correct one for the metal. Report the best value for the radius of the metal.

Data:

| Metal | Unit Cell Edge Length | Measured density |
|---|---|---|
| Chromium | 288.4 pm | 7.19 $g/cm^3$ |
| Iron | 286.6 pm | 7.86 $g/cm^3$ |
| Silver | 407 pm | 10.6 $g/cm^3$ |

| Metal | Calculated Density $g/cm^3$ | Calculated Atomic Radius pm |
|---|---|---|
| ______________ | | |
| if simple cubic | | |
| if body-centered cubic | | |
| if face centered cubic | | |

Based on the above calculations, the metal has a ______________ lattice and the atoms have a radius of ______________ pm.

11. Silver crystallizes in the face-centered cubic system. If the edge of the unit cell is 4.07 Å, what is the apparent radius of a silver atom?

    (A) 1.76 Å (B) 3.52 Å (C) 2.04 Å (D) 2.88 Å (E) 1.44 Å

12. What is the theoretical density of a metallic substance (AW = 210) if the atoms have a radius of $1.5 \times 10^{-8}$ cm and are packed in a simple cubic structure?

    (A) 7.0 $g/cm^3$ (B) 9.0 $g/cm^3$ (C) 11 $g/cm^3$ (D) 13 $g/cm^3$
    (E) 15 $g/cm^3$

13. A crystalline compound contains atoms X and Y in a cubic lattice. A unit cell contains X atoms at the corners *and* face-centers, and Y atoms at the body-center. What is the empirical formula of this compound?

    (A) XY (B) $X_2Y$ (C) $XY_2$ (D) $X_4Y$ (E) $X_2Y_3$

---

## ENERGETICS AND CHANGES OF STATE

Increasing the temperature of a substance increases the kinetic energy of the constituent particles, resulting in a decrease in the ordering of these particles. The energy involved for a substance remaining in a given state is

$$\Delta H = mC\Delta T$$

where $\Delta H$ is the energy change in joules, $m$ the sample mass in grams, $\Delta T = T_{final} - T_{init.}$ with $T$ given in °C, and $C$ is a proportionality constant (the *specific heat*) given in units of J/g-°C and strongly characteristic of the substance. For a given substance there is some particular temperature at which the constituent particles have sufficient energy to break some or all of the orienting forces, resulting in a change of phase. Transformations of the type solid→liquid→gas are disordering processes requiring energy [*endothermic*, $\Delta H$ (+)] ; the reverse of these are ordering [*exothermic*, $\Delta H$ (–)]. Each of these changes of state requires (or liberates) a specific amount of energy for a given quantity of the substance (usually per gram or per mole) undergoing the transformation.

The amount of energy required (or liberated) to change a substance from one set of conditions to another may be calculated by taking the sum of the $\Delta H$ values of the individual steps.

A number of the problems in this section deal with thermal changes in water. The following data will be of use:

| | | |
|---|---|---|
| Specific heat of ice | 2.1 J/g-°C | 0.50 cal/g-°C |
| Specific heat of water | 4.2 J/g-°C | 1.0 cal/g-°C |

| | | |
|---|---|---|
| Specific heat of steam | 2.0 J/g-°C | 0.48 cal/g-°C |
| Heat of fusion | 335 J/g | 80.0 cal/g |
| Heat of vaporization | 2.26 kJ/g | 540. cal/g |

## Example Problem Involving Energetics and Changes of State

### 10-C. Heat Required to Convert Frozen Mercury to Mercury Vapor

The heat of fusion of mercury is 11.7 J/g (m.p. = -38.87°C) and the heat of vaporization is 296 J/g (b.p. = 356.6°C). The specific heats (all in J/g-°C) are: 0.137 (solid), 0.138 (liquid), and 0.104 (gas). Given 1.00 lb of mercury (33.4 ml) at -78°C, how many joules are required to change this mass of mercury to gaseous mercury at 500.°C?

(A) $1.74 \times 10^5$ J (B) 383 J (C) $3.83 \times 10^5$ J (D) $3.89 \times 10^5$ J
(E) $2.15 \times 10^3$ J

### *Solution*

The total change is best represented by a series of thermochemical steps (significant figures underscored):

| Step | $\Delta H$ (in joules) | |
|---|---|---|
| $Hg_{(s,\,-78°)} \rightarrow Hg_{(s,\,-39°)}$ | $(454\text{ g})\left(0.137 \frac{\text{J}}{\text{g-°C}}\right)(39°) =$ | $\underline{24}26$ J |
| $Hg_{(s,\,-39°)} \rightarrow Hg_{(\ell,\,-39°)}$ | $(454\text{ g})(11.7\text{ J/g}) =$ | $\underline{531}2$ J |
| $Hg_{(\ell,\,-39°)} \rightarrow Hg_{(\ell,\,357°)}$ | $(454\text{ g})\left(0.138 \frac{\text{J}}{\text{g-°C}}\right)(396°) =$ | $\underline{248}10$ J |
| $Hg_{(\ell,\,357°)} \rightarrow Hg_{(g,\,357°)}$ | $(454\text{ g})(296\text{ J/g}) =$ | $\underline{134}384$ J |
| $Hg_{(g,\,357°)} \rightarrow Hg_{(g,\,500°)}$ | $(454\text{ g})\left(0.104 \frac{\text{J}}{\text{g-°C}}\right)(143°) =$ | $\underline{6752}$ J |
| $Hg_{(s,\,-78°)} \rightarrow Hg_{(g,\,500°)}$ | | $\underline{173},684$ J |

### *Answer*

| | |
|---|---|
| Calculator | 173,684 J |
| Significant Figures Required | 3 |
| Answer Reported | $1.74 \times 10^5$ J [A] |

## Exercises on Energetics and Changes of State

**For Questions 14 and 15, consider the following diagram for carbon dioxide:**

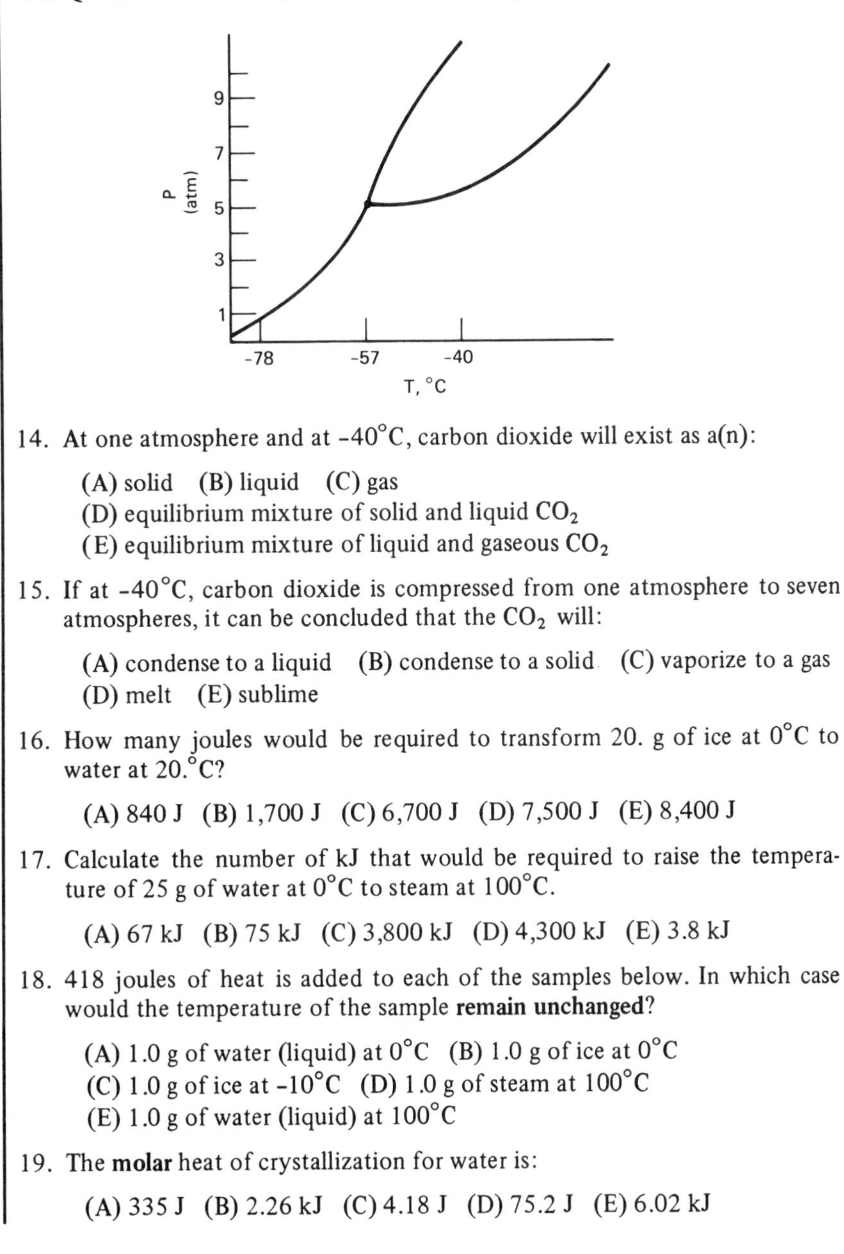


14. At one atmosphere and at -40°C, carbon dioxide will exist as a(n):

    (A) solid (B) liquid (C) gas
    (D) equilibrium mixture of solid and liquid $CO_2$
    (E) equilibrium mixture of liquid and gaseous $CO_2$

15. If at -40°C, carbon dioxide is compressed from one atmosphere to seven atmospheres, it can be concluded that the $CO_2$ will:

    (A) condense to a liquid (B) condense to a solid (C) vaporize to a gas
    (D) melt (E) sublime

16. How many joules would be required to transform 20. g of ice at 0°C to water at 20.°C?

    (A) 840 J (B) 1,700 J (C) 6,700 J (D) 7,500 J (E) 8,400 J

17. Calculate the number of kJ that would be required to raise the temperature of 25 g of water at 0°C to steam at 100°C.

    (A) 67 kJ (B) 75 kJ (C) 3,800 kJ (D) 4,300 kJ (E) 3.8 kJ

18. 418 joules of heat is added to each of the samples below. In which case would the temperature of the sample **remain unchanged**?

    (A) 1.0 g of water (liquid) at 0°C (B) 1.0 g of ice at 0°C
    (C) 1.0 g of ice at -10°C (D) 1.0 g of steam at 100°C
    (E) 1.0 g of water (liquid) at 100°C

19. The **molar** heat of crystallization for water is:

    (A) 335 J (B) 2.26 kJ (C) 4.18 J (D) 75.2 J (E) 6.02 kJ

## GENERAL EXERCISES ON CHAPTER 10—LIQUIDS, SOLIDS, AND CHANGES OF STATE

20. How many joules are required to raise the temperature of 7.5 g of water from 67.2 to 71.2°C?

    (A) 17 J (B) 130 J (C) 31 J (D) 4.2 J (E) 300 J

21. The number of joules required to raise the temperature of the sample of water in Question 20 from 67.2°F to 71.2°F would be:

    (A) the same as required in Question 20
    (B) less than required in Question 20
    (C) more than required in Question 20
    (D) 4.2 J
    (E) there is insufficient information to tell

22. Which of the following indicates a substance with relatively strong intermolecular forces?

    (A) a relatively high vapor pressure at room temperature
    (B) a relatively low melting point
    (C) a relatively high molar heat of vaporization
    (D) very close agreement between calculated and observed variables of the ideal gas law
    (E) a relatively low heat of fusion

23. The boiling point of HF is higher than that of HCl because:

    (A) HF is lighter than HCl and therefore requires more kinetic energy to vaporize
    (B) the London (dispersion) forces are greater in HF
    (C) hydrogen bonding in HF is greater than in HCl
    (D) the vapor pressure of HF is so high that the molecules are unable to evaporate
    (E) all of the above are correct

24. All of the following are involved in linking water molecules together in the liquid and solid states EXCEPT:

    (A) the polarity of the H—O bonds within each water molecule
    (B) dipole-dipole forces (C) hydrogen bonding (D) dispersion forces
    (E) attraction between hydrogen molecules and oxygen molecules

25. A compound containing cations $A^{n+}$ and anions $XO_4^-$ crystallizes in a cubic lattice. The $A^{n+}$ ions are situated on the unit cell corners; the $XO_4^-$ ions are centered on the face-centers and at the body-center. What is the charge on the cation, $A^{n+}$?

    (A) +4 (B) -1 (C) +1 (D) +2 (E) +3

26. Gold exists in a face-centered cubic lattice. Each unit cell contains ______ gold atoms.

    (A) 1 (B) 2 (C) 3 (D) 4 (E) 5

27. Gold crystallizes in the *fcc* system. The unit cell edgelength is found to be 4.08 Å. What is the atomic radius of Au?

    (A) 144 pm (B) 1.77 Å (C) 0.204 nm (D) 1.00 inch (E) 2.00 feet

28. Cooling nitrogen to <63 K causes the element to crystallize. The intermolecular forces of attraction that are responsible for holding the molecules in their lattice sites are called:

    (A) dispersion forces (B) single bonds (C) double bonds
    (D) triple bonds (E) dipole-dipole attractions

29. Which of the following statements is true?

    (A) Covalent bonds are weaker than van der Waals forces between molecules.
    (B) Covalent bonds have about the same strength as dipole-dipole attractions and, indeed, this is another name for a covalent bond.
    (C) Hydrogen bonds between water molecules are stronger than the covalent bonds in water molecules.
    (D) Dipole-dipole attractions between $SO_2$ molecules are weaker than the covalent bonds in $SO_2$ molecules.
    (E) Dispersion forces between molecules are stronger than most ionic bonds.

30. For a given set of *P* and *T* conditions, which of the following gases would be expected to deviate most from the ideal gas law?

    (A) He (B) $H_2$ (C) $CH_4$ (D) $O_2$ (E) $NH_3$

31. From the listed substances and their boiling points, pick the one which has the largest heat of vaporization.

    (A) $H_2O$, 100°C (B) HCN, 26°C (C) $N_2$, 77 K (D) $H_2$, 20 K
    (E) HF, 18°C

32. Which of the following is expected to have the lowest boiling point?

    (A) $O_2$ (B) HF (C) $Cl_2$ (D) $NH_3$ (E) HCl

33. Which of the following has the highest boiling point?

    (A) $O_2$ (B) Ar (C) He (D) HF (E) HCl

34. Which of the following is TRUE?

    (A) The freezing point of a substance is **always** lower than its melting point.
    (B) In a series of related substances, the strength of dispersion forces increases as the molecular weight increases.

(C) Substances with low critical temperatures possess strong intermolecular forces of attraction.
(D) The vapor pressure of a liquid depends on the number of moles of liquid in the container.
(E) "Hydrogen bonding" is always an intramolecular force.

35. Which of the following substances has the largest intermolecular forces of attraction?

(A) $H_2O$ (B) $H_2S$ (C) $H_2Se$ (D) $H_2Te$ (E) $H_2$

36. Sodium metal crystallizes in the body-centered cubic lattice. Each sodium in this structure has ______ nearest-neighbor atoms.

(A) 2 (B) 4 (C) 6 (D) 8 (E) 12

37. The melting point of KCl is less than that of NaF by some 200°C. The primary reason for this is that:

(A) potassium and chloride ions have greater charges than $Na^+$ and $F^-$
(B) the internuclear distance in KCl is greater than that in NaF
(C) the difference in electronegativity between K and Cl is greater than that between Na and F
(D) KCl molecules are more polar than NaF molecules
(E) the formula weight of KCl is greater than that of NaF

38. A sample of benzene at 80°C and 100 torr is cooled at constant volume. What will exist at 20°C? (The vapor pressure of benzene at 20°C is 75 torr.)

(A) benzene vapor at 100 torr (B) benzene liquid at 0 torr
(C) benzene vapor at 75 torr (D) benzene vapor at 83 torr
(E) benzene liquid and vapor at 75 torr

39. Which of the following changes is exothermic?

(A) conversion of 1 g of water at 100°C to steam at 100°C
(B) conversion of 1 g of water to ice at 0°C
(C) sublimation of 1 g of ice at -10°C
(D) changing the temperature of 1 g of ice at -10°C to 0°C
(E) all of the above changes are exothermic

40. At the normal boiling point of a substance, the vapor pressure of the substance:

(A) does not have a definite value since vapor pressure varies with temperature
(B) is equal to 760 torr (C) differs for different substances
(D) all of the above are correct (E) none of the above is correct

41. How many calories would be required to change 10.0 g of water at 70°C to steam at 100°C?

(A) 800 cal (B) 5,100 cal (C) 5,400 cal (D) 5,700 cal (E) 11,000 cal

42. Iron crystallizes in a *bcc* lattice with a unit-celledge length of 2.8664 Å. What is the radius of the iron atom?

    (A) 1.43 Å (B) 1.01 Å (C) 1.38 Å (D) 1.12 Å (E) 1.24 Å

43. Given the data in Question 42, what is the density of crystalline iron?

    (A) 8.95 $g/cm^3$ (B) 7.88 $g/cm^3$ (C) 5.32 $g/cm^3$ (D) 3.94 $g/cm^3$ (E) 15.7 $g/cm^3$

44. What would have been the density of iron if it had crystallized as a primitive cubic solid with a cube edge of 2.8664 Å? (Compare with Question 43.)

    (A) 8.95 $g/cm^3$ (B) 7.88 $g/cm^3$ (C) 5.32 $g/cm^3$ (D) 3.94 $g/cm^3$ (E) 15.7 $g/cm^3$

45. Progressive heating of a solid at its melting point does not result in a temperature increase because the thermal energy being absorbed is used to:

    (A) increase the average kinetic energy of the solid particles
    (B) expand the mercury in the thermometer being used
    (C) overcome the interunit forces of attraction in the solid
    (D) decrease the average potential energy of the solid particles
    (E) vaporize the liquid formed from melting

46. The boiling point of fluorine is about the same as that of:

    (A) He (B) Ne (C) Ar (D) Kr (E) Xe

47. Consider the following processes. In which are covalent bonds broken?

    (A) melting of sodium chloride (B) $CO_{2\,(s)} \rightarrow CO_{2\,(g)}$
    (C) vaporization of water (D) $NH_{3\,(\ell)} \rightarrow NH_{3\,(g)}$
    (E) $C_{(s,\,diamond)} \rightarrow C_{(g)}$

48. 400. calories raise the temperature of 100 g of ethyl alcohol ($C_2H_5OH$) from 20.1°C to 27.0°C. Its specific heat in J/g-°C is:

    (A) 47 (B) 7.1 (C) 120 (D) 2.4 (E) 1.7

49. Which of the following would be expected to form a solid that is brittle, rather high-melting, and a non-conductor of electricity?

    (A) Cu (B) $SO_2$ (C) Fe (D) $Br_2$ (E) $ZnF_2$

50. What final temperature results upon mixing 400. g of water at 20°C with 100. g of water at 70°C?

    (A) 60°C (B) 20°C (C) 70°C (D) 30°C (E) 90°C

51. In the following phase diagram, the melting point of the solid

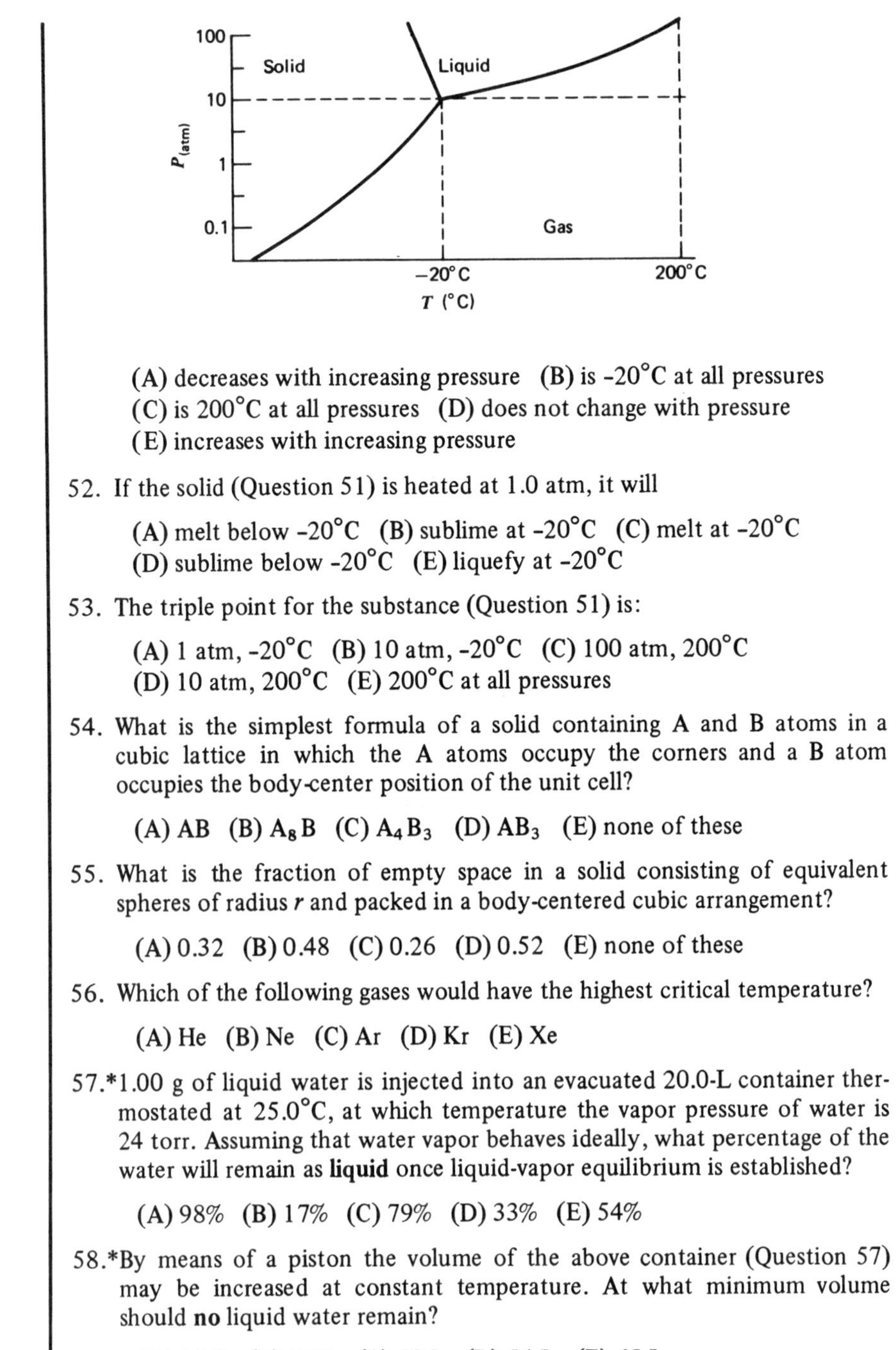


(A) decreases with increasing pressure (B) is -20°C at all pressures
(C) is 200°C at all pressures (D) does not change with pressure
(E) increases with increasing pressure

52. If the solid (Question 51) is heated at 1.0 atm, it will

(A) melt below -20°C (B) sublime at -20°C (C) melt at -20°C
(D) sublime below -20°C (E) liquefy at -20°C

53. The triple point for the substance (Question 51) is:

(A) 1 atm, -20°C (B) 10 atm, -20°C (C) 100 atm, 200°C
(D) 10 atm, 200°C (E) 200°C at all pressures

54. What is the simplest formula of a solid containing A and B atoms in a cubic lattice in which the A atoms occupy the corners and a B atom occupies the body-center position of the unit cell?

(A) AB (B) $A_8B$ (C) $A_4B_3$ (D) $AB_3$ (E) none of these

55. What is the fraction of empty space in a solid consisting of equivalent spheres of radius *r* and packed in a body-centered cubic arrangement?

(A) 0.32 (B) 0.48 (C) 0.26 (D) 0.52 (E) none of these

56. Which of the following gases would have the highest critical temperature?

(A) He (B) Ne (C) Ar (D) Kr (E) Xe

57.*1.00 g of liquid water is injected into an evacuated 20.0-L container thermostated at 25.0°C, at which temperature the vapor pressure of water is 24 torr. Assuming that water vapor behaves ideally, what percentage of the water will remain as **liquid** once liquid-vapor equilibrium is established?

(A) 98% (B) 17% (C) 79% (D) 33% (E) 54%

58.*By means of a piston the volume of the above container (Question 57) may be increased at constant temperature. At what minimum volume should **no** liquid water remain?

(A) 21 L (B) 32 L (C) 43 L (D) 54 L (E) 65 L

59. On the atomic scale, crystalline solids are:

(A) compressible and disordered (B) compressible and ordered
(C) incompressible and ordered (D) incompressible and disordered
(E) incomprehensible

60. The equilibrium partial pressure of water vapor over a 0.10-mole sample of liquid water is 25 torr at 25°C. If the amount of liquid water is increased to 0.20 mole, the equilibrium partial pressure would be:

(A) 25 torr (B) 12.5 torr (C) 50 torr
(D) cannot be determined without knowing the volume of the container
(E) none of these

61. For a given substance the vapor pressure increases when:

(A) the temperature increases
(B) the lid of the container is opened to the atmosphere
(C) the atmospheric pressure is increased
(D) the number of moles of the substance present is increased
(E) none of these

62. Liquids vaporize when:

(A) attractive forces disappear between atoms in the molecule
(B) the heat of vaporization is high
(C) the thermal motion overcomes the attractive forces between molecules
(D) thermal motion stops long enough for the molecule to escape
(E) bonds are broken within the molecule

63. Which of the following substances has the highest melting point?

(A) $CO_2$ (B) $H_2O$ (C) NaCl (D) BaO (E) MgO

64. Calculate the number of joules that would be liberated when 15 g of water at 0.0°C is converted to ice at -25°C.

(A) $3.8 \times 10^2$ J (B) $7.5 \times 10^2$ J (C) $5.0 \times 10^3$ J (D) $5.8 \times 10^3$ J
(E) $3.5 \times 10^4$ J

65. Both solid and liquid $CCl_4$ are molecular in nature. Compared to solid $CCl_4$, liquid $CCl_4$:

(A) is much more compressible and more highly ordered
(B) contains very widely separated and mobile molecules
(C) contains molecules joined by covalent bonds
(D) is a disordered array of molecules with somewhat restricted mobility
(E) is a state in which intermolecular forces are negligible

66. Calcium crystallizes in a face-centered cubic lattice. The edge of the unit cell is 5.57 Å. What is the radius of the calcium atom?

(A) 1.97 Å (B) 2.79 Å (C) 1.39 Å (D) 0.985 Å (E) 4.01 Å

67. Using information from Question 66, how many calcium atoms are there per unit cell?

    (A) 1 (B) 2 (C) 4 (D) 6 (E) 14

68. Using information from Question 66, calculate the density of calcium.

    (A) 0.995 g/cm³ (B) 6.85 g/cm³ (C) 4.36 g/cm³ (D) 1.54 g/cm³ (E) 3.61 g/cm³

69. What is the volume, in cm³, of the unit cell that is made up of atoms of $1.00 \times 10^{-8}$-cm radius arranged in a body-centered cubic lattice?

    (A) $8.00 \times 10^{-24}$ cm³ (B) $1.23 \times 10^{-23}$ cm³ (C) $2.25 \times 10^{-23}$ cm³ (D) $1.00 \times 10^{-24}$ cm³ (E) $1.53 \times 10^{-24}$ cm³

70. An ionic solid, $A_2B_3$, has a cubic unit cell. The arrangement of the ions could be:

    (A) $A^{3+}$ at the corners; $B^{2-}$ at the body center
    (B) $A^{3+}$ at the body center; $B^{2-}$ at the corners
    (C) $A^{3+}$ at the face centers; $B^{2-}$ at the corners
    (D) $A^{3+}$ at the corners **and** face centers; $B^{2-}$ at the body center
    (E) $A^{3+}$ at the corners **and** the body center; $B^{2-}$ at the face centers

71. Consider equivalent spheres of radius *r* packed in a face-centered cubic lattice. What percentage of the solid is **empty** space?

    (A) 26% (B) 32% (C) 45% (D) 18% (E) 51%

72.*Consider liquid mercury at 25°C (density 13.6 g/mL). If a mercury atom can be assumed to be spherical with a radius of 1.44 Å (1 Angstrom, Å, = $1 \times 10^{-8}$ cm), what is the fraction of **empty** space in the liquid?

    (A) 0.49 (B) 0.39 (C) 0.26 (D) 0.62 (E) 0.31

---

## ANSWERS TO CHAPTER 10 PROBLEMS

| | | | | | |
|---|---|---|---|---|---|
| 1. C | 13. D | 25. A | 37. B | 49. E | 61. A |
| 2. C | 14. C | 26. D | 38. E | 50. D | 62. C |
| 3. C | 15. A | 27. A | 39. B | 51. A | 63. E |
| 4. B | 16. E | 28. A | 40. B | 52. D | 64. D |
| 5. E | 17. A | 29. D | 41. D | 53. B | 65. D |
| 6. A | 18. E | 30. E | 42. E | 54. A | 66. A |
| 7. A | 19. E | 31. A | 43. B | 55. A | 67. C |
| 8. A | 20. B | 32. A | 44. D | 56. E | 68. D |
| 9. C | 21. B | 33. D | 45. C | 57. E | 69. B |
| 10. D | 22. C | 34. B | 46. C | 58. C | 70. E |
| 11. E | 23. C | 35. A | 47. E | 59. C | 71. A |
| 12. D | 24. E | 36. D | 48. D | 60. A | 72. A |

# 11
# *Solutions*

Much of chemistry is solution chemistry, *i.e.*, chemistry in which the chemical reactivity of a solute is directly proportional to its effective concentration (activity) in a particular solvent. Molarity, $M$, (Ch. 2) and normality, $N$, (Ch. 8) have already been introduced. In this chapter not only will we continue to deal with these important concentration units, but also additional units will be introduced which are useful in dealing with certain physical properties of solutions that describe how the solvent is affected by the concentrations of solute species.

Solutes that do not dissociate into ions are called non-electrolytes—they exist as molecules in solution. On the other hand, species that exist exclusively as ions in solution, *i.e.*, dissociate 100%, are termed strong electrolytes. Strong electrolytes in water include strong acids, strong bases, and most salts. Species that exist in solution partially as ions and partially as molecules are called weak electrolytes. The vast majority of acids and bases are weak electrolytes, and the total number of solute particles (molecules and ions) produced by a weak electrolyte depends upon the *percent dissociation* (or *percent ionization*) in solution. The terms strong, weak, and non-electrolyte arise because these types of substances produce solutions which are respectively very good, poor, and non-conductors of electricity.

The *percent dissociation* of a weak electrolyte indicates the fraction of the molecules that have dissociated to form ions; it is dependent upon the temperature and the makeup concentration, increasing as the solution is diluted. In the case of 0.100 $M$ acetic acid, the $HC_2H_3O_2$ is 1.32% dissociated at 25°C. This means that the reaction,

$$HC_2H_3O_2 + H_2O \rightarrow H_2O^+ + C_2H_3O_2^-$$

has proceeded only to a small extent, *i.e.*, (0.0132)(0.100 $M$) = 0.00132 moles/liter of the acetic acid has dissociated. This gives a $H_3O^+$ concentration of [0.00132]† and also an acetate ion concentration† of [0.00132] and leaves as the actual concentration of acetic acid, 0.100 - 0.001 = 0.099 *M*.

## CONCENTRATION UNITS

Molarity, in particular, is used in two somewhat different contexts. Until now, the use has been in terms of *solution makeup, e.g.*, the designation 1.0 *M* HCl has been taken as meaning a solution prepared by dissolving 36.5 g of HCl (1.0 mole) in water and diluting this to one liter. In actuality, such a solution is not 1.0 *M* in HCl molecules--it contains **no** detectable HCl molecules. The HCl has entirely reacted with the solvent (water) forming $H_3O^+$ (frequently abbreviated as $H^+_{(aq)}$) and $Cl^-$ ions. Nevertheless, the solution is still referred to as 1.0 *M* HCl, *i.e.*, the concentration is reported based on the makeup of the solution.

A second way of using molarity is to refer to the actual concentration of a molecule or ion in solution. Since the HCl in the aforementioned solution is present **entirely** as ions, it is said to be a *strong electrolyte* and, hence, the solution would be 1.0 *M* in $H_3O^+$ and also 1.0 *M* in $Cl^-$. Other examples of strong electrolytes in aqueous solution are (i) most common salts, *e.g.*, NaCl, $CaCl_2$, and $Na_2SO_4$; (ii) a few acids such as $HClO_4$, $HNO_3$, and $H_2SO_4$ (first proton only); and (iii) a few bases such as NaOH, KOH, and $Ca(OH)_2$. Other compounds are either not completely *dissociated* into ions in solution, *weak electrolytes*, or remain as molecules when dissolved, *nonelectrolytes*. In the case of the 0.100 *M* acetic acid mentioned in the previous paragraph, note carefully that **the makeup concentration (C),** 0.100 *M* $HC_2H_3O_2$, **is the sum of the concentrations of dissociated and undissociated** acetic acid. Many acids and bases are weak electrolytes, and a calculation must be carried out if actual concentrations of the molecules and ions present are desired.

Molarity is a volumetric unit; it does not specify the amount of solvent present. Three other units that do explicitly define both the quantities of solute and solvent present are:

% by weight (%) $$\% \equiv \frac{\text{mass of solute}}{\text{total mass of } \textit{solution}} \times 100$$

molality (*m*) $$m \equiv \frac{\text{moles of solute}}{\text{kilograms of } \textit{solvent}}$$

mole fraction (*X*) $$X_i \equiv \frac{\text{moles of solute (i)}}{\text{total moles of all components}}$$

---

†Square brackets [ ] are used to denote actual molar concentrations of a molecule or ion.

The link between these three concentration units and molarity involves the bulk solution density, grams of *solution*/mL of *solution*.

## Example Problems Involving Concentration Units

### 11-A. Dissociation of a Weak Electrolyte

In addition to solvent molecules, the major species present ($\geqslant$0.1 *M*) in a 1 *M* solution of the weak electrolyte phosphoric acid is (are):

(A) $H^+$ and $H_2PO_4^-$ (B) $H_3PO_4$
(C) $H_3O^+$, $H_2PO_4^-$, $HPO_4^{2-}$, and $PO_4^{3-}$ (D) $H_3O^+$ and $PO_4^{3-}$
(E) $H_3PO_4$, $H_3O^+$, $H_2PO_4^-$, $HPO_4^{2-}$, and $PO_4^{3-}$

***Solution***

The fact that $H_3PO_4$ is a weak electrolyte implies that the first dissociation, $H_3PO_4 + H_2O \rightarrow H_3O^+ + H_2PO_4^-$, does not occur to a large extent and that further dissociations occur even to lesser extents. In a solution of a weak electrolyte one usually finds most of the substance undissociated; hence $H_3PO_4$ would be present in relatively large amounts with small quantities of $H_3O^+$ and $H_2PO_4^-$. These two ions apparently are present at concentrations below the 0.1 *M* level since there is no answer giving just phosphoric acid and these two ions. Indeed, all of the species listed in the answers are present, but some of them occur only at very small concentrations.

***Answer*** $H_3PO_4$ [B]

### 11-B. Concentration of an Ion in a Solution Containing Several Solutes

What is the concentration of sodium ion in a solution made by mixing 100. mL of 0.100 *M* $Na_2SO_4$, 100. mL of 0.250 *M* NaCl, and 200. mL of 0.0250 *M* $Na_3PO_4$? (Assume additive volumes.)

(A) 0.100 *M* (B) 0.150 *M* (C) 0.120 *M* (D) 0.0600 *M* (E) 0.0400 *M*

***Solution***

The molarity of the sodium ion would be the total number of moles of sodium ion present per liter of solution. Since all of these compounds are strong electrolytes, *i.e.*, they occur as sodium ions and the respective anion, the sodium sulfate would contribute 0.0200 moles of $Na^+$; the sodium chloride, 0.0250 moles of $Na^+$; and the sodium phosphate, 0.0150 moles of $Na^+$. Therefore,

$$[Na^+] = \frac{(0.0200 + 0.0250 + 0.0150)\text{ moles Na}^+}{(0.100 + 0.100 + 0.200)\text{ liters}}$$

$$= 0.150\,M\,Na^+$$ [B]

## 11-C. Percent by Weight to Mole Percent

An aqueous solution of ethylene glycol, $C_2H_4(OH)_2$ (a molecular solute, FW = 62.07), has a density of 1.0125 g/cc and consists of 10.00% ethylene glycol by weight. What percent of the molecules present are **water** molecules?

(A) 93.15% (B) 55.49% (C) 96.87% (D) 34.08% (E) 89.11%

### *Solution*

% by molecule is the same numerically as mole %; mole % simply is the mole fraction times 100.

$$\left.\begin{array}{l}\text{density}\\ \%\ \text{glycol}\end{array}\right\} \rightarrow \left.\begin{array}{c}\text{g glycol} \rightarrow \text{moles glycol}\\ \downarrow\\ \text{g water} \rightarrow \text{moles water}\end{array}\right\} \rightarrow \text{mole \% } H_2O$$

The composition of 1 mL of solution is:

$$\left(1\ \cancel{\text{mL soln}}\right)\left(1.0125\ \frac{\cancel{\text{g soln}}}{\cancel{\text{mL soln}}}\right)\left(\frac{10.00\ \text{g glycol}}{100\ \cancel{\text{g soln}}}\right) = 0.1013\ \text{g glycol}$$

$$\left(1\ \cancel{\text{mL soln}}\right)\left(1.0125\ \frac{\text{g soln}}{\cancel{\text{mL soln}}}\right) - 0.1013\ \text{g glycol} = 0.9112\ \text{g}\ H_2O$$

These correspond to 0.001632 mole of glycol and 0.05059 mole of water. Mole % water (and hence % of molecules) is

$$\left(\frac{0.05059}{0.001632 + 0.05059}\right) \times 100 = 96.87\%$$

### *Answer*

All data were given to at least 4 significant figures, none of which were lost in subtraction; hence, the answer should be reported to 4 significant figures, *i.e.*, 96.87% of the molecules are water molecules. **[C]**

## Exercises on Solution Concentrations

1. Barium hydroxide is a strong electrolyte. The $[OH^-]$ in 0.040 $M$ $Ba(OH)_2$ is:

   (A) 0.100 $M$ (B) 0.040 $M$ (C) 0.080 $M$ (D) 0.020 $M$ (E) 0.120 $M$

**The following information is to be used for Questions 2 to 5.** A solution is prepared by dissolving 25.0 g of compound AB (FW = 125) in enough water to make 100.0 mL of solution. The density of the solution is 1.15 g/mL.

2. The % AB by weight in the solution is:

   (A) 25.0% (B) 28.3% (C) 50.8% (D) 20.0% (E) 21.7%

## DRILL ON SOLUTION CONCENTRATIONS

***Student's Name*** ______________

For each solute (i) complete the following table where %i = the percent of the solute in the solution, $M_i$ = molarity, $m_i$ = molality, and $X_i$ = mole fraction of 'i'. You may be asked to turn in this sheet.

| Solute (i) | g solute | g $H_2O$ | Volume of Solution | Density of Solution | % i | $M_i$ | $m_i$ | $X_i$ |
|---|---|---|---|---|---|---|---|---|
| $HC_2H_3O_2$ | 5.00 g | 95.0 g | | 1.007 g/cm$^3$ | | | | |
| $KNO_3$ | | | 1.00 L | 1.120 g/cm$^3$ | | | | 0.03764 |
| $C_{12}H_{22}O_{11}$ | 54.0 g | 36.0 g | 69.8 mL | | | | | |
| $H_2SO_4$ | 34.875 g | | 100. mL | | 50.0% | | | |
| ______ | | | | | | | | |

3. The mole fraction of AB in the solution is:

   (A) 0.038 (B) 0.040 (C) 26.0 (D) 0.415 (E) 0.062

4. The molality of AB is:

   (A) 1.0 $M$ (B) 2.2 $m$ (C) 2.0 $M$ (D) 0.20 $m$ (E) 0.11 $m$

5. The molarity of AB in the solution is:

   (A) 1.00 $M$ (B) 0.200 $M$ (C) 0.100 $M$ (D) 2.00 $M$ (E) 0.000200 $M$

6. An aqueous sulfuric acid solution has a density of 1.25 g/mL and contains 34.0% sulfuric acid by weight. Calculate the weight of sulfuric acid in 200. mL of the solution.

   (A) 85.0 g (B) 7.35 g (C) 5440 g (D) 4.71 g (E) 165 g

7. 200. mL of 1.40 $M$ $Na_2SO_4$ is added to 100. mL of 1.80 $M$ NaCl. If we assume the volumes are additive, the final concentration of $Na^+$ is:

   (A) 1.53 $M$ (B) 2.13 $M$ (C) 1.80 $M$ (D) 2.80 $M$ (E) 2.47 $M$

---

## COLLIGATIVE PROPERTIES

Many physical properties of solutions depend primarily on the relative **numbers** of solute particles (molecules or ions) and solvent particles rather than on the chemical nature of the solute. Such properties are referred to as *colligative* properties and include freezing point depression ($\Delta T_f$), boiling point elevation ($\Delta T_b$), vapor pressure lowering ($\Delta p$), and osmotic pressure ($\Pi$). Experimental measurement of $\Delta p$, $\Delta T_f$, $\Delta T_b$, or $\Pi$ for dilute solutions† allows the number of solute particles in solution to be determined. This, in combination with knowledge of how the solution was prepared, permits calculations of the molecular weight of a molecular solute, the number of ions per formula unit of an ionic substance, or the percent ionization for a substance that partially dissociates into ions. When dealing with these properties it is convenient to use concentration units that directly express the relative numbers of solute and solvent particles. Most commonly used are *molality (m)*, moles of solute per 1000 g of solvent, and *mole fraction ($X_i$)*, moles of component $i$ per total number of moles.

In dilute solution boiling point elevation with nonvolatile solutes and freezing point depression obey the relationship,

$$\Delta T = K_{solvent} m_{solute}$$

---

†The equations relating the magnitudes of the coligative properties are really *limiting laws, i.e.*, they agree more accurately with the experimental measurements as the solute concentration decreases.

where $\Delta T$ is the **absolute** value of the difference between the melting (or boiling) points of the pure solvent and the solution, $K$ is a constant characteristic of the solvent, and $m$ is the **total** molality of solute particles (molecules or ions). For nonvolatile solutes in liquid solution the vapor pressure is given by *Raoult's law,*

$$p_{soln} = p^{\circ} X_{solvent}$$

where at a given temperature, $p_{soln}$ is the equilibrium vapor pressure of the solvent over the solution, $p^{\circ}$ is the vapor pressure of the pure solvent at this temperature, and $X_{solvent}$ is the mole fraction of solvent. In terms of mole fraction of the *solute*, Raoult's Law may also be expressed by the equation,

$$\Delta p = p^{\circ} X_{solute}$$

which states that the vapor pressure **lowering ($\Delta p$)** is directly proportional to the mole fraction of the solute.

*Osmosis* is the process whereby solvent molecules (but **not solute** species) pass through semipermeable membranes (cellophane, parchment paper, cell walls, *etc.*). Thus, if a sugar solution is separated from pure water (or a less concentrated solution) by a suitable membrane, water molecules pass through the membrane in both directions, but a greater number of water molecules pass from the pure water (or less concentrated solution) region into the solution region. This causes the liquid level to rise relative to the level of the less concentrated solution; this 'lifted' water column causes a hydrostatic pressure difference and makes it more difficult for water molecules to move from the less concentrated region to the more concentrated region. Eventually, a point is reached where the number of solvent molecules moving through the membrane is the same in each direction. The pressure exerted by the 'lifted' solution column is called the *osmotic presure,* $\Pi$.

For dilute solutions $\Pi$ has been shown to be the same as the pressure that would be exerted by the solute **if it were** a gaseous substance at the same concentration and temperature. Thus, by analogy with the ideal gas law, $P = \frac{n}{V} RT$, we have

$$\Pi = \frac{n}{V} RT, = RTM_c$$

where $M_c$ has been used as the *colligative molarity*, the **total** number of moles of solute **particles** per liter. Thus, a 0.10 M NaCl solution has $M_c = 0.20$ M and the osmotic pressure at 25°C would be

$$\Pi = \left(0.0821 \frac{L \cdot atm}{mol \cdot K}\right) (298 \text{ K}) \left(0.20 \frac{mol}{L}\right) = 4.89 \text{ atm}$$

Osmosis produces a big effect. Recalling that one atmosphere is equivalent to a column of mercury (density = 13.6 g/cm$^3$) 76.0 cm high, the osmotic pressure of 0.10 M NaCl would be equivalent to a column of water, or a dilute aqueous solution with density ~1.00 g/cm$^3$, over 160 feet high! This contributes to the movement of ground water to the leaves and branches of a tall tree. Osmotic pressure measurements are very important in determining molecular weights of large molecules such as proteins, whose solutions necessarily have very low molarities of solute particles.

## Example Problems Involving Colligative Properties

### 11-D. Molecular Formula from % Composition and Freezing Point Depression Data

A compound analyzes, by weight: P, 26.72%; N, 12.09%; Cl, 61.17%. In a given experiment, 1.008 g of this compound was dissolved in 11.38 mL of benzene ($d$ = 0.879 g/mL); the solution was observed to freeze at 4.37°C. Previous measurements showed a freezing point of 5.48°C for the solvent used ($C_6H_6$; $K_f$ = 5.12°/molal). What is the molecular formula of the compound in the benzene solvent?

(A) $PNCl_2$ (B) $P_2N_2Cl_4$ (C) $P_3N_4Cl_9$ (D) $P_4N_4Cl_8$ (E) $P_4N_2Cl_{10}$

***Solution***

From % composition → empirical formula;
From solution composition → g compound/kg benzene;
From freezing point data → molality ($=\Delta T_f/K_f$);
Then, $MW_{cpd}$ = g cpd/kg benzene/mole cpd/kg benzene, which can be compared with the empirical formula.

The empirical formula calculated from the % composition data is ($PNCl_2$). (See Chapter 2, particularly Example 2-H.) This eliminates answers (C) and (E).

$$\frac{1.008 \text{ g cpd}}{(11.38 \cancel{\text{mL}})(0.879 \cancel{\text{g benzene}}/\cancel{\text{mL}})} \times \frac{10^3 \cancel{\text{g benzene}}}{\text{kg benzene}} = 101 \frac{\text{g cpd}}{\text{kg benzene}}$$

$$\frac{1.11^\circ\cancel{\text{C}}}{5.12^\circ\cancel{\text{C}}/\text{molal}} = 0.217 \frac{\text{mole cpd}}{\text{kg benzene}}$$

$$MW_{cpd} = \frac{101 \text{ g cpd}/\cancel{\text{kg benzene}}}{0.217 \text{ mole cpd}/\cancel{\text{kg benzene}}} = 465 \text{ g cpd/mole cpd}$$

The molecular weight is four times the formula weight of $PNCl_2$ (116); hence the molecular formula must correspond to the tetramer $[PNCl_2]_4$. **[D]**

### 11-E. Number of Ions in a Formula Unit from Freezing Point Data

An aqueous solution containing 0.250 moles of X, a strong electrolyte, in 500.

g of water freezes at -2.79°C. How many ions are formed per formula unit of X? The molal freezing point depression constant for water is 1.86°C/$m$.

(A) 1 (B) 2 (C) 3 (D) 4 (E) 1-1/2

### *Solution*

The freezing point of the solution depends on the total molality of the ions, $m_{total}$. $m_{total}$ is merely the molality of X, $m_X$, times the number of ions derived from each X.

From makeup: $m_x = \dfrac{0.250 \text{ moles X}}{0.500 \text{ kg } H_2O} = 0.500\ m \text{ X}$

From f.p. experiment: $m_{total} = \dfrac{2.79°C}{1.86°C/m} = 1.50\ m \text{ ions}$

Then, $\dfrac{1.50 \text{ moles ions/kg } H_2O}{0.500 \text{ moles X/kg } H_2O} = 3 \dfrac{\text{moles of ions}}{\text{mole of X}}$

or, 3 ions/X unit **[C]**

## Exercises on Colligative Properties

**The following information is to be used for Questions 8 to 12**: Given two solutions of a nondissociating and nonvolatile solute (FW = 40). The solution makeup is as follows:

Solution A: 20 g of solute + 1200 g $H_2O$
Solution B: 30 g of solute + 900 g $H_2O$

For the solvent, water, $K_f$ = 1.86°C and $K_b$ = 0.51°C.

8. Which of the following is CORRECT?

   (A) The vapor pressure above Solution A is **greater** than that above Solution B.
   (B) The boiling point of Solution A is **higher** than that of Solution B.
   (C) The freezing point of Solution A is **lower** than that of Solution B.
   (D) The mole fraction of solute in Solution B is 0.025.
   (E) The molality of solute in Solution B is 0.050.

9. The freezing point of Solution A is:

   (A) -1.86°C (B) -0.51°C (C) -0.78°C (D) 0.00°C (E) -0.42°C

10. If Solution A has a density of 1.10 g/mL, what is the molarity of the solute in this solution?

    (A) 0.38 $M$ (B) 0.50 $M$ (C) 0.018 $M$ (D) 0.45 $M$ (E) 0.10 $M$

11. If the solute in Solution A is completely dissociated into ions, *viz.*,

$$\text{Solute} \rightarrow X^+ + Y^-$$

## DRILL ON COLLIGATIVE PROPERTIES

***Student's Name*** ____________________

Complete the following table by calculating the formula weight of the solute (FW), the normal freezing point of the solution ($T_f$), the normal boiling point of the solution ($T_b$), the decrease in the vapor pressure of the solution ($\Delta p$) at 25°C, and the osmotic pressure at 25°C for each of the following aqueous solutions. The vapor pressure of water at 25°C is 23.8 torr. 'NE' stands for an unknown non-volatile non-electrolyte. You may be asked to turn in this sheet.

| Aqueous Solution | Solution Density, $g/cm^3$ | FW of solute g/mole | $T_f$, °C | $T_b$, °C | $\Delta p$, torr | $\Pi$, atm |
|---|---|---|---|---|---|---|
| 5.00% 'NE' | 1.01 $g/cm^3$ | | −1.63°C | | | |
| 5.00% AX (10.0% ionized as $A^+$ & $X^-$) | 1.01 $g/cm^3$ | 100 | | | | |
| 7.00% $CaCl_2$ | 1.06 $g/cm^3$ | 111 | | | | |
| 10.0% 'NE' | 1.02 $g/cm^3$ | | | | 0.021 torr | |
| 1.00% 'NE' | 1.02 $g/cm^3$ | | | | | 0.0245 atm |
| ______ | | | | | | |

then the boiling point of Solution A would be expected to be:

(A) 99.48°C (B) 0.50°C (C) 100.00°C (D) 99.57°C (E) 100.43°C

12. If half of Solution A is mixed with all of Solution B, the freezing point of the new solution would be:

(A) +1.24°C (B) 0.00°C (C) -0.78°C (D) -1.24°C (E) -1.34°C

---

## GENERAL EXERCISES ON CHAPTER 11—SOLUTIONS

13. In a solution containing 0.100 mole of acetic acid/kg of $H_2O$, the acid is 1.32% ionized. What would be the expected freezing point of the solution?

(A) -0.19°C (B) -0.37°C (C) -0.93°C (D) -0.01°C (E) -0.57°C

14. A salt solution approximating ocean water would have the following **ionic** molalities: chloride, 0.566 *m*; sodium, 0.485 *m*; magnesium, 0.054 *m*; sulfate, 0.029 *m*; calcium, 0.011 *m*; potassium, 0.011 *m*; bicarbonate, 0.002 *m*. At 20°C the vapor pressure of pure water is 17.5 torr. The vapor pressure of this "ocean water" solution would be about ______ % less than that of the pure water at 20°C.

(A) 1 (B) 2 (C) 3 (D) 4 (E) 5

**Questions 15 to 18 refer to a solution prepared by dissolving 85.0 g of $NH_3$ in sufficient water to produce 1.00 L of solution. This solution has a density of 0.960 g/mL.**

15. The percent ammonia by weight in the solution is:

(A) 8.50% (B) 8.85% (C) 9.71% (D) 7.83% (E) 9.29%

16. The mole fraction of ammonia in the solution is:

(A) 0.093 (B) 0.103 (C) 0.086 (D) 0.096 (E) 0.085

17. The molality of ammonia in the solution is:

(A) 5.21 (B) 5.00 (C) 5.71 (D) 4.61 (E) 4.78

18. The molarity of ammonia in the solution is:

(A) 5.21 (B) 5.00 (C) 5.71 (D) 4.61 (E) 4.78

19. A compound that forms an aqueous solution that is a good conductor of electricity is always:

(A) covalent (B) ionic (C) an electrolyte (D) an acid (E) a base

20. The chloride ion concentration in a calcium chloride solution is 2.0 *M*. The solution could be correctly labeled as:

(A) 2.0 *M* $Ca^{2+}$ (B) 2.0 *M* $CaCl_2$ (C) 4.0 *M* $CaCl_2$ (D) 1.0 *M* $CaCl_2$
(E) 4.0 *M* $Cl^-$

21. Which of the following is a weak electrolyte in aqueous solution?

    (A) ethyl alcohol (B) $H_2SO_4$ (C) HCl (D) sodium chloride
    (E) acetic acid

22. The $[H_3O^+]$ in a 0.0100 *M* solution of an acid is $1.00 \times 10^{-3}$ *M*. The percent dissociation of the acid is:

    (A) 1.0% (B) 5.0% (C) 10% (D) 20% (E) 30%

23. If you mix 200 mL of 0.050 *M* calcium hydroxide with 200 mL of 0.30 *M* HCl, what will be the concentration of calcium ion in the resulting solution?

    (A) $1.0 \times 10^{-2}$ *M* (B) $2.5 \times 10^{-2}$ *M* (C) $5.0 \times 10^{-2}$ *M*
    (D) $7.5 \times 10^{-2}$ *M* (E) $1.0 \times 10^{-3}$ *M*

24. If a 0.600 *m* aqueous solution of a weak acid, HA, has a freezing point of -1.28°C, what is the percent ionization of HA in this solution? ($K_f$ = 1.86°C/*m* for water)

    (A) 0.0% (B) 1.3% (C) 8.8% (D) 15% (E) 29%

25. One gram (1.00 g) of "cream of tartar" ($KHC_4H_4O_6$, FW = 188.2) dissolves in 8820 mL of alcohol (density of alcohol at room temperature is 0.810 g/cc). What is the % (by weight) of "cream of tartar" in this saturated alcoholic solution?

    (A) 0.000602% (B) 0.000744% (C) 0.00918% (D) 0.0113%

    (E) 0.0140%

26. One kilogram (1.00 kg) of "tartar emetic" ($KSbC_4H_4O_7 \cdot 0.5H_2O$, FW, 333.93) dissolves in water to give 12.0 L of solution. It is desired to measure out enough solution to contain 30.0 g of **antimony**. How many milliliters of the solution should be taken?

    (A) 2.50 (B) 250 (C) 360 (D) 987 (E) 1000

**Use the following information for Questions 27-29.**

The following experimental data were obtained for pure solvent S and solutions of solute X (MW = 100.) in this solvent. Solute X neither associates nor dissociates in solvent S.

| Experiment | Grams S | Grams X | Freezing-pt, °C |
|---|---|---|---|
| I. | 100.0 | 0.000 | 16.605 |
| II. | 100.0 | 1.000 | 16.200 |
| III. | 100.0 | 1.750 | 15.907 |
| IV. | 100.0 | 2.200 | 15.710 |
| V. | 100.0 | 4.200 | 14.960 |

27. What is the molality of X for the solution of Experiment V?

(A) 0.0420 *m* (B) 0.420 *m* (C) 4.20 *m* (D) 0.0403 *m* (E) 0.104 *m*

28. For Solutions II-V, one would need to know the (__α__) in order to determine the mole fraction of X; one would need to know (__β__) in order to calculate the molarity of X. (Fill in the blanks α and β.)

| | α | β |
|---|---|---|
| (A) | MW of X; | density of S |
| (B) | MW of S; | density of X |
| (C) | MW of S; | density of S |
| (D) | MW of X; | density of the solution |
| (E) | MW of S; | density of the solution |

29. From these data, what is the apparent cryoscopic constant, $K_f$, characteristic of solvent S?

(A) 1.86°C/molal (B) 2.98°C/molal (C) 3.73°C/molal
(D) 4.01°C/molal (E) 5.12°C/molal

30. Dissolving a solute such as NaCl in a solvent at 25°C results in:

(A) an increase in the kinetic energy of the solvent molecules
(B) a decrease in the kinetic energy of the solvent molecules
(C) a decrease in the boiling point of the liquid
(D) an increase in the melting point of the liquid
(E) a decrease in the vapor pressure over the solution

31. Which of the following solutions would have the lowest freezing point?

(A) 0.1 *m* $Ca(NO_3)_2$ (B) 0.1 *m* $KNO_3$ (C) 0.1 *m* HCl
(D) 0.2 *m* sucrose (sugar) (E) 0.1 *m* sucrose (sugar)

32. Which of the following nonelectrolytes would produce the solution having the lowest freezing point, if 10.0 g of the respective compound were dissolved in 1000 g of water?

(A) alcohol, $C_2H_6O$ (B) glycerin, $C_3H_8O_3$ (C) glucose, $C_6H_{12}O_6$
(D) methanol, $CH_4O$ (E) **all** produce the same effect

33. To what volume would you dilute one liter of 0.2 *M* sulfuric acid so that the final solution would be 0.001 *M*?

(A) 40 L (B) 60 L (C) 80 L (D) 120 L (E) 200 L

34. A 1.00 *M* solution of HCl is to be prepared by diluting "constant boiling HCl" to a total volume of 0.500 L. If the constant boiling HCl is 20.22% HCl by weight and has a density of 1.100 g/cc, then the approximate volume of the constant boiling acid that would be required is:

(A) 164 mL (B) 82 mL (C) 90 mL (D) 8 mL (E) 2 mL

35. A solution made from water and a nondissociating solute contains 10.0 g of solute X and 800 g of water. The freezing point of the solution is -0.31°C. What is the molecular weight of X? ($K_f$ for water = 1.86°C/*m*)

    (A) 40 (B) 50 (C) 80 (D) 75 (E) 60

36. Barium phosphate is insoluble in water. If equal volumes of 0.50 *M* $Ba(ClO_4)_2$ and 0.20 *M* $K_3PO_4$ are mixed, what is the concentration of $Ba^{2+}$ ions in the resulting solution?

    (A) 0.20 *M* (B) 0.00 *M* (C) 0.37 *M* (D) 0.10 *M* (E) 0.25 *M*

37. $Ag_2CO_3$ and $Ag_3PO_4$ are both insoluble in water. If equal volumes of 0.800 *M* $AgNO_3$, 0.150 *M* $Na_2CO_3$, and 0.050 *M* $K_3PO_4$ are mixed, the concentration of $Ag^+$ remaining in the resulting solution is:

    (A) 0.000 *M* (B) 0.117 *M* (C) 0.175 *M* (D) 0.351 *M* (E) 0.600 *M*

38. A 0.20 *m* aqueous solution of HY, a weak acid, is 20% ionized. What is the freezing point of this solution? ($K_f$ = 1.86°C/*m* for water)

    (A) -0.45°C (B) -0.31°C (C) -0.61°C (D) -0.53°C (E) -0.37°C

39. In a 0.200 *M* solution the weak base XOH is 5.0% dissociated. What volume of this solution contains the same quantity of $OH^-$ as 1.00 liter of $5.00 \times 10^{-2}$ *M* NaOH?

    (A) 0.20 L (B) 2.0 L (C) 1.0 L (D) 5.0 L (E) 10.0 L

40. 28.2 g of $C_{20}H_{42}$, a nonvolatile solute, was dissolved in 500. g of pure benzene ($C_6H_6$). What would be the vapor pressure of benzene over this solution at 25°C? (The vapor pressure of pure benzene at this temperature is 93.4 torr.)

    (A) 0 torr, since the solute is nonvolatile (B) 92.0 torr (C) 93.4 torr (D) 94.8 torr (E) 760 torr

41. A solution is prepared by dissolving 0.500 mole of $Al_2(SO_4)_3 \cdot 18H_2O$ (FW = 666) in 838 g of water. In terms of $Al_2(SO_4)_3$, the solution has the concentration:

    (A) 0.500 *m* (B) 0.600 *m* (C) 0.500 *M* (D) 0.600 *M* (E) 0.400 *m*

42. The percentage $Al_2(SO_4)_3$ by weight for the above solution (Question 41) is:

    (A) 17.1% (B) 14.6% (C) 33.3% (D) 16.5% (E) 28.4%

43. If the above solution (Questions 41 and 42) has a density of 1.160 g/mL, what is the molarity of the **aluminum ion** in the solution?

    (A) 0.254 *M* (B) 0.495 *M* (C) 0.509 *M* (D) 1.02 *M* (E) 0.990 *M*

44. If the solute behaves ideally as a strong electrolyte, the solution (Questions 41 to 43) should boil at ______ °C above the boiling point of water. ($K_b$ for water = 0.51°C/*m*)

    (A) 0.51 (B) 1.02 (C) 1.28 (D) 2.55 (E) 0.26

45. If 40.0 mL of this solution (Questions 41 to 44) is diluted with water to 250. mL in a volumetric flask, the sulfate ion concentration in the resulting solution would be:

    (A) 0.0792 *M* (B) 0.0990 *M* (C) 0.158 *M* (D) 0.238 *M* (E) 0.0800 *M*

46. Assuming that the volumes are additive, how much water would need to be **added** to 30.0 mL of 12.0 *M* HCl to prepare 2.0 *M* HCl?

    (A) 180 mL (B) 150 mL (C) 120 mL (D) 90 mL (E) 60 mL

47. If 28.6 g of "washing soda," $Na_2CO_3 \cdot 10H_2O$ (FW = 286) is dissolved in 130.g of water, what is the molality of $Na_2CO_3$ in the resulting solution?

    (A) 0.676 *m* $Na_2CO_3$ (B) 0.220 *m* $Na_2CO_3$ (C) 0.893 *m* $Na_2CO_3$
    (D) 0.459 *m* $Na_2CO_3$ (E) 0.769 *m* $Na_2CO_3$

48. An aqueous solution of urea, $(NH_2)_2CO$, contains 10.% urea by weight. How much water is present in one kilogram of this solution?

    (A) 100 g (B) 200 g (C) 400 g (D) 600 g (E) 900 g

49. Water and methanol ($CH_3OH$) are miscible in all proportions. If 16 g of methanol are mixed with 27 g of water, what is the mole fraction of methanol in the solution?

    (A) 0.50 (B) 1.0 (C) 0.30 (D) 0.40 (E) 0.25

50. An aqueous solution contains 19.04 g of magnesium chloride in 500. mL of solution. The $[Cl^-]$ in this solution is:

    (A) 0.400 *M* (B) 4.00 *M* (C) 8.00 *M* (D) 0.800 *M* (E) 2.00 *M*

51. The number of grams of sodium ion in 250 mL of 0.40 *M* sodium carbonate is:

    (A) 106 g (B) 1.2 g (C) 10.6 g (D) 4.6 g (E) 2.3 g

52. If 150. mL of 12.0 *M* HCl is diluted with water to a total volume of 650. mL, the molar concentration of $H_3O^+$ is:

    (A) 18.0 *M* (B) 2.77 *M* (C) 12.0 *M* (D) 3.56 *M* (E) 1.80 *M*

53. If you pour 300 mL of 0.10 *M* KOH into a beaker that contains 100 mL of 0.20 *M* HCl, the resulting solution will be:

    (A) 0.20 *M* with respect to KCl (B) 0.10 *M* with respect to KCl
    (C) 0.025 *M* with respect to KOH (D) 0.01 *M* with respect to HCl
    (E) 0.04 *M* with respect to KOH

54. Equal volumes of 0.80 *M* $HNO_3$ solution and 0.40 *M* NaOH solution are mixed. Which of the following is the true statement about the concentrations in the final solution?

    (A) $[NO_3^-]$ = 0.80 *M*; $[Na^+]$ = 0.40 *M*; $[OH^-]$ = 0.40 *M*; $[H_3O^+]$ = 0.80 *M*
    (B) $[NO_3^-]$ = 0.40 *M*; $[Na^+]$ = 0.20 *M*; $[H_3O^+]$ = 0.20 *M*

(C) $[NO_3^-] = 0.40\,M$; $[Na^+] = 0.20\,M$; $[OH^-] = 0.20\,M$
(D) $[NO_3^-] = 0.40\,M$; $[Na^+] = 0.40\,M$
(E) $[NO_3^-] = 0.20\,M$; $[Na^+] = 0.20\,M$; $[H_3O^+] = 0.20\,M$

55. Equal volumes of 0.60 *M* $Ba(OH)_2$ and 0.20 *M* HCl are mixed. Which of the following is a true statement about the molar concentration in the final solution?

(A) $[Ba^{2+}] = 0.30\,M$; $[Cl^-] = 0.10\,M$; $[OH^-] = 0.50\,M$
(B) $[Ba^{2+}] = 0.60\,M$; $[Cl^-] = 0.20\,M$; $[OH^-] = 0.50\,M$
(C) $[Ba^{2+}] = 0.30\,M$; $[Cl^-] = 0.10\,M$; $[OH^-] = 0.10\,M$
(D) $[Ba^{2+}] = 0.60\,M$; $[Cl^-] = 0.20\,M$; $[H_3O^+] = 0.20\,M$
(E) $[Ba^{2+}] = 0.30\,M$; $[Cl^-] = 0.20\,M$; $[OH^-] = 0.20\,M$

56. What would be the freezing point of a 1.00 molal solution of a weak acid that is 10% dissociated? ($K_f$ for water = 1.86°C)

(A) -2.05°C (B) -2.23°C (C) -1.86°C (D) 0.00°C (E) -1.67°C

57. The freezing point of an aqueous solution of a nonelectrolyte is -0.093°C. The molality of the solute is:

(A) 0.093 *m* (B) 1.86 *m* (C) 1.0 *m* (D) 0.50 *m* (E) 0.050 *m*

58. If the dissolution of an ionic solid in water is endothermic, then it can be concluded that:

(A) the solubility of the solid will increase with increasing temperature
(B) the solubility will decrease with increasing temperature
(C) the solution process releases heat
(D) the solubility of the solid is independent of temperature
(E) the hydration energy exceeds the lattice energy

59. The freezing point of a 0.050 *m* solution of a nonelectrolyte in water would be ($K_f$ for water = 1.86°C):

(A) -1.86°C (B) -0.93°C (C) -0.093°C (D) -0.186°C (E) -3.72°C

60. The freezing point of a 0.200 *m* solution of AB, assuming AB is 5.0% dissociated into $A^+$ and $B^-$ ions, would be:

(A) -0.372°C (B) -0.186°C (C) -0.205°C (D) 0.000°C
(E) -0.391°C

61. The freezing point of a solution prepared by dissolving 3.00 g of compound X (a nonelectrolyte) in 100 g of water is -0.279°C. The molecular weight of compound X is:

(A) 100 (B) 200 (C) 300 (D) 175 (E) 95.0

62. In a given experiment it was observed that by dissolving 3.101 g of solute X in 100. g of $CCl_4$ the vapor pressure of the latter was lowered by 1.85%. The molecular weight of X is:

(A) 124 (B) 300 (C) 254 (D) 191 (E) 458

63. A compound containing phosphorus, nitrogen, and chlorine analyzes 61.17% Cl and 26.72% P, by weight. 1.2952 g of this compound was dissolved in 15.00 mL benzene ($d$ = 0.879 g/mL), producing a solution freezing at 4.03°C. The benzene used as a solvent had a freezing point of 5.48°C. What is the molecular formula of the compound? ($K_f$ for benzene = 5.12°C)

    (A) $P_2N_2Cl_2$ (B) $P_3N_3Cl_6$ (C) $PNCl_2$ (D) $P_4N_4Cl_8$ (E) $(PNCl_2)_6$

64. If 0.720 g of an unknown acid, HB, is completely neutralized by 40.0 mL of 0.200 $M$ NaOH, the formula weight of the acid is:

    (A) 22.5 (B) 30.0 (C) 45.0 (D) 90.0 (E) 180

65. A standard solution of $Ba(OH)_2$ is 0.250 $M$. What volume of 0.200 $M$ nitric acid would be required to neutralize 10.0 mL of the barium hydroxide solution?

    (A) 10.0 mL (B) 12.0 mL (C) 20.0 mL (D) 25.0 mL (E) 30.0 mL

66. A solution containing 6.00 millimoles of $H_3PO_4$ is to be converted so that it contains 6.00 millimoles of $HPO_4^{2-}$. This is to be accomplished by addition of 0.250 $M$ NaOH. How many mL of the NaOH solution will be required?

    (A) 12.0 (B) 24.0 (C) 36.0 (D) 48.0 (E) 3.00

67. The **major** ions present in solution after mixing 0.10 $M$ $H_3PO_4$ with an equal volume of 0.20 $M$ NaOH are:

    (A) $Na^+$ and $HPO_4^{2-}$ (B) $H_3O^+$ and $HPO_4^{2-}$ (C) $OH^-$ and $Na^+$
    (D) $OH^-$, $Na^+$ and $H_2PO_4^-$ (E) $Na^+$ and $H_2PO_4^-$

68. Equal volumes of 0.10 $M$ $H_3PO_4$ and 0.15 $M$ NaOH are mixed. What are the approximate concentrations of the phosphorus-containing species in the resulting solution?

    (A) $[H_2PO_4^-] = 0.050\ M$, $[HPO_4^{2-}] = 0.050\ M$
    (B) $[H_3PO_4] = [H_2PO_4^-] = [HPO_4^{2-}] = [PO_4^{3-}] = 0.025\ M$
    (C) $[HPO_4^{2-}] = 0.075\ M$, $[PO_4^{3-}] = 0.025\ M$
    (D) $[HPO_4^{2-}] = 0.025\ M$, $[H_2PO_4^-] = 0.025\ M$
    (E) $[H_2PO_4^-] = 0.025\ M$, $[HPO_4^{2-}] = 0.050\ M$, $[PO_4^{3-}] = 0.025\ M$

69. A 0.010 $M$ solution of HF is 20.% ionized. Calculate the actual [HF] in solution.

    (A) 0.010 $M$ (B) 0.0020 $M$ (C) 0.0080 $M$ (D) 0.012 $M$ (E) 0.20 $M$

70.*Assuming ideal-solution behavior, how many grams of sucrose (MW = 342, a nonelectrolyte) should be added to 100 g of water in order to produce a solution with a 105.0°C **difference** between the freezing and boiling temperatures? ($K_f$ = 1.86°C/$m$; $K_b$ = 0.51°C/$m$)

    (A) 72 g (B) $4.3 \times 10^2$ g (C) 19 g (D) 3.4 g (E) 29 g

71. A solution is prepared by dissolving 5.00 g of gaseous HCl (FW = 36.5) in

100.0 g of water. Exactly 5.00 g of NaOH (FW = 40.0) is then dissolved in the same solution. What would be the freezing point of this final solution, assuming that it behaves ideally? ($K_f$ for water = 1.86°C)

(A) −0.118°C (B) −4.85°C (C) −4.99°C (D) −5.06°C
(E) −9.71°C

72. What is the molecular weight of a nonvolatile molecular solute whose 1.68% aqueous solution is found to boil 0.026°C higher than pure water at the same atmospheric pressure?

(A) 48 (B) 85 (C) 170 (D) 340 (E) 400

73. What is the osmotic pressure at 0.0°C of a solution containing 1.00 g/L of sucrose, $C_{12}H_{22}O_{11}$?

(A) 0.066 atm (B) 0.11 atm (C) 0.24 atm (D) 0.29 atm
(E) 0.48 atm

74. What is the osmotic pressure at 25°C of a solution containing 0.050 mol/L of $Al(NO_3)_3$?

(A) 0.20 atm (B) 1.3 atm (C) 2.8 atm (D) 3.7 atm
(E) 4.9 atm

75. What is the osmotic pressure at 25°C of a solution containing $1.0 \times 10^{-3}$ mol/L of *each* of the following salts: $NaClO_4$, $Mg(ClO_4)_2$, and $Al(ClO_4)_3$?

(A) 0.073 atm (B) 0.22 atm (C) 0.30 atm (D) 0.33 atm
(E) 0.60 atm

76. The osmotic pressure at 20°C of a solution containing 10.0 g/L of a molecular protein extracted from peanuts was found to be equivalent to a 10.83-cm column height of supported aqueous solution (density = 1.020 g/cm$^3$). What is the molecular weight of the protein? (density of mercury = 13.6 g/cm$^3$)

(A) 8,500 g/mol (B) 15,000 g/mol (C) 22,500 g/mol
(D) 30,000 g/mol (E) 55,000 g/mol

77. Calculate the vapor pressure lowering at 29°C caused by dissolving 70.0 g of $C_{12}H_{22}O_{11}$ in 100.00 g of water. The vapor pressure of water at this temperature is 30.0 millimeters of mercury.

(A) 0.555 torr (B) 1.07 torr (C) 2.00 torr (D) 2.64 torr
(E) 3.55 torr

78. Pure cyclohexane ($C_6H_{12}$, FW = 84.16) freezes at 6.55°C and has a molal freezing point depression constant, $K_f$, of 20.0°C/molal. A solution of a molecular unknown X, made by dissolving 5.00 g of X in 100.0 g of cyclohexane, has a freezing point of 3.00°C. What is the molecular weight of X?

(A) 8.88 (B) 153 (C) 282 (D) 333 (E) 747

79. Solutions were prepared by dissolving 10.0 g of the compounds listed as answers in 100. g of benzene at 25.0 C. Which solution would have the **lowest benzene** vapor pressure at this temperature?

(A) $CH_3OH$ (B) $C_2H_5OH$ (C) $CCl_4$ (D) $C_5H_{12}$
(E) each of the solutions would have the same vapor pressure of benzene

---

## ANSWERS TO CHAPTER 11 PROBLEMS

1. C
2. E
3. A
4. B
5. D
6. A
7. E
8. A
9. C
10. D
11. E
12. D
13. A
14. B
15. B
16. A
17. C
18. B
19. C
20. D
21. E
22. C
23. B
24. D
25. E
26. D
27. B
28. E
29. D
30. E
31. A
32. D
33. E
34. B
35. D
36. D
37. B
38. A
39. D
40. B
41. A
42. B
43. E
44. C
45. D
46. B
47. A
48. E
49. E
50. D
51. D
52. B
53. C
54. B
55. A
56. A
57. E
58. A
59. C
60. E
61. B
62. C
63. B
64. D
65. D
66. D
67. A
68. D
69. C
70. A
71. C
72. D
73. A
74. E
75. B
76. C
77. B
78. C
79. A

# 12
# Chemical Thermodynamics

## ENERGY, ENTHALPY, AND CALORIMETRY

Consider the chemical process indicated by the balanced equation, A → B, with $A_{(amount,\ conditions)}$ representing the **initial state** [**reactant(s)**] and $B_{(amount,\ conditions)}$ representing the **final state** [**product(s)**]. Both A and B possess definite *internal energies* ($E$) and the energy **change** ($\Delta E$), defined arbitrarily as $E_f - E_i = E_{prod} - E_{react} = E_B - E_A$, is the same regardless of the path by which this process is carried out, *i.e.*, energy, like volume, is a *state function.* The *first law of thermodynamics* (*law of energy conservation*) then relates two outward manifestations of energy in transit, $q$ (*heat*) and $w$ (*work*), to the internal energy change, $\Delta E$, or

$$\Delta E = E_B - E_A = q - w \qquad \textbf{[Eqn. 1]}$$

Signs are important: $q$ is positive if heat is absorbed in the process, negative if heat is evolved; $w$ is positive if work is done on the surroundings as a result of the process, negative if the surroundings do work on the chemical system. Likewise, the internal energy may increase ($\Delta E > 0$) or decrease ($\Delta E < 0$). Equation 1 then states that the net change in internal energy is equal to the thermal energy absorbed minus the work energy expended.

The basic energy unit is the joule, J, but the calorie and kilocalorie are still frequently used:

$$1\ \text{J} = 0.2390\ \text{cal} = 2.390 \times 10^{-4}\ \text{kcal}$$

A feeling for the quantity of energy represented by the joule may be had by noting that approximately 4.2 joules are required to raise the temperature of 1 g of liquid water by 1 degree centigrade. Thus, the specific heat capacity of liquid water is 4.184 J/g-°C or 1.00 cal/g-°C (See Ch. 1, pp. 25–26).

A common form of work is expansion (or pressure-volume) work.

Thus in our example, if B has a larger volume than A, then the surroundings must have been forced back in the process, and the *chemical system* A → B must have done work on its surroundings ($w > 0$). For the case in which a volume change ($\Delta V$, again $V_f - V_i$) occurs against, or due to, a **constant** opposing pressure, the work term is:

$$w = P_{opp}(\Delta V) = \frac{\text{force}_{(opp)}}{\text{area}} \times \text{volume change}$$

$$= \text{force} \times \text{distance}$$

Clearly, if $V_f > V_i$, expansion has occurred and $w$ is positive; if $V_f < V_i$, $w < 0$, the work done by the system is negative. Commonly, *P-V* work has the units, *liter-atm* (1 L-atm = 24.2 cal = 101 J).

Experimentally, $\Delta E$ may be determined in a bomb calorimeter where no volume change can occur; in this case no *P-V* work can occur and $\Delta E = q_v$, the **heat absorbed at constant volume** (which may be positive or negative). A typical bomb calorimeter can be described schematically by:

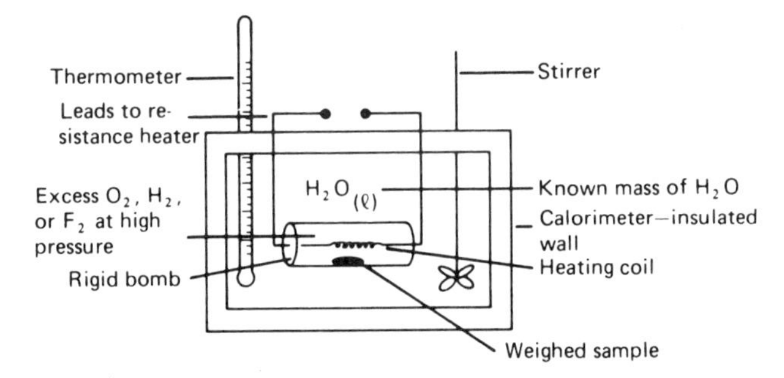


Bomb calorimetry involves reaction of a known mass of a sample with an excess of either $O_{2\,(g)}$, $H_{2\,(g)}$, or $F_{2\,(g)}$. In the case of $O_{2\,(g)}$, the reaction that occurs is called *combustion* and is initiated by electrically heating the reactants. The weighed sample completely reacts with the oxygen in the bomb, **liberating** heat that is **absorbed by** the surrounding water in the calorimeter. The calorimeter water undergoes a temperature rise, $\Delta t$, that may be observed on the thermometer in the surrounding water. Thus, $\Delta E_{reaction} = q_v = -q_{surroundings} \cong -m_{H_2O}C_{H_2O}\Delta t_{H_2O}$. Since the mass, specific heat capacity, and $\Delta t$ of the surrounding $H_2O$ are known or determined, $\Delta E$ for the reaction may be calculated. $\Delta E_{reaction}$ is, of course, dependent upon the mass of sample reacting—an *extensive property*. The value of $\Delta E_{rxn}$ is always reported as an *intensive property*, $\Delta \bar{E}$, where

$$\Delta \bar{E} = \frac{\Delta E_{rxn}}{\text{moles of sample}}$$

and the bar over the $\Delta E$ ($\Delta \bar{E}$) indicates that this is the energy change *per mole*.

The First Law of Thermodynamics says that energy is conserved and that $\Delta E_{total} = 0$; thus, in a bomb calorimeter

$$\Delta E_{total} = \Delta E_{system} + \Delta E_{surroundings} = 0$$

For no change in volume (no work), $\Delta E_{surr} = q_{surr}$ and

$$\Delta E_{syst} + q_{surr} = 0$$

Therefore,

$$\Delta E_{syst} = -q_{surr} = -q_{cal} = -C_{cal}\Delta T_{cal}$$

where we have called the surroundings the calorimeter assembly ('cal') and expressed $q_{cal}$ as the product of the heat capacity of the calorimeter ($C_{cal}$, energy per degree) and the temperature change in the calorimeter, $\Delta T_{cal}$. Alternately, therefore, a bomb calorimeter may be calibrated, under identical conditions, by electrically heating the bomb with a known number of joules, and noting the correspondence between that number of joules and the observed temperature rise. For example, a given calorimeter may in this way be determined to have a *calorimeter heat capacity*, $C_{cal}$, of 2.475 kJ/°C. Then, in a subsequent experiment an observed $\Delta t$ of 3.501 degrees would correspond to a combustion liberating (3.501°C) × (2.475 kJ/°C) = 8.665 kJ; if this amount of heat was liberated by 0.05000 mole of sample, $\Delta E_{comb}$ would be:

$$\Delta \bar{E}_{comb} = \frac{-8.665 \text{ kJ}}{0.05000 \text{ mole}} = -173.3 \text{ kJ/mol}$$

where the minus sign indicates that since the surroundings gained the heat, the reaction must have liberated the heat, *i.e.*, the reaction is *exothermic*.

Frequently, however, chemical processes are carried out in, say, an open beaker exposed to a constant (atmospheric) pressure. It is thus convenient to define the *energy content at constant pressure* as the *enthalpy* ($H$). Therefore, $H = E + PV$ and $\Delta H = \Delta E + \Delta(PV)$. With the restrictions of only expansion-type work and a constant pressure,

$$\Delta H = q - \cancel{w} + \cancel{P_{opp}(\Delta V)} = q_p$$

This states that "**the enthalpy change for a process is the heat absorbed** (positive or negative) **at constant pressure**" or

$$\Delta H = \Delta E + P_{opp}\Delta V \qquad \textbf{[Eqn. 2]}$$

meaning that "**the change in enthalpy is equal to the change in internal energy plus the $P$-$V$ work involved**". $\Delta H$ values are more commonly employed than $\Delta E$ values and are readily found in tables.

For processes involving **only** condensed phases, volume changes are quite small, and $\Delta H \cong \Delta E$; however, when there is a net change in gaseous species, the $P\Delta V$ term becomes significant. From the ideal gas law (at constant temperature and pressure)

$$P\Delta V = \Delta n_{gas}RT$$

and

$$\Delta H = \Delta E + \Delta n_{gas}RT \qquad \text{[Eqn. 3]}$$

where $\Delta n_{gas}$ = (moles of gaseous products–moles of gaseous reactants), $T$ is the absolute temperature, K; and $R$ is the gas constant in **units consistent with** $\Delta H$ and $\Delta E$; $R$ = 1.99 cal/mole-K = $1.99 \times 10^{-3}$ kcal/mole-K = 8.314 J/mole-K = $8.314 \times 10^{-3}$ kJ/mole-K. Thus, Eqn. 3 can be used to convert between $\Delta H$ and $\Delta E$.

## Example Problems Involving Energy, Enthalpy, and Calorimetry

### 12-A. Internal Energy Change

A gas absorbs 300 J of heat energy and is compressed from 20.0 to 10.0 L by an opposing pressure of 2.00 atm. What is $\Delta E$ for this process?

(A) -1720 J (B) -2320 J (C) +1720 J (D) +300 J (E) +2320 J

***Solution***

$\Delta E = q - w$

$q$ = +300 J (heat **absorbed** by the system is **positive**)

$w = P_{opp}\,\Delta V = P_{opp}(V_f - V_i)$

= 2.00 atm(10.0 L - 20.0 L)(101 J/L-atm)

= -2020 J (work **done** by the system is **negative**)

$\Delta E$ = 300 J - (-2020 J) = +2320 J

***Answer***

The internal energy of the system has been **increased** by 2320 J due to (a) the **absorption** of 300 J of thermal energy **(heat)** and (b) the **absorption** of 2020 J of work energy. **[E]**

### 12-B. Conversion of $\Delta \bar{H}$ to $\Delta \bar{E}$

If $\Delta \bar{H}$ for the **sublimation** ($I_{2\,(s)} \rightarrow I_{2\,(g)}$) of one mole of iodine is 62.4 kJ at 25°C and 1.00 atm, what is $\Delta \bar{E}$ for this process?

(A) 29.7 kJ (B) 59.9 kJ (C) 62.3 kJ (D) 64.8 kJ (E) 92.5 kJ

***Solution***

$$\Delta \bar{H} = \Delta \bar{E} + P_{opp}\Delta V = \Delta \bar{E} + \Delta n_{gas}RT$$

Therefore, $\Delta \bar{E} = \Delta \bar{H} - \Delta n_{gas}RT$ (from the chemical equation

$\Delta n_{gas}$ = 1 - 0 = 1 mole).

$$\Delta\bar{E} = 62.4 \text{ kJ} - (1 \cancel{\text{mole}})(8.31 \times 10^{-3} \frac{\text{kJ}}{\cancel{\text{mole}}\text{-}\cancel{\text{K}}})(298 \cancel{\text{K}})$$

$$\Delta\bar{E} = 59.9 \text{ kJ}$$ **[B]**

## 12-C. Calorimetry

A certain bomb calorimeter has a heat capacity of 1611 J/°C. A temperature rise of 3.577°C is observed when the calorimeter is used in the quantitative combustion of a 1.000 g sample of powdered tantalum in excess oxygen to produce $Ta_2O_5$. Calculate $\Delta\bar{E}_{comb}$ in kJ/mole $Ta_2O_5$.

$$2\,Ta_{(s)} + 5/2\,O_{2\,(g)} \rightarrow Ta_2O_{5\,(s)}$$

(A) 5763 (B) 2595 (C) 2086 (D) 1678 (E) 1043

### *Solution*

Since the calorimeter heat capacity is 1611 J/°C, the observed temperature rise corresponds to (3.577°C)(1611 J/°C) = 5,763 J, or 5.763 kJ. $\Delta E$ = -5.763 kJ, since the reaction is exothermic, causing the temperature to rise. In order to convert to $\Delta\bar{E}$, $\Delta E$ must be divided by the number of moles of $Ta_2O_5$ that corresponds to 1.000 g of Ta:

$$(1.000 \text{ g Ta})\left(\frac{1 \text{ mole Ta}}{180.95 \text{ g Ta}}\right)\left(\frac{1 \text{ mole } Ta_2O_5}{2 \text{ mole Ta}}\right) = 0.002763 \text{ mole } Ta_2O_5$$

$$\text{Thus, } \Delta\bar{E} = \frac{-5.763 \text{ kJ}}{0.002763 \text{ mole } Ta_2O_5} = -2{,}086 \text{ kJ/mole } Ta_2O_5$$ **[C]**

## Exercises on Energy, Enthalpy, and Calorimetry

1. At constant pressure, $q$ equals:

   (A) $\Delta H$ (B) $\Delta E$ (C) $\Delta T$ (D) $P\Delta V$ (E) $w$

2. The term adiabatic refers to a process in which there is no heat transfer between the system and the surroundings. For such a process, it follows that

   (A) $q = w$ (B) $q = 0$ (C) $\Delta E = q$ (D) $\Delta E = w$ (E) $P\Delta V = 0$

3. Calculate $\Delta E$ for a system that does 300. J of work on the surroundings when 200. J of heat are lost by the system.

   (A) -500. J (B) 100. J (C) -100. J (D) 500. J (E) 300. J

4. Given the reaction at 25°C:

   $$4\,Ag_{(s)} + 2\,H_2S_{(g)} + O_{2\,(g)} \rightarrow 2\,Ag_2S_{(s)} + 2\,H_2O_{(\ell)}$$

   $\Delta E = -Z$ kJ; then $\Delta H$ equals:

(A) $-Z - 3(8.31)(298)$ kJ (B) $-Z + (101)(298)$ kJ
(C) $-Z - 3(8.31 \times 10^{-3})(298)$ kJ (D) $+Z + 3(8.31 \times 10^{-3})(298)$ kJ
(E) $+Z - 3(8.31 \times 10^{-3})(298)$ kJ

5. When 2.00 moles of an ideal gas expands at room temperature from 5.00 to 10.00 L against a constant pressure of 1.00 atm, the work done by the gas is:

   (A) 505 J (B) 1010 J (C) -5.00 L-atm (D) 253 J (E) 101 J

6. When one mole of $Fe_2O_{3(s)}$ is formed by combustion of iron in oxygen $\Delta\bar{H} = -824.2$ kJ. How many kJ will be released if 0.20 g of iron are converted to $Fe_2O_3$?

   (A) 1.5 (B) 160 (C) 3.0 (D) 8.2 (E) 16

7. The combustion of 0.100 g of ethene causes a temperature rise of 2.00°C in a bomb calorimeter that has a heat capacity of 2.510 kJ/°C. What is the internal energy change **per mole** of ethene for the combustion? Ethene = $C_2H_{4(g)}$.

   (A) -1410 kJ/mol (B) -5.02 kJ/mol (C) -251 kJ/mol
   (D) +251 kJ/mol (E) +179 kJ/mol

8. A 0.25-mole sample of crystalline ammonium nitrate was dissolved in 500. mL of pure water at 21.0°C. As the solute dissolved, the temperature of the solution dropped to a minimum value of 18.0°C. Assume that 500. mL of solution having a density of 1.0 g/cm$^3$, and with a specific heat capacity of 1.0 cal/g-°C is produced. What is the value of $\Delta\bar{H}_{soln}$ for ammonium nitrate?

   (A) +1.5 kcal/mole (B) -1.5 kcal/mole (C) +6.0 kcal/mole
   (D) +0.38 kcal/mole (E) -9.0 kcal/mole

9. How many grams of water must undergo the process,

$$H_2O_{(g,110°C)} \rightarrow H_2O_{(\ell,90°C)}$$

   in order that 100. kJ of heat will be liberated? (For water, the molar heat of vaporization at 373 K = 40.7 kJ/mole; the average molar heat capacity of steam = 36.3 J/mole-K; the specific heat capacity of the liquid = 4.184 J/g-°C.)

   (A) 18 g (B) 43 g (C) 65 g (D) 72 g (E) 95 g

## THERMOCHEMISTRY AND HESS'S LAW

The thermodynamic functions $\Delta E$ and $\Delta H$ are *state functions, i.e.*, the difference in $E$ (or $H$) between some initial and some final state is a constant, and is **totally independent of the route** by which the process, State 1 → State 2, is carried out. Moreover, only differences ($\Delta E$ and $\Delta H$) may be experimentally determined (not the absolute values of $E$ and $H$). It is then convenient for reference to designate a particular form and particular conditions for elements, compounds, and solutes as *standard states.* For solids and liquids, the standard state means "the pure substance at an **external** pressure of 1.00 atm"; for gases it means, "at a **partial** pressure of 1.00 atm"; and for a solute in a liquid solvent, "at a concentration of 1.00 $M$". Also, for an **elementary** substance ($Br_2$, $O_2$, $S_8$, C, *etc.*) the standard state refers to the most stable form of that element at 1.00 atm (*e.g.*, for the element carbon, graphite is more stable than diamond and hence is designated as the standard state). Thermodynamic data for such substances are said to be "standard" and have a superscript (°) appended to the symbol of the thermodynamic function, *e.g.*, $\Delta H°$. Temperature is **not** part of the standard state definition and hence needs to be specified. [Tabulated data are frequently at 298 K (25°C).]

Thermodynamic data are useless unless the transformation or chemical reaction referred to is **clearly specified**. Frequently tabulated for compounds are the *standard, molar enthalpies* at 25°C *of formation* ($\Delta \bar{H}^\circ_{f(298)}$) and *of combustion* ($\Delta \bar{H}^\circ_{comb(298)}$). An elemental substance at 298 K is defined as having $\Delta \bar{H}^\circ_f \equiv 0$. For a compound, $\Delta \bar{H}^\circ_{f(298)}$ refers to the heat absorbed or liberated when **one mole** (*i.e.*, "molar") is formed from the constituent **elements**, both the reactants and product being in their standard states at 298 K. Thus, for $NH_4BrO_3$ (a solid at 298 K), $\Delta \bar{H}^\circ_{f(298)}$ refers to the reaction

$$1/2\ N_{2\,(g)} + 2\ H_{2\,(g)} + 1/2\ Br_{2\,(\ell)} + 3/2\ O_{2\,(g)} \rightarrow NH_4BrO_{3\,(s)}$$

and $\Delta H^\circ_{rxn(298)} = \Delta \bar{H}^\circ_{f(298)}$. Heats of combustion refer to the enthalpy change that takes place when a given compound or element reacts with the stoichiometric amount of oxygen to form specified oxides. $\Delta H^\circ_{rxn}$ values for the following reactions are, respectively, the *molar heats of combustion* of hydrogen, the hydrocarbon, $C_8H_{18}$, and carbon monoxide:

$$H_{2\,(g)} + 1/2\ O_{2\,(g)} \rightarrow H_2O_{(\ell)} \qquad \Delta H^\circ = -\ 286\ \text{KJ}$$

$$C_8H_{18\,(\ell)} + 25/2\ O_{2\,(g)} \rightarrow 9\ H_2O_{(\ell)} + 8\ CO_{2\,(g)} \qquad \Delta H^\circ = -\ 5{,}494\ \text{KJ}$$

$$CO_{(g)} + 1/2\ O_{2\,(g)} \rightarrow CO_{2\,(g)} \qquad \Delta H^\circ = -\ 283\ \text{KJ}$$

It should be noted that for elements $\Delta H^\circ_{comb} = \Delta H^\circ_f$ of a specified oxide.

$\Delta H^\circ_{(298)}$ for a process depends only upon the identity of the products and reactants, not on the pathway selected, nor on the number of steps involved in a pathway. The fact that the enthalpy change in a reaction is the same regardless of whether the reactants are converted into products directly in a **single** step, or in a **series** of stepwise reactions, permits calculation of an unknown enthalpy change by the summation of appropriate reactions to give the desired change—*Hess's law.* The application of Hess's law takes two forms: (a) when using **only** heats of **formation**, the enthalpy change for **any** reaction may be obtained by subtracting the sum of the heats of formation of the reactants from the sum of the heats of formation of the products (the elements involved in the formation reactions cancel), *i.e.*,

$$\Delta H^\circ_{(any\ reaction)} = \sum_{prod} n\, \Delta\bar{H}^\circ_f - \sum_{reac} n\, \Delta\bar{H}^\circ_f$$

where $n$ refers to the number of moles of a substance in the balanced equation; and (b) when using $\Delta H$ values **other than** heats of formation, *if a chemical process can be obtained by summing together other processes, then $\Delta H$ for the process thus obtained will be the algebraic sum of the $\Delta H$ values for the other equations used.* In using Hess's law it must be remembered that (i) if an equation is reversed, the sign of $\Delta H$ changes; (ii) if an equation is multiplied by a factor $x$, then the associated $\Delta H$ value must also be multiplied by the same factor; and (iii) all species other than those desired in the final equation must cancel if the Hess's-law cycle has been correctly conceived. Thus, $\Delta\bar{H}^\circ_{(298)}$ for the reduction of carbon dioxide with hydrogen to carbon monoxide and liquid water is readily obtained by combining two of the equations above:

$$H_{2\,(g)} + 1/2\ O_{2\,(g)} \rightarrow H_2O_{(\ell)} \qquad \Delta\bar{H}^\circ = -286\ KJ$$

$$CO_{2\,(g)} \rightarrow CO_{(g)} + 1/2\ O_{2\,(g)} \qquad \Delta\bar{H}^\circ = +283\ KJ$$

$$CO_{2\,(g)} + H_{2\,(g)} \rightarrow H_2O_{(\ell)} + CO_{(g)} \qquad \Delta\bar{H}^\circ = -\ 3\ KJ$$

Many standard enthalpies of formation are values that are calculated from experimentally determined heats of combustion in a bomb calorimeter ($\Delta\bar{E}^\circ_{comb}$). For example, the $\Delta\bar{H}^\circ_f$ value for the solid hydrocarbon $C_xH_y$ can be calculated from the following experimentally determined quantities:

$$C_{(s)} + O_{2\,(g)} \rightarrow CO_{2\,(g)}, \qquad \Delta\bar{H}^\circ_{comb} = \Delta\bar{H}^\circ_{f(CO_2)}$$

$$H_{2\,(g)} + 1/2\ O_{2\,(g)} \rightarrow H_2O_{(\ell)} \qquad \Delta\bar{H}^\circ_{comb} = \Delta\bar{H}^\circ_{f(H_2O)}$$

$$C_xH_{y(s)} + \left(x + \frac{y}{4}\right)O_{2(g)} \rightarrow xCO_{2(g)} + \frac{y}{2}H_2O_{(\ell)} \quad \Delta\bar{H}^\circ_{comb\ (C_xH_y)}$$

Using these data in Hess's law, we can say:

$$\Delta\bar{H}^\circ_{comb\,(C_xH_y)} = x\,\Delta\bar{H}^\circ_{f(CO_2)} + \frac{y}{2}\Delta\bar{H}^\circ_{f(H_2O)} - (x + \frac{y}{4})(0) - \Delta\bar{H}^\circ_{f(C_xH_y)}$$

Since the first four terms in this equation are experimentally determined or defined, the last term, $\Delta\bar{H}^\circ_{f(C_xH_y)}$, may be calculated and then added to the tables of thermochemical data.

## Example Problems Involving Thermochemistry and Hess's Law

### 12-D. Standard Molar Heat of Formation from Heat of Reaction and Tabulated Data

For the reaction in a bomb calorimeter,

$$NH_{3(g)} + 3\,F_{2(g)} \rightarrow NF_{3(g)} + 3\,HF_{(g)}$$

$\Delta E^\circ = -881.2$ kJ. From tabulated data (*cf.* Appendix), find $\Delta H^\circ_f$ for gaseous $NF_3$.

(A) -114 kJ/mol (B) +22 kJ/mol (C) -1741 kJ/mol (D) -1198 kJ/mol
(E) -656 kJ/mol

***Solution***

Since $\Delta n_{gas} = 0$, then $\Delta E^\circ = \Delta H^\circ$ for the **reaction**. For any reaction (see text),

$$\Delta H^\circ_{rxn} = \sum_{prod} n\,\Delta\bar{H}^\circ_f - \sum_{reac} n\,\Delta\bar{H}^\circ_f$$

Therefore,

$$-881.2 = (1)\Delta\bar{H}^\circ_{f(NF_3)} + (3)\Delta\bar{H}^\circ_{f(HF)} - (1)\Delta\bar{H}^\circ_{f(NH_3)} - (3)\Delta\bar{H}^\circ_{f(F_2)}$$

$$-881.2 = \Delta\bar{H}^\circ_{f(NF_3)} + (3)(-271) - (-46.11) - (3)(0)$$

from which $\Delta\bar{H}^\circ_{f(NF_3)}$ is calculated as equal to -114 kJ/mole [A]

### 12-E. Summation of Heats of Combustion to Obtain $\Delta H^\circ$ for Another Reaction

For the liquids $C_2H_5OH$ and $C_2H_4O$ the standard enthalpies of combustion are, respectively, -327.6 and -279.0 kcal/mole. Using Hess's law find $\Delta H^\circ$ for the partial oxidation,

$$2\,C_2H_5OH_{(\ell)} + O_{2(g)} \rightarrow 2\,C_2H_4O_{(\ell)} + 2\,H_2O_{(\ell)}$$

(A) +97.2 kcal (B) -97.2 kcal (C) -48.6 kcal (D) -376.2 kcal
(E) -230.4 kcal

***Solution***

The reactions to which the $\Delta H^\circ_{comb}$ values refer are:

$$C_2H_5OH_{(\ell)} + 3\,O_{2(g)} \rightarrow 2\,CO_{2(g)} + 3\,H_2O_{(\ell)} \qquad \Delta H^\circ = -327.6 \text{ kcal}$$

$$C_2H_4O_{(\ell)} + 5/2\,O_{2(g)} \rightarrow 2\,CO_{2(g)} + 2\,H_2O_{(\ell)}$$

$\Delta H^\circ$ = -279.0 kcal

By doubling the first reaction and adding this to twice the reverse of the second reaction, the desired reaction is obtained:

| | | | | $\Delta H^\circ$, kcal |
|---|---|---|---|---|
| $2\,C_2H_5OH_{(\ell)}$ | $+ 6\,O_{2\,(g)}$ | $\rightarrow 4\,CO_{2\,(g)}$ | $+ 6\,H_2O_{(\ell)}$ | 2(-327.6) |
| $4\,CO_{2\,(g)}$ | $+ 4\,H_2O_{(\ell)}$ | $\rightarrow 5\,O_{2\,(g)}$ | $+ 2\,C_2H_4O_{(\ell)}$ | 2(+279.0) |

$2\,C_2H_5OH_{(\ell)} + O_{2\,(g)} \rightarrow 2\,H_2O_{(\ell)} + 2\,C_2H_4O_{(\ell)}$ -97.2 **[B]**

This exercise illustrates an important point–the incorrect answer (A) will be obtained if you indiscriminantly use heats of combustion in the equation,

$$\Delta H^\circ_{rxn} = \sum_{prod} n\,\Delta \overline{H}^\circ_f - \sum_{reac} n\,\Delta \overline{H}^\circ_f$$

This relationship can be **used only** with heats of **formation, not** with heats of combustion.

### 12-F. Bond Energy from Standard Molar Heats of Formation

Using the answer to Exercise 12-D and from tabulated data (*cf.* Appendix), find the average N—F bond energy, $D_{N—F}$, in kJ per mole of bonds.

(A) 38 (B) 114 (C) 222 (D) 235 (E) 275

#### *Solution*

A valuable use of bulk thermochemical data is that it may be used to estimate the properties of molecules, here the strength of a chemical bond. The average N–F bond energy, as calculated from data concerning $NF_{3\,(g)}$, refers to one-third of the enthalpy change for the reaction,

$$NF_{3\,(g)} \rightarrow N_{(g)} + 3\,F_{(g)}$$

*i.e.*, $D_{N-F} = \Delta \overline{H}^\circ_{rxn}/3.$

$$\Delta H^\circ_{rxn} = \Delta \overline{H}^\circ_{f(N)} + 3\cdot\Delta \overline{H}^\circ_{f(F)} - \Delta \overline{H}^\circ_{f(NF_3)}$$
$$= 473 + 3(79) - (-114) = 824 \text{ kJ}$$

Since in this reaction three N–F bonds are broken, the average $D_{N-F}$ = 824/3 = 275 kJ is taken. **[E]**

## Exercises on Thermochemistry and Hess's Law

In this set of exercises, at times it is necessary to select some data from the thermochemical table in the Appendix. When additional data must be looked up, the exercise is marked with a double dagger (‡). When doing problems in any text, be sure to use the data tabulated **in that text**—values tabulated frequently vary somewhat from one text to another.

10. In which of the following reactions would the heat absorbed at 25°C and 1 atm be $\Delta\overline{H}_f^\circ$ for alcohol ($C_2H_5OH$)?

    (A) $2\,C_{(s)} + 3\,H_{2\,(g)} + O_{(g)} \rightarrow C_2H_5OH_{(\ell)}$
    (B) $2\,C_{(g)} + 6\,H_{(g)} + O_{(g)} \rightarrow C_2H_5OH_{(\ell)}$
    (C) $2\,C_{(s)} + 3\,H_{2\,(g)} + 1/2\,O_{2\,(g)} \rightarrow C_2H_5OH_{(\ell)}$
    (D) $4\,C_{(s)} + 6\,H_{2\,(g)} + O_{2\,(g)} \rightarrow 2\,C_2H_5OH_{(\ell)}$
    (E) $2\,CO_{2\,(g)} + 3\,H_2O_{(\ell)} \rightarrow C_2H_5OH_{(\ell)} + 3\,O_{2\,(g)}$

11. $\Delta H$ for the reaction, $SiH_{4\,(g)} + 2\,O_{2\,(g)} \rightarrow SiO_{2\,(s)} + 2\,H_2O_{(\ell)}$, is -1517 kJ. What would be the value of $\Delta H$ if the water formed is in the gaseous state? $\Delta\overline{H}$ for $H_2O_{(\ell)} \rightarrow H_2O_{(g)}$ is + 44.0 kJ/mole.

    (A) -1561 kJ (B) -1517 kJ (C) -1473 kJ (D) -1429 kJ (E) -1605 kJ

12. If $\Delta H^\circ$ for the reaction,

$$4\,NH_{3\,(g)} + 5\,O_{2\,(g)} \rightarrow 4\,NO_{(g)} + 6\,H_2O_{(g)}$$

    is -905.4 kJ at 25°C, calculate $\Delta\overline{H}_f^\circ$ for $NO_{(g)}$ in kilojoules. (‡)

    (A) -390.4 (B) -96.7 (C) +90.3 (D) 361.1 (E) 905.4

13. Given the following two reactions:

    $2\,Na_{(s)} + 2\,HCl_{(g)} \rightarrow 2\,NaCl_{(s)} + H_{2\,(g)}$ $\Delta H^\circ = -152.34$ kcal

    $H_{2\,(g)} + Cl_{2\,(g)} \rightarrow 2\,HCl_{(g)}$ $\Delta H^\circ = -\ 44.12$ kcal

    What is $\Delta\overline{H}_f^\circ$, in kcal/mole, for $NaCl_{(s)}$?

    (A) -54.11 (B) -98.23 (C) -108.22 (D) -120.29 (E) -196.46

14. Calculate $\Delta H_{298}^\circ$ for the reaction:

$$Cu_{(s)} + 2\,H_2SO_{4\,(\ell)} \rightarrow CuSO_{4\,(s)} + SO_{2\,(g)} + 2\,H_2O_{(\ell)}$$

    in kJ per mole of Cu. (‡)

    (A) -2167.94 (B) -825.96 (C) -540.11 (D) -11.87 (E) +273.80

15. Given the following thermochemical reactions:

    $XO_{2\,(s)} + CO_{(g)} \rightarrow XO_{(s)} + CO_{2\,(g)}$ $\Delta H = -20.0$ kJ
    $X_3O_{4\,(s)} + CO_{(g)} \rightarrow 3\,XO_{(s)} + CO_{2\,(g)}$ $\Delta H = +\ 6.0$ kJ
    $3\,X_2O_{3\,(s)} + CO_{(g)} \rightarrow 2\,X_3O_{4\,(s)} + CO_{2\,(g)}$ $\Delta H = -12.0$ kJ

    Find $\Delta H$ in kJ for the reaction:

$$2\,XO_{2\,(s)} + CO_{(g)} \rightarrow X_2O_{3\,(s)} + CO_{2\,(g)}$$

    (A) -40.0 (B) -28.0 (C) -26.0 (D) -18.0 (E) 0.0

16. What is the strength of the H—Br bond in kJ/mole? (‡)

    (A) 36.4 (B) 201 (C) 293 (D) 366 (E) 431

# DRILL ON THERMOCHEMISTRY AND HESS'S LAW

***Student's Name*** ______________________

Use the Tables in the back of this book to complete this table. You may be asked to turn in this sheet.

| Process | $\Delta H°/kJ$ | $\Delta E°/kJ$ | w/J | q/J |
|---|---|---|---|---|
| Evaporation of 8.0 g $Br_2$ at 25°C | | | | |
| Sublimation of 1.0 g of $C_{(diamond)}$ at 25°C | | | | |
| $H_2S_{(g)} + \frac{3}{2} O_{2(g)} \rightarrow SO_{2(g)} + H_2O_{(l)}$ (For 10.0 g of $H_2S$ at 25°C) | | | | |
| $CaO_{(g)} + H_2O_{(l)} \rightarrow Ca(OH)_{2(s)}$ (For 14.0 g of CaO at 25°C) | | | | |
| ______________________ | | | | |
| ______________________ | | | | |

17. How many kJ will be liberated when 8.00 g of oxygen gas reacts with excess iron to form $Fe_2O_3$ (25°C, $P_{O_2}$ = 1 atm)? (‡)

 (A) 824.2 (B) 206.1 (C) 274.5 (D) 328.9 (E) 137.4

---

## ENTROPY, FREE ENERGY, AND REACTION SPONTANEITY

The term *spontaneous* can be defined as meaning "proceeding from, or acting by, *internal* impulse, energy, or natural law, without *external* force; self-acting." It does not imply anything about the time required by, or the rate of, the process. One of the most powerful features of thermodynamics is that it is able to define the criteria for reaction spontaneity under given conditions.

What processes, or reactions, are spontaneous? Thermodynamics defines the two factors that must be considered in answering this question:

(a) The driving force toward a state of **minimum energy**—an exothermic process is **more likely** to be spontaneous than an endothermic one (*i.e.*, at constant pressure, if $\Delta H<0$, the process has a favorable tendency to occur).

(b) The driving force toward a state of **maximum *entropy*,** $S$, (maximum **probability, disorder,** or **disarray**)—those processes that show an increase in entropy ($\Delta S>0$) are more probable than those of $\Delta S<0$.

The interconnection between these two driving forces is a **state function,** the *free energy change* ($\Delta G$).

The *Second Law of Thermodynamics* may be formulated as

$$\Delta S_{total} = (\Delta S_{system} + \Delta S_{surroundings}) \geqslant 0 \qquad \textbf{[Eqn. 1]}$$

The impact of which is that no process spontaneously can occur for which $\Delta S_{total}<0$. The increase in entropy (disorder) of the surroundings becomes larger as the amount of heat absorbed by the surroundings ($\Delta E_{surr}$) increases; for a given $\Delta E_{surr}$, the **increase** in disorder is less, the larger the thermal disorder already present (as measured by the magnitude of $T$). $\Delta S_{surr}$ can therefore be taken as equal to $\Delta E_{surr}/T$, which equals $-\Delta H_{syst}/T$. Insertion of this term into Eqn. 1, multiplying by $-T$, and by **defining** $-T\Delta S_{total} \equiv \Delta G_{system}$, we have the change in free energy,

$$\Delta G_{system} = \Delta H_{syst} - T\Delta S_{syst} \leqslant 0 \qquad \textbf{[Eqn. 2]}$$

For standard-state conditions, this may be written

$$\Delta G^\circ = \Delta H^\circ - T\Delta S^\circ \qquad \textbf{[Eqn. 2']}$$

Equations 2 and 2′ are valid under conditions of constant temperature and pressure; under these conditions a process is spontaneous if $\Delta G<0$, at equilibrium (no net change occurring) if $\Delta G = 0$.

The *Third Law of Thermodynamics* sets a lower limit of zero on the *absolute entropy*, $\bar{S}$, for a pure, crystalline solid at absolute zero (*i.e.*, the degree of disorder is zero at 0 K).* Thus, for a process at a given temperature, $T$,

$$\Delta S_T = S_{final} - S_{initial} = \sum_{prod} n\bar{S} - \sum_{reac} n\bar{S}$$

The **absolute** values of $H$ (undetermined) and $S$ (determinable) are functions of $T$; the **differences** are, however, essentially independent of temperature.† On the other hand, Eqns. 2 and 2′ show that $\Delta G$ has a strong temperature dependence. Thus, if a reaction having $\Delta H>0$ and $\Delta S>0$ is nonspontaneous ($\Delta G>0$) at a low temperature, there is a higher temperature ($T$) at which the system will equilibrate ($\Delta G = 0$), and above this $T$, the reaction will become spontaneous ($\Delta G<0$).

Standard molar free energies of formation ($\Delta\bar{G}_f^\circ$) are often tabulated. These may be used (as with $\Delta\bar{H}_f^\circ$ values) to find $\Delta G_{rxn}^\circ$ by summation in a Hess's law cycle (also, $\Delta\bar{G}_f^\circ$ for *elements* is zero at 298 K).

$\Delta G_{rxn}^\circ$ has several significant properties:

1. At constant $T$ and $P$ and at standard-state conditions, $\Delta G_{rxn}^\circ$ can be thought of as a quantity that gives the net balance between the important driving forces $\Delta H_{rxn}^\circ$ and the product of the absolute temperature and $\Delta S_{rxn}^\circ$, *i.e.*,

$$\Delta G_T^\circ = \Delta H_T^\circ - T\Delta S_T^\circ$$

If a reaction has $\Delta G_T^\circ > 0$, this reaction will not produce products in their standard-state quantities at temperature $T$. Such a reaction will lead to a final mixture that has constant composition (an *equilibrium mixture*; see Chapter 14) that will contain relatively greater amounts

---

*The absolute entropy ($\bar{S}_T$) for a substance may be determined from the measurement of its molar heat capacity at constant pressure ($\bar{C}_p$) as a function of $T$, taking into account the enthalpy changes for any phase changes that occur between 0 K and $T$. The tabulated values of absolute entropy are usually given in cal/mole-K. (Note that in using Eqns. 2 and 2′, $\Delta\bar{G}$ and $\Delta\bar{H}$ are usually tabulated in kcal/mole.) $C_p$ is measured in cal/mole-K and is merely the (formula weight) × (specific heat).

†This means that $H$ and $S$ for both products and reactants vary with $T$ in an almost parallel way over small temperature ranges, such that the differences between those for products and those for reactants ($\Delta H$ and $\Delta S$) are approximately constant.

of reactant species (original) than product species. Such a reaction is said to be *non-spontaneous.* Conversely, if $\Delta G^\circ_T < 0$, the reaction will lead to a final mixture that is richer in products than in original reactants; such a reaction is said to be *spontaneous.* Finally, if $\Delta G^\circ = 0$, no **net** reaction will occur–the system will be at *equilibrium* and **all** species, reactants and products, could be in their standard-state concentrations.

2. The standard free energy change for a process, $\Delta G^\circ_T$, is a measure of the maximum useful work when the reaction is carried out in such a way that both reactants and products are in their standard states initially.

$$\Delta G^\circ_T = -w^\circ_{max}$$

If $\Delta G^\circ_T = -100$ kJ, this means that $w_{max}$ will be positive–the reaction releases energy sufficient to perform, at most, 100 kJ of work.

It should be remembered that like $\Delta H^\circ$, $\Delta G^\circ$ is a derived quantity that can be applied to the system alone (not system **plus** surroundings). Furthermore, conclusions based on the sign of $\Delta G^\circ$ are valid if the initial $T$ and $P$ are the same as the final $T$ and $P$, although both may have changed during the reaction. Example problems 12-F and 12-G illustrate the use of $\Delta G^\circ_{rxn}$.

## Example Problems Involving Entropy, Free Energy, and Reaction Spontaneity

### 12-G. Entropy Change in a Reaction

For the reaction at 25°C,

$$2\ SO_{2\,(g)} + O_{2\,(g)} = 2\ SO_{3\,(g)}$$

the entropy change, $\Delta S^\circ$, in J/K, is: (‡)

(A) -188.0 J/K (B) -195.6 J/K (C) +60.1 J/K (D) +17.1 J/K
(E) -239.6 J/K

#### *Solution*

Since the reaction involves compacting a total of three moles of gases into two moles of a single gas, intuitively one surmises that the reaction involves a certain "ordering," *i.e.*, $\Delta S_{rxn} < 0$ [$\Delta S_{rxn}$ generally has the same sign as the change in number of moles of gas; here $\Delta n_{gas} = -1$]. This alone eliminates answers (C) and (D), and from the thermodynamic data tabulated in the Appendix,

$$\Delta S^\circ_{rxn} = \sum_{prod} n\bar{S}^\circ - \sum_{react} n\bar{S}^\circ$$

$= (2)(256.6) - (1)(205.03) - (2)(248.1)$ J/K

$= -188.0$ J/K [A]

## 12-H. Variation of $\Delta G$ with Temperature

The reaction, $CaCO_{3\,(s)} \rightarrow CaO_{(s)} + CO_{2\,(g)}$, is endothermic. It has a $\Delta G^\circ$ of +31.1 kcal at 25°C, meaning that at 25°C $CaCO_3$ will not decompose to yield $CO_2$ and CaO at a $CO_2$ partial pressure of 1.00 atm. Estimate the minimum temperature above which the decomposition would become spontaneous with $P_{CO_2} = 1.00$ atm. (‡)

(A) ~200°C (B) ~400°C (C) ~600°C (D) ~800°C (E) ~1000°C

### *Solution*

From the relationship, $\Delta G^\circ = \Delta H^\circ - T\Delta S^\circ$, it is seen that there is a temperature above which $\Delta G^\circ$ becomes negative, provided that $\Delta S^\circ > 0$. This temperature will be the one at which $\Delta G^\circ = 0$. Assuming that $\Delta H^\circ$ and $\Delta S^\circ$ remain constant as the $CaCO_3$ is heated and since $\Delta S^\circ$ is positive, as the temperature increases $\Delta G^\circ$ becomes less positive, then zero, and then negative:

$$\Delta G^\circ = 0 = \Delta H^\circ - T\Delta S^\circ$$

From the tabulated data, $\Delta H^\circ = +42.5$ kcal and $\Delta S^\circ = +38.4$ cal/K. Then

$$T \cong \frac{\Delta H^\circ}{\Delta S^\circ} = \frac{42.5 \text{ kcal}}{0.0384 \text{ kcal/K}} \cong 1107 \text{ K (or } 834^\circ\text{C)}$$

### *Answer*

Thus, the decomposition of calcium carbonate should become feasible upon heating to around 800°C. (Actually this temperature should be an overestimate for a reaction carried out in an open container since the opposing pressure of $CO_2$ in air is only about 0.2 torr.) [D]

## 12-I. Conclusions Drawn from $\Delta G^\circ$ for a Reaction

At 25°C, 2.04 mole of $H_{2\,(g)}$ at 1 atmosphere, 2.04 mole of $O_{2\,(g)}$ at 1 atmosphere, and one drop (~0.05 g) of $H_2O_{(\ell)}$ are placed in a 50-L container. Qualitatively, what would be the final equilibrium result of the process that occurs?

| | $P_{H_2}$ | $P_{O_2}$ | wt. $H_2O$ |
|---|---|---|---|
| (A) | >1 atm | >1 atm | <0.05 g |
| (B) | ~1 atm | ~1 atm | ~0.05 g |
| (C) | ~0 atm | ~0 atm | >0.05 g |
| (D) | ~2 atm | ~1.5 atm | ~0 g |
| (E) | <<1 atm | <1 atm | >0.05 g |

### *Solution*

For the reaction at 25°C, $H_{2\,(g)} + 1/2\ O_{2\,(g)} \rightarrow H_2O_{(\ell)}$, the entropy change is negative, so that $\Delta G^\circ_{rxn}$ will be more positive than $\Delta H^\circ_{rxn}$:

$$\Delta G^\circ_{rxn} = \Delta \bar{G}^\circ_{f(H_2O)} - \Delta \bar{G}^\circ_{f(H_2)} - (1/2)\Delta \bar{G}^\circ_{f(O_2)}$$

$$= -237.2 - 0 - (1/2)(0)$$

$$= -237.2 \text{ kJ}$$

or alternatively,

$$\Delta H^\circ_{rxn} = -285.83 \text{ kJ}$$

$$\Delta S^\circ_{rxn} = 69.91 - 130.57 - (1/2)(205.03) = -163.18 \text{ J/K}$$

$$\Delta G^\circ_{rxn} = \Delta H^\circ_{rxn} - T\Delta S^\circ_{rxn} = -285.83 - 298\frac{(163.18)}{1000} = -237.2 \text{ kJ}$$

Since $\Delta G^\circ_{rxn}$ has a large negative value, the reaction,

$$H_{2\,(g)} + 1/2\ O_{2\,(g)} \rightarrow H_2O_{(\ell)},$$

should go far to the right. If it is assumed that this means that the reaction goes essentially to completion, the $H_2$ will be depleted and its partial pressure will approach zero; half of the $O_2$ will be used and its partial pressure will approach 0.5 atmosphere; an additional amount of liquid $H_2O$, approaching 2 moles or 36 g, will be formed. The reaction can occur spontaneously—the initial pressure is ~1500 torr; after equilibrium is established at 25°C, the pressure will be ~400 torr (0.5 atm of excess $O_2$ plus the vapor pressure of $H_2O_{(\ell)}$ at 25°C).

[E]

## Exercises on Entropy, Free Energy, and Reaction Spontaneity

18. It is found that a given temperature $\Delta G^\circ = -394$ kJ/mole for the reaction:

$$C_{(s)} + O_{2\,(g)} = CO_{2\,(g)}$$

At this temperature which statement is TRUE for the reaction?

(A) The system is at equilibrium.
(B) Gaseous $CO_2$ will be formed spontaneously.
(C) Gaseous $CO_2$ will decompose spontaneously.
(D) The process, as written, is not possible.
(E) The process will proceed very rapidly.

19. For which of the following reaction systems would $\Delta S_{syst} < 0$?

(A) $2\ H_2O_{(g)} \rightleftharpoons 2\ H_{2\,(g)} + O_{2\,(g)}$ (B) $H_2O_{(s)} \rightleftharpoons H_2O_{(\ell)}$
(C) $CaCO_{3\,(s)} \rightleftharpoons CaO_{(s)} + CO_{2\,(g)}$ (D) $2\ NH_{3\,(g)} \rightleftharpoons N_{2\,(g)} + 3\ H_{2\,(g)}$
(E) none of the above

20. Calculate $\Delta G^\circ_{298}$ for the reaction: $2\ HI_{(g)} = H_{2\,(g)} + I_{2\,(g)}$. (‡)

(A) 17.7 kJ (B) 9.5 kJ (C) 6.5 kJ (D) 7.9 kJ (E) 15.9 kJ

21. From the thermodynamic data in the Appendix and information given in Q-12, p. 282, calculate $\Delta G^\circ_{298}$ for the reaction: (‡)

$$4\ NH_{3\,(g)} + 5\ O_{2\,(g)} \rightarrow 4\ NO_{(g)} + 6\ H_2O_{(g)}$$

(A) -905 kJ (B) -1,085 kJ (C) -726 kJ (D) -959 kJ (E) -2.47 kJ

22. To which of the following does "$\Delta \overline{G}^\circ_f$ of $NH_4Br_{(s)}$" apply at 25°C?

(A) $1/2\ N_{2\,(g)} + 2\ H_{2\,(g)} + 1/2\ Br_{2\,(g)} \rightarrow NH_4Br_{(s)}$
(B) $NH_{3\,(g)} + HBr_{(g)} \rightarrow NH_4Br_{(s)}$
(C) $NH_{3\,(aq)} + HBr_{(aq)} \rightarrow NH_4Br_{(s)}$
(D) $1/2\ N_{2\,(g)} + 2\ H_{2\,(g)} + 1/2\ Br_{2\,(\ell)} \rightarrow NH_4Br_{(s)}$
(E) $1/2\ N_{2\,(g)} + 2\ H_{2\,(g)} + 1/2\ Br_{2\,(s)} \rightarrow NH_4Br_{(s)}$

**Use the following table in answering Questions 23 through 25. Processes labeled I to IV are characterized by sign of $\Delta H$ and $\Delta S$.**

| *Process* | $\Delta H$ | $\Delta S$ |
|---|---|---|
| I | + | + |
| II | + | - |
| III | - | - |
| IV | - | + |

23. Which processes are **definitely** spontaneous, and which are **possibly spontaneous** (depending on magnitudes) at constant *P* and *T*?

(A) definitely I and II; possibly IV (B) definitely IV; possibly I and III
(C) definitely III; possibly II and IV
(D) definitely IV; possibly II and III
(E) definitely III and IV; possibly II

24. Which process is improbable at low *T*, but becomes **more** probable at a higher *T* (assuming $\Delta H$ and $\Delta S$ are independent of *T*)?

(A) I (B) II (C) III (D) IV (E) cannot tell from data

25. Which process should be **nonspontaneous** at **all** values of *T* (same assumptions as in Question 24 above)?

(A) I (B) II (C) III (D) IV (E) cannot tell from data

26. At 25°C what is the value of $\Delta S^\circ_{rxn}$ in cal/K for

$$N_{2\,(g)} + 3\ H_{2\,(g)} \rightleftharpoons 2\ NH_{3\,(g)}?\ (‡)$$

(A) -47.45 (B) -31.01 (C) 14.96 (D) -93.42 (E) 122.95

27. Assuming that $\Delta H^\circ_{rxn}$ and $\Delta S^\circ_{rxn}$ are independent of $T$ for the reaction in Question 26, what is $\Delta G^\circ_{rxn}$ in kcal at 100°C?

 (A) -17.29 (B) -4.33 (C) +0.41 (D) +8.65 (E) -10.81

## GENERAL EXERCISES ON CHAPTER 12—CHEMICAL THERMODYNAMICS

28. A gas absorbs 300. cal of heat and is compressed from 20.0 L to 10.0 L by an opposing pressure of 2.00 atm. The internal energy change, in calories, for the gas is:

 (A) +784 (B) +320 (C) +184 (D) -184 (E) -784

29. From tabulated enthalpies of **combustion** find the enthalpy change at standard-state conditions for the reaction

$$C_2H_{2\,(g)} + 2\,H_{2\,(g)} \rightarrow C_2H_{6\,(g)} \quad (\ddagger)$$

 (A) +335.7 kJ (B) -335.7 kJ (C) -49.8 kJ (D) -807.5 kJ
 (E) +521.7 kJ

30. The structure of $C_2H_2$ is H—C≡C—H and that for $C_2H_6$ is H—C(H)(H)—C(H)(H)—H.

 The enthalpy change ($\Delta H$) for the reaction given in Question 29 could be estimated from bond energies ($D$) to be approximately:

 (A) $2\,D_{H-H} + 2\,D_{C\equiv C} + 2\,D_{C-H} - D_{C-C} - 4\,D_{C-H}$
 (B) $-\,D_{H-H} - D_{C\equiv C} + 6\,D_{C-H} + D_{C-C} - 2\,D_{C-H}$
 (C) $2\,D_{H-H} + D_{C\equiv C} - 4\,D_{C-H} - D_{C-C}$
 (D) $D_{H-H} + 2\,D_{C\equiv C} - 6\,D_{C-H} - D_{C-C}$
 (E) $2\,D_{H-H} + D_{C\equiv C} - 6\,D_{C-H} - 2\,D_{C-C}$

31. Calculate ΔE for evaporation of 1.00 g of $Br_2$ at 25°C and 1 atm. (‡)

 (A) 28.5 kJ (B) 178 kJ (C) 193 J (D) 33.4 kJ (E) 44.9 cal

32. Silver tableware is tarnished by foods such as eggs that contain certain sulfur compounds, *e.g.*,

$$4\,Ag_{(s)} + 2\,H_2S_{(g)} + O_{2\,(g)} \rightarrow 2\,Ag_2S_{(s)} + 2\,H_2O_{(\ell)}$$

 The value of $\Delta H^\circ_{298}$ (in kJ) for this reaction is: (‡)

 (A) -244.30 kJ (B) -297.82 kJ (C) -339.07 kJ (D) -488.61 kJ
 (E) -595.66 kJ

33. Consider the following experimental set-up:

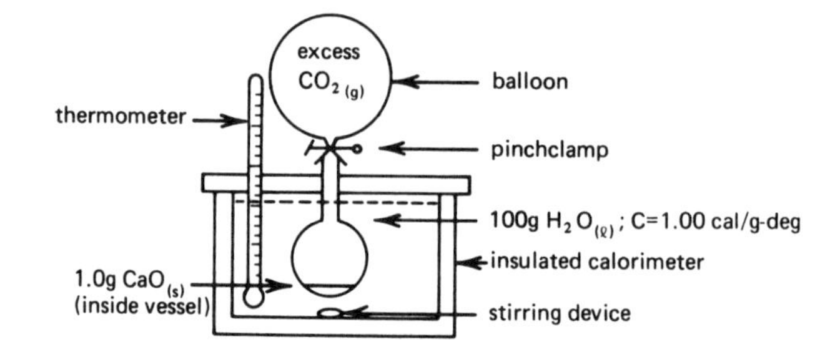


The pinchclamp is opened, and the CaO reacts quantitatively with the $CO_2$. If the pre-reaction temperature of the water is taken as 25.0°C, what maximum temperature could be reached by the water as a result of the heat of reaction?

$$CaO_{(s)} + CO_{2\,(g)} = CaCO_{3\,(s)} \qquad \Delta H = -42 \text{ kcal.}$$

(A) 35.0° (B) 32.5° (C) 30.0° (D) 28.5° (E) 27.0°

34.*A certain bomb calorimeter has a heat capacity of 1610.8 J/°C. A temperature rise of 3.577°C is observed when the calorimeter is used in the quantitative combustion of a 1.000 g sample of powdered tantalum in excess oxygen to produce $Ta_2O_5$. Find $\Delta \overline{H}_f^\circ$ of tantalum(V) oxide at 298 K.

(A) -2,085 kJ/mol (B) -2,079 kJ/mol (C) -2,091 kJ/mol
(D) -1,043 kJ/mol (E) -1,040 kJ/mol

35.*Determine the enthalpy of formation in kcal/mole for chloroethane, $C_2H_5Cl_{(g)}$, from the following data:

| | $\Delta H^\circ$, (kcal) |
|---|---|
| $H_{2\,(g)} + 1/2\, O_{2\,(g)} \rightarrow H_2O_{(\ell)}$ | - 68.3 |
| $C_{(s)} + O_{2\,(g)} \rightarrow CO_{2\,(g)}$ | - 94.1 |
| $C_2H_5Cl_{(g)} \rightarrow C_2H_{4\,(g)} + HCl_{(g)}$ | + 17.2 |
| $2\,CO_{2\,(g)} + 2\,H_2O_{(\ell)} \rightarrow C_2H_{4\,(g)} + 3\,O_{2\,(g)}$ | +337.2 |
| $H_{2\,(g)} + Cl_{2\,(g)} \rightarrow 2\,HCl_{(g)}$ | - 44.2 |

(A) -26.9 (B) -39.3 (C) -60.3 (D) -135.5 (E) -580.6

36. In which of the following reactions would $\Delta H$ be most nearly equal to $\Delta E$?

(A) $H_{2\,(g)} + 1/2\, O_{2\,(g)} \rightarrow H_2O_{(g)}$
(B) $Cl_{2\,(g)} + F_{2\,(g)} \rightarrow 2\,ClF_{(g)}$
(C) $H_2O_{(\ell)} \rightarrow H_2O_{(g)}$
(D) $2\,SO_{3\,(g)} \rightarrow 2\,SO_{2\,(g)} + O_{2\,(g)}$
(E) $Ba(NO_3)_{2\,(s)} \rightarrow Ba(NO_2)_{2\,(s)} + O_{2\,(g)}$

37. To which of the following reactions would the term "$\Delta \overline{H}_f^\circ$ of $NH_4I_{(s)}$" apply at 25°C?

(A) $1/2\, N_{2\,(g)} + 2\,H_{2\,(g)} + I_{(s)} \rightarrow NH_4I_{(s)}$

(B) $1/2\ N_{2\,(g)} + 2\ H_{2\,(g)} + 1/2\ I_{2\,(g)} \rightarrow NH_4I_{(s)}$
(C) $NH_{3\,(g)} + HI_{(g)} \rightarrow NH_4I_{(s)}$
(D) $NH_{3\,(aq)} + HI_{(aq)} \rightarrow NH_4I_{(s)}$
(E) none of the above

38. Which of the following statements is FALSE?

(A) For an irreversible process, entropy is not conserved.
(B) The enthalpy of a perfect crystal at 0 K is zero.
(C) Energy can be transferred as heat or work.
(D) For reactions involving no gases, $\Delta H \cong \Delta E$.
(E) $P\Delta V$ can have units of joules.

39. For an endothermic reaction at constant pressure:

(A) $\Delta H < 0$ (B) $\Delta H > 0$ (C) $\Delta G < 0$ (D) $\Delta S > 0$ (E) $\Delta S < 0$

40. For a system at constant volume, $\Delta E$ **must** equal:

(A) $\Delta H$ (B) $q$ (C) $\Delta H + w$ (D) $P\Delta V$ (E) $q + w$

41. Which of the following must be true for a spontaneous process?

(A) $\Delta G > 0$ (B) $\Delta S_{syst} > 0$ (C) $T\Delta S_{syst} < \Delta H$ (D) $\Delta H < 0$
(E) $\Delta S_{total} > 0$

42. For a spontaneous endothermic process

(A) $\Delta G > 0$ (B) $\Delta G = 0$ (C) $\Delta S > 0$ (D) $\Delta H < 0$ (E) $S° = 0$

43. In a laboratory experiment, a student dissolved 5.30 g of lithium chloride in 100. mL of water, giving 105 g of solution with a specific heat capacity of 0.950 cal/g-°C. The dissolution caused the temperature to change from 24.0°C to a maximum of 35.0°C. What is the molar enthalpy of solution ($\Delta\bar{H}^\circ_{soln}$) of lithium chloride?

(A) +46.0 kJ/mole (B) +20.5 kJ/mole (C) -36.7 kJ/mole
(D) -20.5 kJ/mole (E) -43.1 kJ/mole

44. What is $\Delta n_{gas}$ for the combustion of one mole acetic acid$_{(\ell)}$, when both reactions and products are at 25°C?

(A) -1/2 (B) +2 (C) +1/2 (D) -1 (E) 0

45. From the $\Delta H°$ values given below, find $\Delta H°$ for the reaction whereby $V_2O_3$ disproportionates into a mixture of $V_2O_5$ and VO. Express your answer in kilocalories per mole of $V_2O_5$ formed.

$2\ V_{(s)} + 3/2\ O_{2\,(g)} \rightarrow V_2O_{3\,(s)}$ $\Delta H° = -290$ kcal
$2\ V_{(s)} + 5/2\ O_{2\,(g)} \rightarrow V_2O_{5\,(s)}$ $\Delta H° = -373$ kcal
$V_{(s)} + 1/2\ O_{2\,(g)} \rightarrow VO_{(s)}$ $\Delta H° = -100$ kcal
$V_{(s)} + O_{2\,(g)} \rightarrow VO_{2\,(s)}$ $\Delta H° = -172$ kcal

(A) -276 (B) -183 (C) +56 (D) +97 (E) +397

46. From the data given in Question 45, find the standard enthalpy change for the reaction:

$$V_2O_{5\,(s)} + 3\,V_{(s)} \rightarrow 5\,VO_{(s)}$$

(A) +1140 kJ (B) -1980 kJ (C) -531 kJ (D) -481 kJ (E) +273 kJ

47. Which of the following statements is FALSE?

(A) According to the third law of thermodynamics, $S = 0$ for a perfect crystal at 0°C.
(B) According to the first law of thermodynamics, energy is conserved.
(C) According to the second law of thermodynamics, the total entropy change is positive for a spontaneous process.
(D) $P\Delta V$ has the units of energy.
(E) For reactions involving only condensed phases, $\Delta H \cong \Delta E$.

48. You have discovered substance $Q$ that reacts by rearrangement to form $P$, *i.e.*, $Q \rightarrow P$. This reaction is endothermic and will occur spontaneously at constant temperature. This information implies that:

(A) $\Delta G > 0$
(B) $P$ has a less orderly structure than does $Q$
(C) the reaction is quite rapid at this temperature
(D) $\Delta H < 0$
(E) $\Delta S_{syst} < 0$

49. Which of the following is the standard **formation** reaction for $NH_4Br_{(s)}$ (25°C)?

(A) $1/2\,N_{2\,(g)} + 2\,H_{2\,(g)} + 1/2\,Br_{2\,(g)} \rightarrow NH_4Br_{(s)}$
(B) $NH_{3\,(g)} + HBr_{(g)} \rightarrow NH_4Br_{(s)}$
(C) $NH_{3\,(aq)} + HBr_{(aq)} \rightarrow NH_4Br_{(s)}$
(D) $1/2\,N_{2\,(g)} + 2\,H_{2\,(g)} + 1/2\,Br_{2\,(\ell)} \rightarrow NH_4Br_{(s)}$
(E) $1/2\,N_{2\,(g)} + 2\,H_{2\,(g)} + 1/2\,Br_{2\,(s)} \rightarrow NH_4Br_{(s)}$

50. For the sublimation of iodine at 25°C, what mass of iodine would require one kilojoule of heat? (‡)

(A) about 250 mg (B) about 4 g (C) about 100 mg (D) about 13 g
(E) about 65 mg

51. Suppose that at constant pressure the reaction of fluorine with excess silane releases 17.54 kJ of heat per gram of fluorine reacting:

$$SiH_{4\,(g)} + 4\,F_{2\,(g)} \rightarrow SiF_{4\,(g)} + 4\,HF_{(g)}$$

What is the standard molar enthalpy of formation ($\Delta\bar{H}_f^\circ$) of gaseous $SiF_4$? (‡)

(A) -2667 kJ/mole (B) -2360 kJ/mole (C) -1550 kJ/mole
(D) -1330 kJ/mole (E) -665 kJ/mole

52. $\Delta \overline{H}^{\circ}_{comb}$ for $C_3H_{8\,(g)}$ is -530.59 kcal/mole at 25°C. What is $\Delta \overline{H}^{\circ}_{f}$ for $C_3H_{8\,(g)}$ in kcal/mole? (‡)

    (A) -368.22 (B) -91.69 (C) -50.56 (D) -24.82 (E) +17.25

53. The heat of combustion of $C_6H_{6\,(\ell)}$ is tabulated as $\Delta \overline{H}^{\circ}_{comb}$ = -3268 kJ/mole at 25°C. What is the **sum** of the coefficients in the balanced equation to which this quantity applies?

    (A) 33 (B) 17.5 (C) 16.5 (D) 35 (E) none of these

54. Given that $\Delta H^{\circ}$ = -300 kJ at 25°C for the process:

$$A_{(g)} + B_{(\ell)} \rightarrow 2\,C_{(g)} + D_{(s)}$$

In order to find $\Delta E^{\circ}$ in kJ, all of the following numerical values could be substituted directly into the equation,

$$\Delta H^{\circ} = \Delta E^{\circ} + \Delta n_{gas} RT$$

EXCEPT:

    (A) -300 (B) +1 (C) $1.99 \times 10^{-3}$ (D) 298 (E) 0.00831

55. A system gives up 0.200 kcal of heat to the surroundings and does 1.67 kJ of work upon the surroundings. The internal energy of the system **(increases/decreases)** by (______) kJ in this process.

    (A) increases; 0.84 (B) decreases; 0.84 (C) decreases; 2.51
    (D) increases; 1.67 (E) decreases; 1.67

56. The reaction of 1 mole of solid calcium oxide with 1 mole of gaseous carbon dioxide to form 1 mole of solid calcium carbonate releases 42.5 kcal of heat. Which of the following reaction mixtures would liberate **42.5 kJ** of heat?

    (A) 56.1 g of CaO + 44.0 g of $CO_2$
    (B) 1.32 g of CaO + 1.04 g of $CO_2$
    (C) 0.418 g of CaO + 0.418 g of $CO_2$
    (D) 235 g of CaO + 184 g of $CO_2$
    (E) 13.4 g of CaO + 10.5 g of $CO_2$

57. For the combustion of one mole of $C_2H_{4\,(g)}$ at 25°C, what is the difference, $\Delta \overline{H} - \Delta \overline{E}$?

    (A) -7.4 kJ (B) -5.0 kJ (C) +4.2 kJ (D) +8.3 kJ (E) -2.5 kJ

58. 1.000 g of ethylene, $C_2H_{4\,(g)}$ (FW = 28.05), is exploded at constant pressure and at 25°C with an excess of air. If all the heat released is absorbed by 1.000 kg of liquid water (specific heat capacity = 1.000 cal/g-°C), what temperature rise would be expected in water? (‡)

    (A) 11.98°C (B) 6.25°C (C) 12.00°C (D) 6.29°C (E) 12.02°C

59. The same quantity of ethylene (see the previous problem) is exploded with an excess of pure oxygen at 20.0 atm in a rigid closed bomb at 25.00°C. If all of the heat released is absorbed by 1.000 kg of liquid water, what temperature rise would be found in the water? (‡)

    (A) 11.98°C (B) 11.92°C (C) 12.04°C (D) 6.25°C (E) 6.21°C

60. Calculate the internal energy change, $\Delta E$, for a system when 300. J of heat are absorbed by the system, the system doubles its volume, has its temperature increased from 273 K to 546 K, and does 200. J of work on the surroundings.

    (A) 27.3 kJ (B) 0.500 kJ (C) 100. J (D) -0.500 kJ (E) -100. J

61. If a system **absorbs** 80. J of heat from the surroundings and simultaneously increases its volume by 1.20 L against a constant pressure of 1.00 atm, what is the internal energy change, in joules, for the system?

    (A) -75 (B) -41 (C) 0 (D) +85 (E) +201

62. When 1 mole of $Fe_2O_{3\,(s)}$ is formed by combustion of $Fe_{(s)}$ in $O_{2\,(g)}$, 824 kJ of heat is liberated. By this reaction, which of the following would liberate 100. kcal of heat?

    (A) the reaction of 0.508 mole of Fe
    (B) the reaction of 0.508 mole of $O_2$
    (C) the reaction of 2.03 mole of Fe
    (D) the reaction of 0.761 mole of $O_2$
    (E) the formation of 0.254 mole of $Fe_2O_3$

63. $\Delta\overline{H}^\circ_{comb}$ for $C_2H_{2\,(g)}$ at 25°C in kcal/mole (use $\Delta\overline{H}^\circ_f$ data) is: (‡)

    (A) -216.6 (B) -256.4 (C) -310.6 (D) -600.4 (E) -621.2

64. Estimate the standard molar heat of formation of hydrogen cyanide, H—C≡N, from the following average bond energies: H—C, 99 kcal/mole; H—H, 104 kcal/mole; N≡N, 226 kcal/mole; and C≡N, 210 kcal/mole. (‡)

    (A) +27 kcal (B) +144 kcal (C) +474 kcal (D) -196 kcal
    (E) -144 kcal

65. From appropriate enthalpies of formation find the N—H bond energy in kilocalories per mole of bonds. (‡)

    (A) 77 (B) 86 (C) 93 (D) 99 (E) 107

66. For the combustion of one mole of propane, $C_3H_{8\,(g)}$, at 298 K to form $CO_{2\,(g)}$ and $H_2O_{(\ell)}$, $\Delta H^\circ$ = -2220. kJ and $\Delta G^\circ$ = -2108 kJ. Using the $\overline{S}^\circ$ values in the table, find the standard molar entropy ($\overline{S}^\circ$) for propane gas. (‡)

    (A) 376 J/mole-K (B) 186 J/mole-K (C) 271 J/mole-K
    (D) 214 J/mole-K (E) 64.9 J/mole-K

67. Calculate $\Delta E$ for a system that absorbs 200. cal of heat at the same time that the surroundings are doing 300. cal of work on the system.

(A) -500. cal (B) -100. cal (C) +100. cal (D) +300. cal (E) +500. cal

68.*The temperature rise in a calorimeter may be correlated with the heating effect in joules by passing a current of *I amperes* through a resistance of *R ohms* for a period of *t seconds.* Given that $V = IR$ (Ohm's law), where $V$ is the *voltage* (joules/coulomb), and that the units of $I$ are coulombs/second. If a 5.00-minute current flow of 1.00 amp through a 100. ohm resistor caused the temperature of a given calorimeter to increase by 5.00°C, then the heat capacity of the calorimeter is:

(A) 6.00 kJ/°C (B) 10.0 kJ/°C (C) 1.50 kJ/°C (D) 50.0 kJ/°C (E) 0.500 kJ/°C

69. Calculate the molar heat of combustion for gasoline ($C_8H_{18}$) at 25°C and 1 atm. (‡)

(A) -226 kJ (B) -454 kJ (C) -5,495 kJ (D) -4,916 kJ (E) -905 kJ

70. Upon dissolving 8.63 g of $NH_4H_2PO_4$ (FW = 115) in 100. mL of water in a calorimeter, it was found that $\Delta T$ = -2.7°C. Assuming 109 g of solution having a specific heat capacity of 1.0 cal/g-°C, calculate $\Delta \overline{H}_{soln}$, the molar enthalpy of solution of $NH_4H_2PO_4$ in kJ/mole.

(A) -11 (B) +11 (C) -8.2 (D) +16 (E) +23

71. Given the following thermochemical reactions:

$MnO_{2\,(s)} + CO_{(g)} \rightarrow MnO_{(s)} + CO_{2\,(g)}$ $\quad \Delta H = -151$ kJ
$Mn_3O_{4\,(s)} + CO_{(g)} \rightarrow 3\,MnO_{(s)} + CO_{2\,(g)}$ $\quad \Delta H = -\ 54$ kJ
$3\,Mn_2O_{3\,(s)} + CO_{(g)} \rightarrow 2\,Mn_3O_{4\,(s)} + CO_{2\,(g)}$ $\quad \Delta H = -142$ kJ

Find $\Delta H$ in kilojoules for the following reaction:

$$2\,MnO_{2\,(s)} + CO_{(g)} \rightarrow Mn_2O_{3\,(s)} + CO_{2\,(g)}$$

(A) -145.2 (B) -218.7 (C) -264.8 (D) -312.5 (E) -384.9

72.*At 200°C, $\Delta G_{rxn}$ = -56.9 kJ and $\Delta H_{rxn}$ = -1.7 kJ. What is the corresponding value of the **total** entropy change $\Delta S_{total}$, in J/K?

(A) +285 (B) +37.2 (C) +120 (D) +87.9 (E) none of these

73. How many kJ would be **liberated** in the thermal decomposition of 0.250 moles of $(NH_4)_2Cr_2O_{7\,(s)}$ at standard-state conditions into $Cr_2O_{3\,(s)}$, $H_2O_{(g)}$, and $N_{2\,(g)}$? (‡)

(A) 75 (B) 107 (C) 167 (D) 298 (E) 427

74. Calculate the bond energy in kJ/mole of C—H bond from the enthalpy of formation of methane ($CH_4$) and the enthalpies of formation of $H_{(g)}$ and $C_{(g)}$. (‡)

(A) 251 (B) 377 (C) 416 (D) 464 (E) 1,664

75. How many kJ will be liberated when a 5.6-g sample of iron is reacted to form $Fe_2O_3$ at standard-state conditions? (‡)

(A) 824 (B) 259 (C) 82.4 (D) 41 (E) 23

76. Given the following thermochemical data:

$Fe_2O_{3\,(s)} + CO_{(g)} \rightarrow 2\,FeO_{(s)} + CO_{2\,(g)}$ $\Delta H = -\ 2.9$ kJ
$Fe_{(s)} + CO_{2\,(g)} \rightarrow FeO_{(s)} + CO_{(g)}$ $\Delta H = +\ 11.3$ kJ

$\Delta H$ in kj for the reaction:

$$Fe_2O_{3\,(s)} + 3\,CO_{(g)} \rightarrow 2\,Fe_{(s)} + 3\,CO_{2\,(g)}$$

will be:

(A) -25.5 kJ (B) -14.2 kJ (C) -8.4 kJ (D) +8.4 kJ (E) -19.7 kJ

77. The heat of combustion of natural gas (almost pure $CH_{4\,(g)}$) is $\Delta \bar{H}^\circ_{comb}$ = -212.8 kcal/mole $CH_4$ at 25°C. Approximately how much heat is given off by the combustion of one cubic foot (28.32 L) of natural gas, measured at 1.00 atm and 25°C?

(A) ~6000 kcal (B) ~150 kcal (C) ~5000 kcal (D) ~250 kcal
(E) ~ 400 kcal

78.*From the information in #77 (above) **and** the tabulated data below, find the standard molar free energy of formation of $CH_4$ (g), at 25°C.

| Substance | $\Delta \bar{H}^\circ_f$, kcal/mole | $\Delta \bar{G}^\circ_f$, kcal/mole | $\bar{S}^\circ$, cal/mole K |
|---|---|---|---|
| $H_2O_{(\ell)}$ | -68.3 | -56.7 | – |
| $CO_{2\,(g)}$ | -94.1 | - 5.8 | – |
| $O_{2\,(g)}$ | 0 | 0 | 49.00 |
| $CH_{4\,(g)}$ | – | – | 44.50 |
| $C_{(s)}$ | 0 | 0 | 1.36 |
| $H_{2\,(g)}$ | 0 | 0 | 31.21 |

(A) +37.1 kcal/mole (B) -12.1 kcal/mole (C) -31.1 kcal/mole
(D) -57.3 kcal/mole (E) -44.5 kcal/mole

79. A gas absorbs 350. calories of heat and is compressed from 20.0 L to 10.0 L at a constant applied pressure of 1.00 atm. The result of this process is that the gas **(increases/decreases)** in internal energy by ______ calories. (Given that 1 L-atm = 24.2 calories.)

(A) decreases; 108 (B) decreases; 340 (C) increases; 340
(D) increases; 360 (E) increases; 592

80. A one-pound bottle of calcium turnings is inadvertently knocked off the shelf into a sink of water. The bottle is broken, and $H_{2(g)}$ and $Ca(OH)_{2(s)}$ are formed. Approximately how many kJ are released by the reaction? (‡)

(A) ~40 kJ (B) ~400 kJ (C) ~4,000 kJ (D) ~40,000 kJ (E) ~400,000 kJ

81. Consider the following experimental set-up:

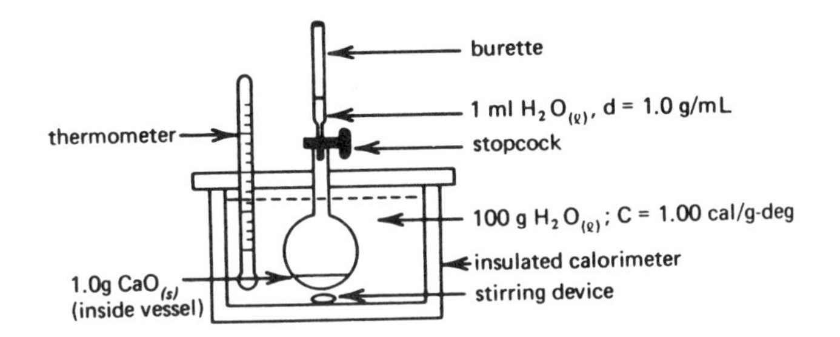


The stopcock is opened, and the CaO reacts with the $H_2O$. If the pre-reaction temperature of the water is taken as 25.0°C, what maximum temperature could be reached by the water as a result of the heat of reaction? (‡)

(A) 27.9° (B) 28.5° (C) 30.0° (D) 31.7° (E) 33.7°

82. If $\Delta H^\circ$ = +162 kcal for the reaction,

$$H_2S_{(g)} \rightarrow 2\,H_{(g)} + S_{(g)}$$

what is the average sulfur-hydrogen bond energy ($D_{S-H}$) in kcal/mole of bonds? [$\Delta\bar{H}_f^\circ$ values; $H_{(g)}$, 52 kcal/mole; $S_{(g)}$, 53 kcal/mole; $H_2S_{(g)}$, -4.93 kcal/mole]

(A) 2.5 (B) 4.9 (C) 50 (D) 81 (E) 76

83. For $Br_{2(\ell)} \rightarrow Br_{2(g)}$, $\Delta\bar{H}^\circ$ is +7.4 kcal/mole at 25°C. How many calories are evolved upon condensation of 1.0 g of bromine at 25°C and 1 atm?

(A) 46 (B) 35 (C) 74 (D) $1.2 \times 10^3$ (E) 39

84. Given the following hypothetical thermochemical equations:

$$(1)\ A_{(s)} + 2\,B_{(g)} \rightarrow 3\,C_{(s)} \quad \Delta H = +39\text{ kJ}$$

$$(2)\ 2\,B_{(g)} + A_{(s)} \rightarrow 3\,D_{(\ell)} \quad \Delta H = -39\text{ kJ}$$

Find $\Delta H$ for $C_{(s)} \rightarrow D_{(\ell)}$

(A) +45 kJ (B) -78 kJ (C) -52 kJ (D) +39 kJ (E) -26 kJ

85. Given the following **unbalanced** equations for three combustions, A, B, and C:

A. $\underline{CH_{4(g)}} + O_{2(g)} \rightarrow CO_{2(g)} + H_2O_{(\ell)}$

B. $\underline{C_2H_{6\,(g)}} + O_{2\,(g)} \rightarrow CO_{2\,(g)} + H_2O_{(\ell)}$

C. $\underline{H_{2\,(g)}} + O_{2\,(g)} \rightarrow H_2O_{(\ell)}$

The values of $\Delta H_{rxn}$ for these reactions, **per mole of the underlined species**, are $\Delta H_A = -212.8$ kcal, $\Delta H_B = -372.8$ kcal, and $\Delta H_C = -68.32$ kcal. From these values, determine $\Delta H_{rxn}$ for the reaction (as written):

$$2\,CH_{4\,(g)} \rightarrow C_2H_{6\,(g)} + H_{2\,(g)}$$

(A) -31.0 kcal (B) +31.0 kcal (C) +15.5 kcal (D) -15.5 kcal
(E) +65.4 kcal

86. If $\Delta S°$ is 80 cal/K for the reaction (25°C)

$$A_{(s)} \rightarrow B_{(g)} + 2\,C_{(g)}$$

what is the standard molar entropy, $\bar{S}°$, of $A_{(s)}$ at this temperature? Given: $\bar{S}°$ of $B_{(g)}$ is 50 cal/mole-K; $\bar{S}°$ of $C_{(g)}$ is 30 cal/mole-K.

(A) 190 cal/mole-K (B) -90 cal/mole-K (C) 30 cal/mole-K
(D) -30 cal/mole-K (E) none of these

87. Given the reaction below, with thermodynamic data (25°C) tabulated beneath each species:

| | $C_3H_{8\,(g)}$ → | $CH_{4\,(g)}$ + | $C_2H_{4\,(g)}$ | |
|---|---|---|---|---|
| $\Delta \bar{H}_f^\circ$ | -24.82 | -17.88 | +12.49 | kcal/mole |
| $\bar{S}°$ | +64.51 | +44.49 | +52.45 | cal/mole-K |

The standard free energy change for this reaction at 298 K is:

(A) -9644 kcal (B) -9.77 kcal (C) 9.77 kcal (D) 19.42 kcal
(E) 9644 kcal

## ANSWERS TO CHAPTER 12 PROBLEMS

| | | | | | |
|---|---|---|---|---|---|
| 1. A | 16. D | 31. B | 46. C | 61. B | 76. A |
| 2. B | 17. E | 32. E | 47. A | 62. D | 77. D |
| 3. A | 18. B | 33. B | 48. B | 63. C | 78. B |
| 4. C | 19. E | 34. C | 49. D | 64. A | 79. E |
| 5. A | 20. E | 35. A | 50. B | 65. C | 80. C |
| 6. A | 21. D | 36. B | 51. C | 66. C | 81. A |
| 7. A | 22. D | 37. E | 52. D | 67. E | 82. D |
| 8. C | 23. B | 38. B | 53. B | 68. A | 83. A |
| 9. B | 24. A | 39. B | 54. C | 69. C | 84. E |
| 10. C | 25. B | 40. B | 55. C | 70. D | 85. C |
| 11. D | 26. A | 41. E | 56. E | 71. B | 86. C |
| 12. C | 27. B | 42. C | 57. B | 72. C | 87. C |
| 13. B | 28. A | 43. C | 58. E | 73. A | |
| 14. D | 29. B | 44. E | 59. A | 74. C | |
| 15. A | 30. C | 45. D | 60. C | 75. D | |

# 13
# *Reaction Rates*

Previous sections have dealt with spontaneity, direction, and extent of chemical reactions but have not indicated how fast a reaction occurs, *i.e.*, how much time would be required for a particular reaction mixture to reach a state of equilibrium. The topic that deals with reactions *rates* and *times* is called *kinetics*.

## CONCEPTS AND RATE LAWS

Four factors influence reaction rates: (1) *nature* of reactants, (2) *concentration* of reactants, (3) *temperature*, and (4) *catalysis*. By choosing for study a specific chemical reaction at a given temperature, factors (1), (3), and (4) are held constant; then the *rate* of the reaction can (and **must**) be studied **experimentally** by measuring the concentration **changes** that occur over a given time interval. A variety of techniques may be used (chemical analysis, pressure measurement, *etc.*, as a function of elapsed **time**) in order to find the concentration **changes** ($\Delta c$) that occur over the time interval ($\Delta t$). The rate of a reaction is always spoken of in positive terms. Thus, for a given **reactant**,

$$\text{Rate of disappearance} = -\frac{\Delta c_{reactant}}{\Delta t}$$

for a **product**,

$$\text{Rate of formation (appearance)} = \frac{\Delta c_{product}}{\Delta t}$$

Most often $\Delta c$ is measured in molarity and, once the rate of disappearance (or formation) of one species in a chemical equation has been **experimentally** determined, the rates with respect to all other species can be calculated from the stoichiometry of the reaction. For example, if $-\Delta c_A/\Delta t$

has been found to be 0.010 $M$/min for the reaction, $aA + bB + \cdots \rightarrow cC + dD + \cdots\cdots$, then $-\Delta c_B/\Delta t = 0.010(b/a)$; $\Delta c_C/\Delta t = 0.010(c/a)$; and $\Delta c_D/\Delta t = 0.010(d/a)$.

For a specific reaction at a given temperature a *rate law* may be found from a series of experiments in which the concentration of **one reactant** is systematically varied while the concentrations of the other reactants are held constant. A *rate law* **is an algebraic equation that relates the rate of the reaction to the concentrations of the reactants.** It takes the form of a constant times the concentration of each reactant raised to a power and for the above reaction the rate law would be:

$$R = k_T[A]^n[B]^m \cdots\cdots$$

where the rate, $R$, refers to $\Delta c/\Delta t$ for any one of the species, $k_T$ is the *specific rate constant* (a function only of $T$), [ ] indicate the use of molar concentrations, and the exponents "$m, n, \cdots\cdots$" are the *reaction orders* with respect to reactants A, B, $\cdots\cdots$. The *overall order* of the reaction is the sum of $m + n + \cdots\cdots$. The exponents (orders) in the rate law **must** be determined experimentally although occasionally one or more of the exponents may equal the coefficient of the species in the balanced chemical equation. This is the case since most reactions are really the sum of a series of steps, each proceeding at its own rate and involving *intermediate species* that are not observed. These steps are referred to as the *reaction mechanism.* In some instances, the rate of the overall, net reaction is limited by one very slow step (the *rate-determining step*) so that **if the rate law can be determined experimentally** and the orders deduced, these orders will be the stoichiometric coefficients of the reactants in this **slow step.** Then knowing the slow step and the net reaction (sum of **all** steps), the mechanism may be **speculated** upon.

## Example Problems Involving Concepts and Rate Laws

### 13-A. Determination of a Rate Law for a Reaction

The following table gives some experimental data at 200°C for a hypothetical reaction, $X + 2Y + 3Z \rightarrow 3M$. These data are used in all of the example problems.

| Experiment Number | Initial Concentrations | | | Initial Rate |
|---|---|---|---|---|
| | $[X]_0$ | $[Y]_0$ | $[Z]_0$ | $\Delta[M]/\Delta t$, $M$/min |
| I | 0.10 | 0.010 | 0.050 | 10 |
| II | 0.10 | 0.020 | 0.10 | 20 |
| III | 0.10 | 0.020 | 0.20 | 20 |
| IV | 0.20 | 0.020 | 0.050 | 80 |
| V | 0.40 | 0.010 | 0.080 | – |

The rate law for the reaction is $\Delta[M]/\Delta t$ =

(A) $k[X][Y]^2[Z]^3$ (B) $k[X][Y][Z]$ (C) $k[X][Y]^2$ (D) $k[X]^2[Y]$
(E) $k[X]^2[Z]$

### *Solution*

The rate law for the reaction takes the following form: $R$ = rate of initial formation of M = $\Delta[M]/\Delta t = k_T[X]_0^p[Y]_0^q[Z]_0^r$. This rate law is valid for each of the experiments. There are two methods of evaluating the exponents and arriving at the specific rate constant, $k$: A (by inspection) and B (analytically).

*Method* A. It is seen that in Exps. II and III [X] and [Y] are held constant while [Z] is doubled; the rate remains unchanged and hence, [Z] does not affect the rate, *i.e.*, the reaction rate is **zero order** in [Z]. (Note that $[Z]^0 = 1$.) Henceforth, [Z] may be ignored in finding the orders with respect to [X] and [Y].

From Exps. I and II it is noted that doubling [Y] doubles the rate; therefore, the rate is first order in [Y]. Then, from Exps. III and IV, the rate quadruples as [X] is doubled (holding [Y] constant). Thus, the rate law is:

$$R \propto [X]^2[Y]^1[Z]^0 = [X]^2[Y]$$

or

$$R = k[X]^2[Y] \qquad \textbf{[D]}$$

By substituting data into the equation from any of the five experiments, the specific rate constant may be evaluated, *i.e.*, $k_{473} = 1.0 \times 10^5\ M^{-2}\text{min}^{-1}$.

*Method* B. First, the rate law for Exp. II when divided by that for Exp. III yields

$$\frac{R_{II}}{R_{III}} = \frac{k(0.10)^p(0.020)^q(0.10)^r}{k(0.10)^p(0.020)^q(0.20)^r} = \frac{20}{20} = 1$$

Therefore, $(1/2)^r = 1, r = 0 \Rightarrow [Z]^0$.

Likewise, using the rate expressions for Exps. I and II,

$$\frac{R_{II}}{R_I} = \frac{(0.020)^q}{(0.010)^q} = 2.0 \quad \text{or} \quad (2.0)^q = 2.0, q = 1 \Rightarrow [Y]^1$$

From Exps. III and IV,

$$\frac{R_{IV}}{R_{III}} = \frac{(0.20)^p}{(0.10)^p} = 4.0, \quad \text{or} \quad (2.0)^p = 4.0, p = 2 \Rightarrow [X]^2$$

Hence, as before, the rate law found is: $R = k[X]^2[Y]$. **[D]**

### 13-B. Rate of Formation of a Product from the Rate Law for a Reaction

What is the value of ($\Delta$[M] /$\Delta t$) for Exp. V?

(A) $1.0 \times 10^2$ $M$/min (B) $1.6 \times 10^2$ $M$/min (C) $1.2 \times 10^2$ $M$/min
(D) $3.2 \times 10^2$ $M$/min (E) $6.4 \times 10^2$ $M$/min

*Solution*

Having found the rate law, $R = k[X]^2[Y]$, and having evaluated $k = 1.0 \times 10^5\ M^{-2}\ \text{min}^{-1}$, then

$$R_V = (1.0 \times 10^5\ M^{-2}\text{min}^{-1})(0.40\ M)^2(0.010\ M) = 1.6 \times 10^2\ M/\text{min} \quad [B]$$

[Note that by comparing Exps. IV to V, [X] is doubled (quadrupling the rate) while [Y] is halved (halving the rate); thus, relative to $R_{IV}$, the rate has been changed by total factor of $4(1/2) = 2$, giving $R_V = 2 \cdot R_{IV} = 1.6 \times 10^2$ $M$/min]

### 13-C. Rate of Disappearance of a Reactant from Rate of Formation of a Product

What is the rate of disappearance of Y in Exp. III?

(A) 20 $M$-min$^{-1}$ (B) 30 $M$-min$^{-1}$ (C) 13 $M$-min$^{-1}$ (D) 7 $M$-min$^{-1}$
(E) 40 $M$-min$^{-1}$

*Solution*

This is simple stoichiometry! In Exp. III, the concentration of the product M has increased by 20 $M$ in one minute's time. In order to produce 20 moles/liter of M, (20 molar M/min) (2 molar Y/3 molar M) = 13 molar Y/min must have been consumed; therefore, $-\Delta[Y]/\Delta t = 13\ M\text{-min}^{-1}$. [C]

## Exercises on Concepts and Rate Laws

1. Factors that influence reaction rates include all of the following, EXCEPT the:

   (A) size of solid reactant particles
   (B) heat of reaction (C) reaction temperature
   (D) concentration of reactants (E) specific rate constant

2. For a reaction of the type A + B $\rightarrow$ C, it is experimentally found that doubling the concentration of B causes the reaction rate to be increased four-fold, but doubling the concentration of A has no effect on the rate. The rate equation is:

   (A) $R = k[A]^2$ (B) $R = k[B]^2$ (C) $R = [C]/[A][B]$
   (D) $R = k[A]$ (E) $R = k[A][B]$

3. For the net reaction: 2A + B → C, the rate law for formation of C is:

   (A) $R = k[A]^2[B]$ (B) $R = k[A][B]$ (C) $R = [C]^2/[A]^2[B]$
   (D) $R = k[A]^2$ (E) cannot tell from above data

**The following data are for Questions 4 through 6 and refer to the reaction: A + 2B + C → D + 2E.**

| Experiment Number | Initial Molar Concentrations × 100: A | B | C | Initial Rate of Formation of D *M*/min |
|---|---|---|---|---|
| I | 2.0 | 2.0 | 2.0 | 2.0 |
| II | 2.0 | 1.0 | 2.0 | 2.0 |
| III | 4.0 | 5.0 | 2.0 | 8.0 |
| IV | 2.0 | 4.0 | 1.0 | 1.0 |

4. The rate law for the above reaction is:

   (A) $k[A][B][C]$ (B) $k[A][B]$ (C) $k[A][B]^2$ (D) $k[A]^2[C]$
   (E) $k[A][C]^2$

5. In Exp. II the rate of formation of E (in *M*/min) is:

   (A) 2.0 (B) 4.0 (C) 6.0 (D) 8.0 (E) 1.0

6. In Exp. III, the rate of disappearance of B (in *M*/min) is:

   (A) 2.0 (B) 4.0 (C) 8.0 (D) 16.0 (E) 32.0

7. The rate of a reaction is given by $k[A][B]$. The reactants are gases. If the volume occupied by the reacting gases is suddenly reduced to one-fourth the original volume, the rate of the reaction (relative to the original rate) will be:

   (A) 8/1 (B) 16/1 (C) 1/8 (D) 1/16 (E) 4/1

**Questions 8 through 10 refer to data taken at 27°C for the reaction $2\ NOCl_{(g)} \rightarrow 2\ NO_{(g)} + Cl_{2\ (g)}$**

| Experiment | $[NOCl]_0$ | Initial Rate |
|---|---|---|
| I | 0.30 *M* | $3.60 \times 10^{-9}$ *M*-sec$^{-1}$ |
| II | 0.60 *M* | $1.44 \times 10^{-8}$ *M*-sec$^{-1}$ |
| III | 0.90 *M* | $3.24 \times 10^{-8}$ *M*-sec$^{-1}$ |

8. What is the rate constant, $k_{300}$, for the above reaction?

   (A) $1.2 \times 10^{-8}\ M^{-1}\ sec^{-1}$ (B) $2.4 \times 10^{-8}\ M^{-1}\ sec^{-1}$
   (C) $4.0 \times 10^{-8}\ M^{-1}\ sec^{-1}$ (D) $3.6 \times 10^{-8}\ M^{-1}\ sec^{-1}$
   (E) $1.2 \times 10^{-8}\ M^{-1}\ sec^{-1}$

## DRILL ON KINETICS AND RATE LAWS

***Student's Name*** ________________

The following rate data is for the reaction: $A_{(g)} + 2\ B_{(g)} + C_{(g)} \rightarrow Y_{(g)} + 2\ Z_{(g)}$. Deduce the Rate Law, calculate the rate constant ($k$) and complete the table. You may be asked to turn in this sheet.

| Exp. | Initial Concentrations [A] | [B] | [C] | Initial Rate of Δ[Y]/Δt | Initial Rate of Disappearance of B, −Δ[B]/Δt | Rate of Formation of Y when [C] has decreased to 0.0300 M |
|---|---|---|---|---|---|---|
| A | 0.100 | 0.200 | 0.0400 | 0.100 *M*/sec | *M*/sec | *M*/sec |
| B | 0.100 | 0.300 | 0.0400 | 0.225 *M*/sec | *M*/sec | *M*/sec |
| C | 0.200 | 0.200 | 0.0400 | 0.100 *M*/sec | *M*/sec | *M*/sec |
| D | 0.400 | 0.200 | 0.0600 | 0.150 *M*/sec | *M*/sec | *M*/sec |
| E | 0.300 | 0.100 | 0.100 | *M*/sec | *M*/sec | *M*/sec |
| Write the rate law for the reaction | | | | | | |
| What is the value of the specific rate constant ($k$)? | | | | | | |

9. The value of the rate constant at 400 K would be:

   (A) greater than $k_{300}$ (B) the same as $k_{300}$ (C) less than $k_{300}$
   (D) indeterminate (E) predictable only if $\Delta G^{\circ}_{rxn}$ is known

10. By what factor would the rate change if the initial concentration of NOCl were increased from 0.30 *M* to 0.40 *M*?

    (A) 1.8 (B) 1.3 (C) 0.56 (D) 0.75 (E) 0.33

---

## TEMPERATURE DEPENDENCE AND THEORY OF REACTION RATES

Reaction rates are dependent upon temperature, with a good rule of thumb being that rates approximately double when $T$ is increased by 10°C. The temperature dependence arises because molecules (or ions) must possess a certain minimal energy to react—the *activation energy*, $E_a$. Increasing the temperature increases the number of molecules (or ions) that possess sufficient energy to undergo reaction. This temperature dependence of the specific rate constant is expressed by the Arrhenius equation:

$$k_T = Ae^{-E_a/RT} \quad \text{or} \quad \ln k_T = \ln A - \frac{E_a}{R}\left(\frac{1}{T}\right)$$

where $A$ is a constant (called the collision frequency factor), $R$ is the gas constant in J/mole-K, and $T$ the temperature in kelvins. The constants $E_a$ and $A$ may be evaluated by carrying out a series of experiments at different temperatures but with the concentrations of the reactants kept constant; a plot of $\ln k_T$ versus $1/T$ is a straight line with a slope of $-E_a/R$ and an intercept of $\ln A$. Or, analytically, from two points taken from the straight line:

$$\log \frac{k_{T_2}}{k_{T_1}} = \frac{E_a}{2.303R}\left(\frac{T_2 - T_1}{T_2 T_1}\right)$$

*Catalysts* alter the reaction mechanism by introducing at least one additional step; the net effect is that at a given temperature,

$$E_{a\,(catalyzed)} < E_{a\,(uncatalyzed)}$$

and the reaction rate is increased, *i.e.*, a greater fraction of the reactants have the necessary activation energy since at a given $T$ the $E_a$ barrier has been lowered by introduction of the catalyzed pathway.

## Example Problems Involving Temperature Dependence and Theory of Reaction Rates

### 13-D. Calculation of the Activation Energy

For the decomposition reaction,

$$N_2O_{5\,(g)} \rightarrow 2\,NO_{2\,(g)} + 1/2\,O_{2\,(g)}$$

the rate is first order in $N_2O_5$, *i.e.*, $-\Delta[N_2O_5]/\Delta t = k[N_2O_5]$. At 0°C, $k_{273} = 7.33 \times 10^{-7}$ sec$^{-1}$; at 65°C, $k_{338} = 4.87 \times 10^{-3}$ sec$^{-1}$. What is the activation energy for the decomposition?

(A) $3.8 \times 10^5$ J/mol (B) $9.1 \times 10^2$ J/mol (C) $5.0 \times 10^4$ J/mol
(D) $2.5 \times 10^4$ J/mol (E) $1.0 \times 10^5$ J/mol

### *Solution*

The Arrhenius equation, $k = Ae^{-E_a/RT}$, relates the rate constant at temperature $T$ to the activation energy, $E_a$. Since both the activation energy and the frequency factor, $A$, are unknown, values of $k$ at two different temperatures are needed to solve for these two unknowns. It is convenient to rewrite the Arrhenius equation in the logarithmic form in terms of two different rate constants and their respective temperatures, eliminating the frequency factor. Upon conversion of the natural log to $\log_{10}$, this gives:

$$\log\frac{k_2}{k_1} = \frac{E_a}{2.303R}\left(\frac{T_2 - T_1}{T_1 T_2}\right)$$

where the subscripts "2" and "1" refer to the higher and lower temperatures, respectively. Substituting the given data, we have:

$$\log\frac{4.87 \times 10^{-3}\ \text{sec}^{-1}}{7.33 \times 10^{-7}\ \text{sec}^{-1}} = \frac{E_a}{(2.303)(8.314\ \text{J/mol-K})}\left(\frac{65\ \text{K}}{(273)(338)\ \text{K}^2}\right)$$

Solving for $E_a$,

$$E_a = \frac{3.822(2.303)(8.314\ \text{J/mol-K})(273 \cdot 338\ \text{K}^2)}{65\ \text{K}} = 1.0 \times 10^5\ \text{J/mol}$$

(or $2.4 \times 10^4$ cal/mol)

This value indicates that 1 mole of $N_2O_5$ molecules must be activated (increased in potential energy) by 100,000 J to form a high-energy activated complex, $N_2O_5^*$, before its molecules can successfully decompose into $NO_2$ and $O_2$ molecules as the final products. Note that $E_a$ will have the energy units of R substituted into this equation.

The logarithmic form of the Arrhenius equation is:

$$\log(k) = \log(A) - \frac{E_a}{2.303R}\left(\frac{1}{T}\right)$$

a form of the equation for a straight line,

$$y = b + m(x)$$

In the original work on this reaction the rate was followed by measuring the increase in pressure with time. The original experimental data were plotted as follows:

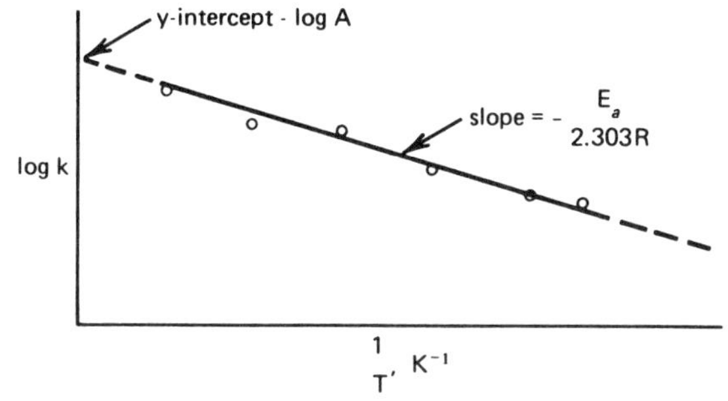


## Exercises on Temperature Dependence and Theory of Reaction Rates

11. With reference to a reaction at room temperature (22°C), what activation energy in kilojoules is implied by the rule of thumb, "the rate of a reaction approximately doubles for every 10°C increase in temperature"?

    (A) 50 kJ/mol (B) 150 kJ/mol (C) 250 kJ/mol (D) 350 kJ/mol
    (E) 450 kJ/mol

12. A catalyst can:

    (A) shift the equilibrium of a reaction
    (B) diminish the enthalpy of a reaction
    (C) make a reaction thermodynamically more feasible
    (D) diminish the activation energy of a reaction
    (E) increase the rate constant of the forward reaction without changing that of the reverse reaction

13. Which of the following statements is FALSE?

    (A) In stepwise reactions the rate-determining step is the slow one.
    (B) It is possible to change the specific rate constant for a reaction by changing the temperature.
    (C) The rates of most reactions change as the reaction proceeds, even if the temperature is held constant.
    (D) The specific rate constant for a reaction is independent of reactant concentrations.
    (E) The rate of a catalyzed reaction is always independent of the concentration of the catalyst.

**For Questions 14 through 17, refer to the potential energy diagram (below) for the reaction:**

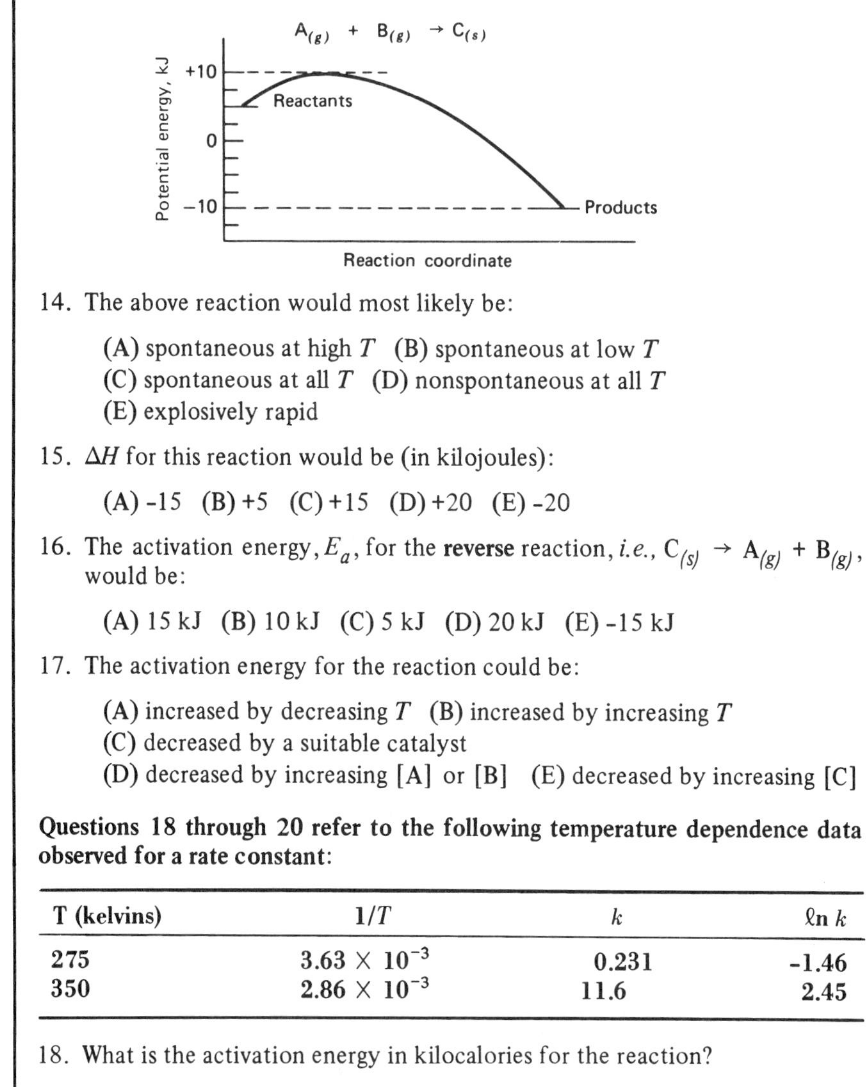


14. The above reaction would most likely be:

    (A) spontaneous at high $T$ (B) spontaneous at low $T$
    (C) spontaneous at all $T$ (D) nonspontaneous at all $T$
    (E) explosively rapid

15. $\Delta H$ for this reaction would be (in kilojoules):

    (A) -15 (B) +5 (C) +15 (D) +20 (E) -20

16. The activation energy, $E_a$, for the **reverse** reaction, *i.e.*, $C_{(s)} \rightarrow A_{(g)} + B_{(g)}$, would be:

    (A) 15 kJ (B) 10 kJ (C) 5 kJ (D) 20 kJ (E) -15 kJ

17. The activation energy for the reaction could be:

    (A) increased by decreasing $T$ (B) increased by increasing $T$
    (C) decreased by a suitable catalyst
    (D) decreased by increasing [A] or [B] (E) decreased by increasing [C]

**Questions 18 through 20 refer to the following temperature dependence data observed for a rate constant:**

| T (kelvins) | $1/T$ | $k$ | $\ln k$ |
|---|---|---|---|
| 275 | $3.63 \times 10^{-3}$ | 0.231 | -1.46 |
| 350 | $2.86 \times 10^{-3}$ | 11.6 | 2.45 |

18. What is the activation energy in kilocalories for the reaction?

    (A) 5 (B) 10 (C) 15 (D) 20 (E) 25

19. What is the value of $\ln A$?

    (A) 12.7 (B) 16.8 (C) 9.3 (D) 4.6 (E) 19.7

20. What is the rate constant at 299 K?

    (A) 2.8 (B) 1.0 (C) 0.46 (D) 0.25 (E) 9.9

## GENERAL EXERCISES ON CHAPTER 13—REACTION RATES

21. Which of the following is TRUE?

    (A) The rate law for a reaction is an algebraic expression relating the forward reaction rate to product concentrations.
    (B) Reactions with high activation energies are usually endothermic.
    (C) The mechanism of a reaction can sometimes be deduced from the rate law.
    (D) Increasing the total pressure in a gas phase reaction increases the fraction of collisions effective in producing reaction.
    (E) None of the above is true.

22. The times listed in the following table are those required for the concentration of $S_2O_8^{2-}$ to decrease by 0.00050 *M* as measured in an "iodine clock" reaction at 23°C. The net reaction is:

$$S_2O_8^{2-} + 2\,I^- \rightarrow I_2 + 2\,SO_4^{2-}$$

| Experiment | $[S_2O_8^{2-}]_0$ | $[I^-]_0$ | Time, sec |
|---|---|---|---|
| I | 0.0400 | 0.0800 | 39 |
| II | 0.0400 | 0.0400 | 78 |
| III | 0.0100 | 0.0800 | 156 |
| IV | 0.0200 | 0.0200 | ? |

Calculate the expected time in seconds for Exp. IV.

(A) 78 (B) 156 (C) 634 (D) 312 (E) 234

**The following data at 300 K are for Questions 23 through 25 and are for the reaction: $A_{(g)} + 2B_{(g)} + 3C_{(g)} \rightarrow X_{(g)} + 4Y_{(s)}$**

| Experiment | $[A]_0$ | $[B]_0$ | $[C]_0$ | Initial Rate, $\frac{\Delta[X]}{\Delta t}$ *M*/min |
|---|---|---|---|---|
| I | 0.10 | 0.10 | 0.10 | $2.0 \times 10^{-2}$ |
| II | 0.050 | 0.20 | 0.10 | $4.0 \times 10^{-2}$ |
| III | 0.10 | 0.20 | 0.10 | $4.0 \times 10^{-2}$ |
| IV | 0.050 | 0.10 | 0.025 | $5.0 \times 10^{-3}$ |
| V | 0.020 | 0.010 | 0.010 | $2.0 \times 10^{-4}$ |

23. The rate law for this reaction is $\Delta[X]/\Delta t$ =

    (A) $k_{300}[A][B]^2[C]^3$ with $k_{300} = 20\,M^{-5}\,\text{min}^{-1}$
    (B) $k_{300}[B]^2[C]$ with $k_{300} = 20\,M^{-2}\,\text{min}^{-1}$

(C) $k_{300}$ [A] [B]$^2$ with $k_{300}$ = 20 $M^{-2}$min$^{-1}$
(D) $k_{300}$ [B] [C] with $k_{300}$ = 2.0 $M^{-2}$min$^{-1}$
(E) $k_{300}$ [B] [C]$^2$ with $k_{300}$ = 20 $M^{-2}$min$^{-1}$

24. In Exp. III the rate of disappearance of C, $-\Delta[C]/\Delta t$, equals:

(A) 0.040 $M$/min (B) 0.013 $M$/min (C) 0.020 $M$/min (D) 0.12 $M$/min
(E) 0.053 $M$/min

25. All of the following are true EXCEPT:

(A) This reaction is zero order in [A].
(B) $k_{305}$ would be greater than $k_{300}$ for this reaction.
(C) The rate-determining step involves a collision of one B molecule with one C molecule.
(D) Thermodynamics predicts that $\Delta S_{rxn}$ would likely be negative.
(E) The activated complex involves an $AB_2$ intermediate.

**The following data are for Questions 26 through 32 and refer to the reaction: A + 2B + 3C → 2Y + Z. All data were taken at 50.0°C.**

| Experiment Number | Initial Molar Concentrations | | | Initial Rate of Formation of Y |
|---|---|---|---|---|
| | [A] | [B] | [C] | |
| I | 0.10 | 0.02 | 0.04 | 10 $M$/hr |
| II | 0.10 | 0.03 | 0.04 | 15 $M$/hr |
| III | 0.20 | 0.02 | 0.08 | 80 $M$/hr |
| IV | 0.20 | 0.02 | 0.16 | 160 $M$/hr |
| V | 0.05 | 0.01 | 0.08 | ? |

26. Doubling [B] would change the rate of formation of Y by factor of:

(A) 1/2 (B) 3 (C) 3/2 (D) 2 (E) 1 (no change)

27. The rate of formation of Z in Exp. III was (in $M$/hr):

(A) 160 (B) 80 (C) 60 (D) 40 (E) 20

28. The rate of disappearance of C in Exp. II was (in $M$/hr):

(A) 45 (B) 10 (C) 30 (D) 7.5 (E) 22.5

29. The rate law derived for the reaction from the above data is:

(A) $k$[A] [B] [C] (B) $k$[A] [B]$^2$ (C) $k$[A]$^2$ [B]$^2$ (D) $k$[A]$^2$ [B] [C]
(E) $k$[A] [B] [C]$^2$

30. The missing rate (Exp. V) in units of $M$/hr should be:

(A) 2.5 (B) 5.0 (C) 10 (D) 7.5 (E) none of these

31. The value of the specific rate constant is:

(A) $1.25 \times 10^{-6}\ M^3\text{hr}^{-1}$ (B) $1.25 \times 10^{4}\ M^{-2}\text{hr}^{-1}$
(C) $1.25 \times 10^{-4}\ M^{-3}\text{hr}^{-1}$ (D) $1.25 \times 10^{6}\ M^{-3}\text{hr}^{-1}$
(E) none of these

32. After a long time (assuming the reverse reaction can be neglected), the concentration of Z in Exp. III would be:

(A) 0.20 $M$ (B) 0.02 $M$ (C) 0.01 $M$ (D) 0.04 $M$ (E) 0.005 $M$

**Use the following data that were obtained at 25°C for Q-33 and Q-34.**

$$A_{(g)} + 2B_{(g)} + 3C_{(g)} \rightarrow D_{(g)} + 4E_{(s)}$$

| Experiment | $[A]_0$ | $[B]_0$ | $[C]_0$ | Initial Rate ($M$/hr) |
|---|---|---|---|---|
| I | 0.40 | 0.15 | 0.20 | 4.20 |
| II | 0.40 | 0.25 | 0.60 | 12.6 |
| III | 1.20 | 0.30 | 0.60 | 113.4 |
| IV | 0.40 | 0.30 | 0.40 | 8.40 |

33. The rate law for this reaction is $\Delta[D]/\Delta t$ =

(A) $k_{298}[A]^2[B][C]$ (B) $k_{298}[A][B]$ (C) $k_{298}[A][B]^2[C]$
(D) $k_{298}[A]^2[C]$ (E) $k_{298}[A]^3[B]^2[C]$

34. At time-zero, the rate for experimental **Run IV** is that [D] is increasing at the rate of 8.40 $M$/hr. What would be the rate in **Run IV**, when the limiting reactant has been **90.0% consumed**?

(A) 0.41 $M$/hr (B) 7.6 $M$/hr (C) 0.84 $M$/hr (D) 0.0042 $M$/hr
(E) 0.11 $M$/hr

35. The rate of the reaction between $BrO_3^-$ and $Br^-$ in acid solution, producing $Br_2$, follows the rate law:

$$\text{rate} = k_T\,[BrO_3^-]\,[Br^-]\,[H^+]^2.$$

The time required for $[BrO_3^-]$ to *decrease by* $1.0 \times 10^{-4}$ $M$ is shown below for a first experimental run:

| Experiment | $[BrO_3^-]_0$ | $[Br^-]_0$ | $[H^+]_0$ | Time Req., sec. |
|---|---|---|---|---|
| I | 0.015 $M$ | 0.096 $M$ | 0.041 $M$ | 100 |
| II | 0.045 $M$ | 0.016 $M$ | 0.123 $M$ | ? |

At the same temperature, approximately how many seconds should theoretically elapse before $[BrO_3^-]$ decreases by the same amount in Exp. II?

(A) ~10 sec (B) ~20 sec (C) ~30 sec (D) ~40 sec (E) ~50 sec

36. Suppose a reaction follows the rate law: Rate = $k[Q]^x$. If at a given temperature the rate is observed to double when [Q] is increased from 0.10 $M$ to 0.15 $M$, then $x$ is about:

(A) 1 (B) 1.7 (C) 2 (D) 2/3 (E) 1.5

37. Consider the reaction A + 2B → C + 3D. If the initial rate of formation of C is $\Delta[C]/\Delta t = 1.0$ mole-$L^{-1}hr^{-1}$, then which of the following corresponds to an initial rate?

(A) $-\Delta[B]/\Delta t = 2.0\ M\text{-}hr^{-1}$ (B) $\Delta[D]/\Delta t = 1/3\ M\text{-}hr^{-1}$
(C) $\Delta[A]/\Delta t = 1.0\ M\text{-}hr^{-1}$ (D) $-\Delta[B]/\Delta t = 1.0\ M\text{-}hr^{-1}$
(E) none of these

38. If the concentration of reactant A is doubled in a reaction that is third order in A, by what factor will the rate of reaction change?

(A) 2 (B) 3 (C) 5 (D) 6 (E) 8

**Questions 39 through 41 refer to data taken at 50.0°C for the reaction: 2A + 3B + 2C → 4D. The data are given in the table below.**

| Experiment | $[A]_0$ | $[B]_0$ | $[C]_0$ | $\Delta[D]/\Delta t$ |
|---|---|---|---|---|
| I | 1.0 | 1.0 | 1.00 | 0.010 $M$-min$^{-1}$ |
| II | 1.0 | 1.0 | 0.50 | 0.0025 $M$-min$^{-1}$ |
| III | 1.0 | 2.0 | 0.50 | 0.0050 $M$-min$^{-1}$ |
| IV | 2.0 | 2.0 | 1.00 | 0.040 $M$-min$^{-1}$ |
| V | 0.50 | 0.50 | 0.50 | ? |

39. The rate law for the above reaction is $\Delta[D]/\Delta t$ =

(A) $k[A][B][C]$ (B) $k[A]^2[B]^3[C]^2$ (C) $k[A][B][C]^2$
(D) $k[B][C]^2$ (E) none of these

40. The rate of appearance of D in Exp. V is:

(A) 0.0050 $M$-min$^{-1}$ (B) 0.00125 $M$-min$^{-1}$ (C) $2.50 \times 10^{-4}$ $M$-min$^{-1}$
(D) $5.00 \times 10^{-4}$ $M$-min$^{-1}$ (E) $6.25 \times 10^{-4}$ $M$-min$^{-1}$

41. The rate of disappearance of B in Exp. II is:

(A) 0.0025 $M$-min$^{-1}$ (B) 0.0019 $M$-min$^{-1}$ (C) 0.0033 $M$-min$^{-1}$
(D) 0.0075 $M$-min$^{-1}$ (E) none of these

42. The reaction A + 2B + C → D occurs by the mechanism:

(1) A + B ⇌ X very rapid equilibrium
(2) X + C → Y slow
(3) Y + B → D very fast

What is the rate law for the reaction?

(A) $R = k[C]$ (B) $R = k[A][B]^2[C]$ (C) $R = k[D]$
(D) $R = k[A][B][C]$ (E) $R = k[A][B]$

**Questions 43–47 deal with the following reversible reaction,**

$$N_2O_{5\,(g)} \underset{(rev)}{\overset{(for)}{\rightleftharpoons}} 2\,NO_{2\,(g)} + 1/2\,O_{2\,(g)}$$

**for which the rate law is**

$$\text{Rate} = k[N_2O_5]$$

**The potential energy diagram for the reactant, intermediate, and products is given below.**

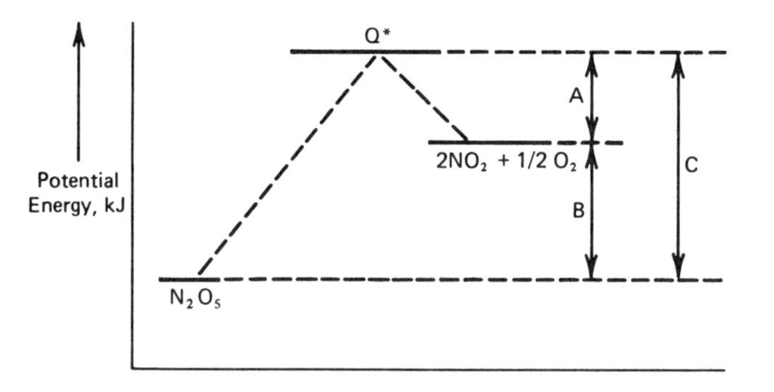


43. The chemical composition represented by Q* is an activated form of:

    (A) $(N_4O_{10})^*$ (B) $(NO_2)^*$ (C) $(N_2O_4)^*$ (D) $(N_2O_5)^*$ (E) $(N_2)^*$

44. Line Segment A has a length equal to:

    (A) $\Delta H_{rxn}$ (B) $E_{a\,(rev)}$ (C) $E_{a\,(for)}$ (D) $k$ (E) none of these

45. Line Segment B has a length equal to:

    (A) $\Delta H_{rxn}$ (B) $E_{a\,(for)}$ (C) $E_{a\,(rev)}$ (D) $k$ (E) none of these

46. Line Segment C has a length equal to:

    (A) $\Delta H_{rxn}$ (B) $E_{a\,(rev)}$ (C) $E_{a\,(for)}$ (D) $k$ (E) none of these

47. If this reaction were catalyzed, the effect would be to:

    (A) decrease Line Segment A only. (B) increase Line Segment B only.
    (C) decrease Line Segment B only. (D) increase Line Segment A only.
    (E) decrease both Line Segments A and C.

48. The reaction 2 A + B → C occurs in one step. A student places 3.0 moles of A and 2.0 moles of B in a 1.0-L flask. When [C] reaches 1.0 *M*, the rate of this reaction, relative to the original rate, will be:

    (A) 1/18 (B) 18/1 (C) 9/1 (D) 1/9 (E) 1/10

49. The rate of a reaction is given by $k$[A] [B]. When the concentrations of both A and B are 0.20 *M*, the rate is $4.0 \times 10^{-6}$ $M\text{-min}^{-1}$. What is the value of the specific rate constant, $k$? (Note the units!)

(A) $1.0 \times 10^{-4}\ M^{-2}\text{min}^{-1}$ (B) $4.0 \times 10^{-1}\ M^{-1}\text{min}^{-1}$
(C) $2.0 \times 10^{-3}\ M^{-2}\text{min}^{-1}$ (D) $3.0 \times 10^{-4}\ M^{-1}\text{min}^{-1}$
(E) $1.0 \times 10^{-4}\ M^{-1}\text{min}^{-1}$

***Questions 50 through 53 refer to the data taken at 50.0°C for the hypothetical reaction: P + 3Q + 2R → 2S + T.**

| Experiment | $[P]_0$ | $[Q]_0$ | $[R]_0$ | Rate ($M$/min) |
|---|---|---|---|---|
| I | 0.25 | 0.30 | 0.10 | 1.30 |
| II | 0.40 | 0.30 | 0.25 | 8.13 |
| III | 0.40 | 0.50 | 0.10 | 2.17 |
| IV | 0.25 | 0.40 | 0.10 | 1.73 |
| V | 0.15 | 0.20 | 0.20 | 3.47 |

50. The rate of formation of T is equal to:

(A) $k[P][Q]^3[R]$ (B) $k[P][Q][R]$ (C) $k[P][Q]^2$ (D) $k[Q][R]^2$
(E) $k[P]^2[Q]^2[R]$

51. The rate of this reaction could be defined as:

(A) $1/2\Delta[S]/\Delta t$ (B) $\Delta[P]/\Delta t$ (C) $-\Delta[Q]/\Delta t$ (D) $1/2\Delta[R]/\Delta t$
(E) $-\Delta[T]/\Delta t$

52. In Exp. V, the rate of disappearance of Q equals:

(A) 10.41 $M$/min (B) 3.47 $M$/min (C) 1.16 $M$/min (D) 1.74 $M$/min
(E) 6.94 $M$/min

53. The value of the specific rate constant is:

(A) $2.3 \times 10^3\ M^{-4}\text{min}^{-1}$ (B) $4.3 \times 10^2\ M^{-2}\text{min}^{-1}$ (C) $58\ M^{-2}\text{min}^{-1}$
(D) $1.7 \times 10^2\ M^{-2}\text{min}^{-1}$ (E) $1.9 \times 10^4\ M^{-5}\text{min}^{-1}$

**For problems 54-56, consider a reaction that has $k_{273} = 7.33 \times 10^{-7}\text{sec}^{-1}$, $k_{338} = 4.87 \times 10^{-3}\text{sec}^{-1}$, and $E_a = 1.039 \times 10^5$ J/mol.**

54.*This reaction is:

(A) a first order reaction (B) a second order reaction
(C) a third order reaction (D) an exothermic reaction
(E) can't tell from the data given

55.*Use the Arrhenius equation and the value of $k_{273}$ to find $k_{373}$.

(A) 0.00999 $\text{sec}^{-1}$ (B) 0.156 $\text{sec}^{-1}$ (C) 0.0831 $\text{sec}^{-1}$ (D) 4.18 $\text{sec}^{-1}$
(E) 32.5 $\text{sec}^{-1}$

56.*Use the Arrhenius equation and the value of $k_{338}$ to find $k_{373}$.

(A) 32.5 $\text{sec}^{-1}$ (B) 4.18 $\text{sec}^{-1}$ (C) 0.0831 $\text{sec}^{-1}$ (D) 0.156 $\text{sec}^{-1}$
(E) 0.00999 $\text{sec}^{-1}$

## ANSWERS TO CHAPTER 13 PROBLEMS

1. B
2. B
3. E
4. D
5. B
6. D
7. B
8. C
9. A
10. A
11. A
12. D
13. E
14. B
15. A
16. D
17. C
18. B
19. B
20. B
21. C
22. D
23. D
24. D
25. E
26. D
27. D
28. E
29. D
30. A
31. D
32. C
33. D
34. A
35. B
36. B
37. A
38. E
39. C
40. E
41. B
42. D
43. D
44. B
45. A
46. C
47. E
48. A
49. E
50. D
51. A
52. A
53. B
54. A
55. B
56. D

# 14
# Chemical Equilibrium

## THE EQUILIBRIUM CONSTANT

Although many reactions proceed essentially to completion (*quantitatively*), there are just as many, or more, that do not. The *equilibrium constant* ($K$) expresses mathematically the extent to which a reaction will proceed at a given temperature. The most important property of $K$ is that, given its value for a specified temperature, the ultimate composition of a system maintained at that temperature can be calculated (*equilibrium conditions*). However, $K$ has several other important properties: (i) if a reaction is written in reverse, the equilibrium constant of the reverse reaction is the reciprocal of that for the forward direction, $K_{rev} = K_{for}^{-1}$; (ii) if an original chemical equation with $K_{orig}$ is multiplied by a factor $n$ to give a final equation, then $K_{fin} = K_{orig}^{n}$; (iii) if several reactions with equilibrium constants $K_i$, $K_j$, $K_k$, $\cdots$ are summed (according to Hess's law, see Ch. 11) so as to give a **new** equation, then $K_{new} = K_i \cdot K_j \cdot K_k \cdots$; (iv) an equilibrium may be approached and established "from either side"—for a given system (as defined by a written, balanced equation), $K$ has the same value regardless of whether the initial mixture consists of only reactants, of only products, or of both reactants and products; and (v) for a given system, the value of $K$ is a function only of temperature—for an *exothermic* reaction, $K$ decreases as $T$ is increased, while for an *endothermic* reaction, $K$ increases as $T$ increases.

In order to decide the **direction** of net chemical change as equilibrium is approached, it is convenient to define a general *reaction quotient* ($Q$) that has the same mathematical form and units as does $K$ but with defined concentrations (those that have been given), rather than equilibrium concentrations. For the reaction,

$$a\,A + b\,B \rightleftharpoons c\,C + d\,D$$

in terms of molar concentrations, $Q_c = [D]^d[C]^c/[A]^a[B]^b$, in terms of partial pressures, $Q_p = p_D^d \cdot p_C^c / p_A^a \cdot p_B^b$. Then if $Q = K$, no net chemical change will occur in the reaction mixture (the system is at equilibrium); if $Q<K$ the reaction will move **to the right**, and if $Q>K$, the reaction will be driven **to the left** in order to establish equilibrium.

In addition to equilibrium concepts, one must keep in mind that (a) a great deal of stoichiometric reasoning goes into equilibrium calculations; (b) the difference between **moles** and **molarity** (moles/L) must be carefully remembered; (c) the equilibrium state is defined in terms of the concentrations that remain at equilibrium, these being calculated from the **original** concentrations plus (or minus) the **concentration changes**; (d) if for a system at equilibrium, a change in one concentration is caused to occur, **all** concentrations then change in reestablishing equilibrium, $K$ remaining constant at constant temperature; and (e) equilibrium changes caused by external perturbations (changes in $P$, $T$, $V$, or concentration) can be **qualitatively** predicted by *Le Chatelier's principle* (see your textbook); more quantitatively predicted by use of the $Q$ vs. $K$ routine; and finally, accurately calculated by using reaction stoichiometry and the $K$-value.

## Example Problems Involving the Equilibrium Constant

### 14-A. The Equilibrium Constant, $K_c$, from Moles of Reactants and Products

1.20 mole of $PCl_{5\,(g)}$ is put into a 3.00-L vessel. When the temperature is raised to 200°C the $PCl_5$ partially decomposes:

$$PCl_{5\,(g)} \rightleftharpoons PCl_{3\,(g)} + Cl_{2\,(g)}$$

At equilibrium 1.00 mole of $PCl_5$ remains. What is the value of $K_c$ for this reaction?

(A) 0.033 (B) 0.067 (C) 0.050 (D) 0.013 (E) 0.040

#### *Solution*

In this problem both stoichiometry and the difference between **moles** and **moles per liter** are important. Using the square bracket notation for molarity and "$n$" for number of moles,

$$K_c = \frac{[PCl_3][Cl_2]}{[PCl_5]} \quad \text{or} \quad K_c = \left(\frac{n_{PCl_3} \cdot n_{Cl_2}}{n_{PCl_5}}\right)\left(\frac{1}{V}\right)$$

| System: | $PCl_{5\,(g)}$ | $\rightleftharpoons$ $PCl_{3\,(g)}$ | + $Cl_{2\,(g)}$ |
|---|---|---|---|
| Initial number of moles: | 1.20 | 0 | 0 |
| Change in number of moles: | -0.20 | +0.20 | +0.20 |
| Number of moles at equilibrium: | 1.00 | 0.20 | 0.20 |
| Equilibrium concentrations: | $\frac{1.00}{3}M$ | $\frac{0.20}{3}M$ | $\frac{0.20}{3}M$ |

$$K_c = \frac{(0.20/3)^2}{(1.00/3)} \quad \text{or} \quad K_c = \frac{(0.20)^2}{1.00} \cdot \frac{1}{3} = 0.013 \qquad \text{[D]}$$

## 14-B. Calculation of an Equilibrium Constant from Constants for Other Reactions

Given the reactions and equilibrium constants at 25°C:

(a) $H_{2\,(g)} + Cl_{2\,(g)} \rightleftharpoons 2\,HCl_{(g)}$ $\quad K_p = 2.5 \times 10^{33}$
(b) $N_{2\,(g)} + 3\,H_{2\,(g)} \rightleftharpoons 2\,NH_{3\,(g)}$ $\quad K_p = 6.0 \times 10^{5}$
(c) $N_{2\,(g)} + 4\,H_{2\,(g)} + Cl_{2\,(g)} \rightleftharpoons 2\,NH_4Cl_{(s)}$ $\quad K_p = 3.9 \times 10^{70}$

Find $K_p$ for the reaction: $NH_{3\,(g)} + HCl_{(g)} \rightleftharpoons NH_4Cl_{(s)}$.

(A) $5.1 \times 10^{15}$ (B) $5.8 \times 10^{109}$ (C) $2.6 \times 10^{31}$ (D) $3.8 \times 10^{-32}$
(E) $2.0 \times 10^{70}$

### *Solution*

It is seen that the desired reaction may be obtained by summing one-half of reaction (c) with the reverse of one-half of equations (a) and (b). After dividing (c) by 2, $K_p$ becomes $\sqrt{3.9 \times 10^{70}}$; reversing equation (a) and dividing by 2 gives $K_p = 1/\sqrt{2.5 \times 10^{33}}$; reversing (b) and dividing by 2 given $K_p = 1/\sqrt{6.0 \times 10^{5}}$. When equations are summed to give a new equation, the equilibrium constant for the reaction obtained is the product of the $K$'s for the stepwise reactions added. Thus,

$$K_p = \left(\frac{3.9 \times 10^{70}}{(2.5 \times 10^{33})(6.0 \times 10^{5})}\right)^{1/2} = 5.1 \times 10^{15} \qquad \text{[A]}$$

## 14-C. Shifting an Equilibrium by Adding Reactants and Products

Given the system: $H_2O_{(g)} + CO_{(g)} \rightleftharpoons CO_{2\,(g)} + H_{2\,(g)}$. At a given high temperature in a 1.00-liter container, an equilibrium mixture contains 0.400 mole each of $CO_2$ and $H_2$ and 0.200 mole each of $H_2O$ and CO. Suddenly, an additional 0.400 mole of both CO and $CO_2$ are added to the mixture. What will be the final [CO] when the new equilibrium is established?

[A] 0.600 $M$ (B) 0.83 $M$ (C) 0.563 $M$ (D) 1.43 $M$ (E) 0.037 $M$

### *Solution*

Since the volume is one liter, the number of moles and the molarity are numerically the same. From the original equilibrium (EQ) concentrations, $K_c$ is readily obtained; the added substances are then summed with those originally present to give the concentrations at the instant of mixing (INST), from which $Q$ may be evaluated and the direction of the equilibrium "shift" ascertained. Simple stoichiometry then allows the new equilibrium (EQ′) concentrations to be expressed in terms of the concentration change (Δ) necessary to reestablish the equilibrium quotient at the value $K_c$.

| | $H_2O_{(g)}$ | + | $CO_{(g)}$ | $\rightleftharpoons$ | $H_{2\,(g)}$ | + | $CO_{2\,(g)}$ | |
|---|---|---|---|---|---|---|---|---|
| SYST: | | | | | | | | |
| EQ: | $0.200\,M$ | | $0.200\,M$ | | $0.400\,M$ | | $0.400\,M$ | ($K_c$ = 4.00) |
| ADD: | 0 | | $0.400\,M$ | | 0 | | $0.400\,M$ | |
| INST: | $0.200\,M$ | | $0.600\,M$ | | $0.400\,M$ | | $0.800\,M$ | ($Q_c$ = 2.67) ("Shift" →) |
| Δ: | $-x\,M$ | | $-x\,M$ | | $+x\,M$ | | $+x\,M$ | |
| EQ: | $0.200-x$ | | $0.600-x$ | | $0.400+x$ | | $0.800+x$ | |

$$4.00 = \frac{(0.800 + x)(0.400 + x)}{(0.200 - x)(0.600 - x)}$$

Multiplying and collecting terms: $3x^2 - 4.40x + 0.160 = 0$

By the quadratic equation, $x = \dfrac{4.40 \pm \sqrt{(4.40)^2 - 4(3)(0.160)}}{6}$

$$x_+ = 1.43 \quad \text{and} \quad x_- = 0.037$$

It is seen that the mathematical root $x_+ = 1.43$ has no chemical significance, since it would require more $H_2O$ and CO to be consumed than were originally present. The valid root, $x_- = 0.037$, then yields

$$[CO]_{EQ'} = (0.600 - x) = 0.563\,M \qquad [C]$$

The other concentrations at EQ′ would be

$$[H_2]_{EQ'} = 0.437\,M$$
$$[CO_2]_{EQ'} = 0.837\,M$$
$$[H_2O]_{EQ'} = 0.163\,M$$

The result may be checked by seeing that these values yield, to within "round-off" error, $K_c = 4.00$.

## Exercises on the Equilibrium Constant

1. Consider the reaction, $2\,A_{(g)} + B_{(g)} \rightleftharpoons C_{(s)} + 2D_{(g)}$. The correct expression for the equilibrium constant (mass action expression) is:

   (A) $\dfrac{[C][D]}{[A][B]}$ (B) $\dfrac{[C][D]^2}{[A]^2[B]}$ (C) $\dfrac{[D]^2}{[A]^2[B]}$ (D) $\dfrac{[C] + [D]^2}{[A]^2 + [B]}$
   (E) $\dfrac{[A]^2[B]}{[D]}$

2. In the system $AB_{(s)} \rightleftharpoons A_{(g)} + B_{(g)}$, doubling the equilibrium concentration of A will cause the equilibrium concentration of B to become:

   (A) twice its original value (B) 1/3 its original value
   (C) 1/4 its original value (D) 1/8 its original value
   (E) none of the above

3. Which of the following equilibria would **not** be affected by volume changes at constant temperature?

   (A) $N_{2(g)} + 3\,H_{2(g)} \rightleftharpoons 2\,NH_{3(g)}$ (B) $H_{2(g)} + Cl_{2(g)} \rightleftharpoons 2\,HCl_{(g)}$
   (C) $2\,NO_{(g)} + O_{2(g)} \rightleftharpoons 2\,NO_{2(g)}$ (D) $2\,NOCl_{(g)} \rightleftharpoons 2\,NO_{(g)} + Cl_{2(g)}$
   (E) $3\,F_{2(g)} + Cl_{2(g)} \rightleftharpoons 2\,ClF_{3(g)}$

4. If the value of the equilibrium constant for

$$2\,X_{(s)} + 3\,Y_{(g)} \rightleftharpoons 2\,Z_{(g)}$$

   is 4000 L/mole at 150 K, the concentration of Z **in equilibrium with** 0.10 **mole** of X and 0.20 *M* Y at this temperature is:

   (A) 8.0 *M* (B) 4.0 *M* (C) 5.7 *M* (D) 6.3 *M* (E) 2.8 *M*

5. For the equilibrium system: $N_2O_{4(g)}$ + Heat $\rightleftharpoons 2\,NO_{2(g)}$, which of the following factors would cause the value of the equilibrium constant to **decrease**?

   (A) decreasing the temperature (B) removing some of the $NO_2$
   (C) adding a catalyst (D) adding some helium gas
   (E) none of the above

6. If the value of the equilibrium constant for

$$A_{(g)} + 2\,B_{(\ell)} \rightleftharpoons 4\,C_{(g)}$$

   is 0.125, the value for the equilibrium constant for

$$4\,C_{(g)} \rightleftharpoons A_{(g)} + 2\,B_{(\ell)}$$

   is:

   (A) 0.125 (B) -0.125 (C) 0.975 (D) 8.00 (E) none of these

7. For the ammonia synthesis (an exothermic reaction)

$$N_{2(g)} + 3\,H_{2(g)} \rightleftharpoons 2\,NH_{3(g)}$$

   what conditions of pressure and temperature would be optimum for the synthesis?

   (A) low *T* and low *P* (B) low *T* and high *P* (C) high *T* and low *P*
   (D) high *T* and high *P* (E) the reaction is independent of *T* and *P*

8. A reaction is of the type $A_{(g)} + B_{(g)} \rightleftharpoons C_{(g)}$. An unmeasured quantity of C is introduced into a 1.0-liter container and is allowed to come to equilibrium. At this time a sample is analyzed and found to be 1.0 *M* in C. If under the conditions of the experiment the equilibrium constant is $1.0 \times 10^6\ M^{-1}$, what is the equilibrium concentration of B?

   (A) $1.0 \times 10^{-6}\ M$ (B) $1.0 \times 10^{-4}\ M$ (C) 0.040 *M*
   (D) $1.0 \times 10^{-3}\ M$ (E) 0.50 *M*

9. For a reaction of the type, $A_{(g)} + B_{(s)} \rightleftharpoons 2\,C_{(g)}$, at equilibrium 3.0 moles of C, 1.0 mole of A, and 2.0 moles of B are present in a 0.50-L vessel.

What is the value of $K_c$?

(A) 9.0 (B) 4.5 (C) 0.67 (D) 6.0 (E) 18

10. For the equilibrium,

$$2\ SO_{2\,(g)} + O_{2\,(g)} \rightleftharpoons 2\ SO_{3\,(g)} + 45\ \text{kcal}$$

the number of moles of sulfur trioxide will increase if:

(A) the temperature of the system is increased (at constant $V$)
(B) the volume of the reaction flask is decreased (at constant temperature).
(C) $O_{2\,(g)}$ is removed from the flask (leaving $T$ and $V$ constant).
(D) the pressure of the system is increased by adding an inert gas (leaving $V$ and $T$ constant).
(E) none of the above.

11. $K_c$ for the above reaction (Question 10) will decrease if:

(A) some $SO_3$ is removed from the reaction flask
(B) oxygen is added to the reaction flask
(C) the temperature is decreased
(D) the volume of the reaction vessel is decreased
(E) the temperature is increased

12. At a given temperature $K_c = 1.69$ for the reaction:

$$H_{2\,(g)} + CO_{2\,(g)} \rightleftharpoons H_2O_{(g)} + CO_{(g)}$$

If 0.40 mole of $H_2$ and 0.40 mole of $CO_2$ are placed in a 1.0-L container at this temperature, the equilibrium concentration of water will be:

(A) 0.80 $M$ (B) 0.20 $M$ (C) 0.23 $M$ (D) 0.25 $M$ (E) 0.40 $M$

13. The equilibrium constant for a reaction of the type,

$$A_{(g)} + B_{(g)} \rightleftharpoons C_{(g)} + D_{(g)}$$

is 4.0. A mixture is prepared in which the concentrations of both A and B are 0.10 $M$ and those of both C and D are 0.30 $M$. What will be the equilibrium concentration of A in the mixture?

(A) 0.10 $M$ (B) 0.20 $M$ (C) 0.033 $M$ (D) 0.13 $M$ (E) 0.33 $M$

14. Consider the equilibrium: $N_{2\,(g)} + 3\ H_{2\,(g)} \rightleftharpoons 2\ NH_{3\,(g)}$. At a certain temperature, an equilibrium mixture in a 4.00-L vessel contains 1.60 moles of $NH_3$, 0.800 moles of $N_2$, and 1.20 moles of $H_2$. What is the value of $K_c$?

(A) 9.00 (B) 29.6 (C) 3.37 (D) 17.1 (E) 7.41

15. If the equilibrium, $P_{4(g)} + 6\ Cl_{2(g)} \rightleftharpoons 4\ PCl_{3(g)}$, is established by adding equal numbers of moles of $P_4$ and $Cl_2$ to an evacuated flask, which of the following must be true at equilibrium?

(A) $[Cl_2] > [PCl_3]$ (B) $[Cl_2] > [P_4]$ (C) $[PCl_3] > [P_4]$
(D) $[P_4] > [Cl_2]$ (E) $[P_4] = [PCl_3]$

16. 2.0 moles of ammonia are introduced into a 1.0-L container in which it partially dissociates at high temperatures:

$$2\ NH_{3\,(g)} \rightleftharpoons 3\ H_{2\,(g)} + N_{2\,(g)}$$

At equilibrium at a particular temperature, 1.0 mole of ammonia remains. $K_c$ is:

(A) 0.42 (B) 0.75 (C) 1.0 (D) 1.5 (E) 1.7

17. Given the reaction: $X_{(g)} + 2\ Y_{(g)} \rightleftharpoons 3\ Z_{(g)}$. At a given temperature 2.00 moles of X and 2.00 moles of Y are placed in a 3.00-L container. After equilibrium is established, the concentration of X is 0.500 *M*. The value of $K_c$ is:

(A) 2.25 (B) 27.0 (C) 0.222 (D) 3.80 (E) 10.6

18. Suppose that the reaction between the cation $M^+$ and a molecule L (frequently called a ligand) has an enormously large equilibrium constant:

$$M^+_{(aq)} + 2\ L_{(aq)} \rightleftharpoons ML^+_{2\,(aq)} \qquad K_c >> 1.$$

The equilibrium is to be established by mixing 10 drops of a solution with $[M^+] = 0.04\ M$ with 10 drops of a solution with $[L] = 0.04\ M$. As $K_c$ approaches infinity, the equilibrium concentration of free $M^+$ will approach:

(A) 0.00 *M* (B) 0.03 *M* (C) 0.02 *M* (D) 0.01 *M* (E) 0.04 *M*

**Questions 19 to 22 refer to the following equilibrium and data:**

$$2\ NO_{(g)} + Cl_{2\,(g)} \rightleftharpoons 2\ ClNO_{(g)}$$

**At a given temperature, 0.300 mole of NO, 0.200 mole of $Cl_2$, and 0.500 mole of ClNO were introduced into a 25.0-L vessel. At equilibrium 0.600 mole of ClNO was present.**

19. The number of moles of $Cl_2$ present at equilibrium is:

(A) 0.200 (B) 0.100 (C) 0.150 (D) 0.250 (E) 0.050

20. If the container volume were increased to 50.0 liters,

(A) the number of ClNO molecules would increase
(B) the number of $Cl_2$ molecules would increase
(C) the concentration of ClNO would increase
(D) the number of moles of ClNO would increase
(E) there would be no "shift" in the equilibrium

21. The units for $K_c$ would be:

(A) L/mole (B) $L^2/mole^2$ (C) mole/L (D) $mole^2/L^2$ (E) no units

22. The numerical value for $K_c$ at this temperature would be

(A) $1.50 \times 10^3$ (B) $3.00 \times 10^2$ (C) $2.50 \times 10^3$
(D) $2.02 \times 10^{-1}$ (E) $9.00 \times 10^2$

23.*An equilibrium mixture of $N_2$, $H_2$, and $NH_3$ in a 3.00-L vessel is found to contain 0.150 mole of $N_2$, 0.600 mole of $H_2$, and 0.300 mole of $NH_3$. Assuming that the temperature and volume remain unchanged, theoretically, how many moles of $N_2$ must be introduced into the vessel to double the equilibrium concentration of $NH_3$?

(A) 12.8 (B) 4.66 (C) 2.34 (D) 1.80 (E) 38.4

24. Solid naphthalene ($C_{10}H_8$, MW = 128) sublimes, so that it may be used to fumigate enclosed spaces with its vapors:

$$C_{10}H_{8(s)} \rightleftharpoons C_{10}H_{8(g)}, \quad K_c = 4.29 \times 10^{-6} \text{ (at 298 K)}$$

Suppose that initially 1.00 g of the solid is placed into a ten-liter container at 25°C. Approximately what percentage of the $C_{10}H_8$ will have sublimed, once equilibrium is established?

(A) ~0.1% (B) ~0.5% (C) ~1% (D) ~5% (E) ~10%

25. Suppose the ten-liter container (Q-24, above) had been completely evacuated prior to the addition of 1.00 g of $C_{10}H_{8(s)}$. At 25°C, approximately what equilibrium pressure would result?

(A) ~0.1 torr (B) ~1 torr (C) ~10 torr (D) ~50 torr (E) ~100 torr

**Questions 26 to 32 refer to the following equilibrium at a given temperature:**

$$2\,N_2O_{(g)} + N_2H_{4(g)} \rightleftharpoons 3\,N_{2(g)} + 2\,H_2O_{(g)}$$

26. If this equilibrium is established by adding equal numbers of moles of $N_2O$ and $N_2H_4$ to a container, then which of the following must be true at equilibrium?

(A) $[N_2H_4] < [N_2O]$ (B) $[N_2] > [H_2O]$ (C) $[N_2O] = [H_2O]$
(D) $[N_2O] = [N_2H_4]$ (E) none of these

27. At equilibrium, $K_c$ =

(A) $\dfrac{(3[N_2])^3(2[H_2O])^2}{(2[N_2O])^2(1[N_2H_4])}$ (B) $\dfrac{[N_2]^3 + [H_2O]^2}{[N_2O]^2 + [N_2H_4]}$

(C) $\dfrac{[N_2O]^2[N_2H_4]}{[N_2]^3[H_2O]^2}$ (D) $\dfrac{[N_2][H_2O]}{[N_2O][N_2H_4]}$

(E) none of these

28. $K_p$ is related to $K_c$ by the equation:

(A) $K_p = K_c/(RT)^2$ (B) $K_p = K_c$ (C) $K_p = K_c(RT)$
(D) $K_p = K_c/(RT)$ (E) $K_p = K_c(RT)^2$

29. If $\Delta H° < 0$ for the reaction as written, then $K$ will:

(A) be negative (B) decrease in value as $T$ is increased
(C) be related to $\Delta H°$ by the relation, $\Delta H° = RT\,\ln K$

(D) have the same value at all temperatures
(E) depend upon the number of moles of $N_2O$ added to the vessel

30. Suppose this equilibrium were established by adding 0.10 mole **each** of $N_2O$ and $N_2H_4$ to a 10-L container. If $x$ is defined as the number of moles of $N_2O$ reacting, then the number of **moles** of $N_2H_4$ **remaining at equilibrium** will be:

    (A) $0.10 - x$ (B) $0.10 - 2x$ (C) $0.10 - 0.5x$ (D) 0.10 (E) $0.050 - x$

31. Again following the suppositions in Question 30, the equilibrium **concentration** of $N_2$ will be, $[N_2] =$

    (A) $10x$ (B) $0.10x$ (C) $0.15x$ (D) $0.67x$ (E) $x$

32. Suppose this system is at equilibrium. If the pressure is increased by halving the volume at constant $T$,

    (A) the number of moles of $N_2O$ will increase
    (B) the equilibrium constant will decrease
    (C) the equilibrium constant will increase
    (D) the number of moles of water will increase
    (E) the molar concentration of $N_2H_4$ will decrease

---

## FREE ENERGY AND THE EQUILIBRIUM CONSTANT

In Chapter 12, the role of the **standard** free energy change for a reaction, $\Delta G^\circ_{rxn}$, was mentioned in relation to reaction spontaneity for standard-state conditions. For the reaction, $A \rightleftharpoons B$,

| If $\Delta G^\circ_T$ is: | Under Standard-State Conditions, the reaction: |
|---|---|
| $> 0$ | is NOT spontaneous to the right;<br>IS spontaneous to the left. |
| $< 0$ | IS spontaneous to the right;<br>is NOT spontaneous to the left. |
| $= 0$ | IS AT EQUILIBRIUM, *i.e.*, the mixture is of constant composition and no net change occurs; all concentrations are constant. |

Remember that standard conditions refer to gases at **one atmosphere** partial pressure, solutes in liquid solution at **one-molar concentration**, and pure solids at a **mole fraction of one**. Standard-state conditions are seldom achieved in practice and are often impossible to achieve experimentally.

For substances $A$ and $B$ (under any conditions), the molar free energies (in absolute terms) may be related to their respective standard molar free energies by equations of the form:

$$\bar{G}_A = \bar{G}^\circ_A + RT\,\ell n\, a_A \quad \text{and} \quad \bar{G}_B = \bar{G}^\circ_B + RT\,\ell n\, a_B$$

In general, this means that the free energy of a substance differs from its $G^\circ$ value by an amount that depends on both the temperature and the *activity* ($a$) of the substance. The activity is a concentration unit that is in accord with the standard-state conventions (partial pressure for gases, molarity for solutes, *etc.*).

Now consider the reaction,

$$x\,A_{(aq)} \rightarrow y\,B_{(aq)}$$

for which $\Delta G_{rxn} = y\bar{G}_B - x\bar{G}_A$. By substitution one can write then,

$$\Delta G = \Delta G^\circ + yRT\,\ell n\, a_B - xRT\,\ell n\, a_A \quad \text{or} \quad \Delta G = \Delta G^\circ + RT\,\ell n\, \frac{a_B^y}{a_A^x}$$

If we recall the reaction quotient,

$$\Delta G = \Delta G^\circ + RT\,\ell n\, Q \qquad \textbf{[Eqn. 3]}$$

Then at equilibrium $\Delta G = 0$ and $Q = K$. Therefore,

$$\Delta G^\circ = -RT\,\ell n\, K \qquad \textbf{[Eqn. 4]}$$

This states that the **standard** free energy change ($\Delta G^\circ$) is related to the equilibrium constant and therefore to the **extent of reaction**! The more positive $\Delta G^\circ$ is, the more unfavorable (in terms of net driving force) is the attainment of product(s) **in the standard state.** Furthermore, Eqn. 4 can be cycled back into Eqn. 3 with the result that

$$\Delta G = RT\,\ell n\, \frac{Q}{K} \qquad \textbf{[Eqn. 5]}$$

by which $Q$ and $K$ can be compared to establish the **direction** of net chemical change. If $Q > K$, $\Delta G > 0$—the reaction **as written** is **not** spontaneous (in the left-to-right direction); however, the **reverse** reaction has $\Delta G < 0$—the **right-to-left** reaction proceeds spontaneously until equilibrium is established and $Q = K$. Note also, that in an initial mixture of reactants, $Q = 0$ and $\Delta G = -\infty$; thus, all reactions must proceed to a finite, but perhaps exceedingly small, extent.

Most tabulated $\Delta\bar{G}^\circ_f$ values have had their origins (either directly or indirectly) in Eqn. 4. Standard free energy changes are calculated from experimental measurements of concentrations at equilibrium.

## Example Problems Involving Free Energy and the Equilibrium Constant

### 14-D. Estimation of the Normal Boiling Point

Given that $\Delta H^\circ_{vap}$ = 5.58 kJ/mole and $\Delta S^\circ_{vap}$ = 72 J/mole-K for nitrogen, these data being obtained on liquid nitrogen at -207°C. Estimate the normal boiling point for $N_2$, in °C, from these data.

(A) +12.9 (B) -273 (C) +77.3 (D) -196 (E) -85.1

***Solution***

The subscript "*vap*" stands for vaporization and the process may be symbolized in chemical terms:

$$N_{2\,(\ell)} \rightarrow N_{2\,(g)}, \text{ for which } K_p = P_{N_2}$$

The normal boiling point is defined as the temperature at which the **equilibrium** vapor pressure of nitrogen reaches 1.00 atm. Thus

$$\Delta G = \Delta G^\circ + RT \ln Q = \Delta G^\circ + RT \ln P_{N_2}$$

Since $\Delta G = 0$ (equilibrium) and $P_{N_2}$ = 1.00 atm, $\Delta G^\circ = 0$. Then, since $\Delta G^\circ \cong \Delta H^\circ - T\Delta S^\circ$,

$$T \cong \frac{\Delta H^\circ}{\Delta S^\circ} = \frac{5.58 \text{ kJ/mole}}{0.072 \text{ kJ/mole-K}} \cong 77 \text{ K} \qquad \text{[D]}$$

$$\cong 77 - 273 = -196°\text{C}$$

(The approximation sign $\cong$ indicates that it has been assumed that both $\Delta H_{vap}$ and $\Delta S_{vap}$ have essentially the same values at the normal boiling point and at -207°C, *i.e.*, both are constant over a relatively small temperature range.)

### 14-E. Equilibrium Constant for a Reaction from $\Delta G^\circ$

For the reaction,

$$2\, SO_{2\,(g)} + O_{2\,(g)} = 2\, SO_{3\,(g)}$$

the equilibrium constant at 25°C is about: (‡)

(A) $1.5 \times 10^{-25}$ atm$^{-1}$ (B) $4.0 \times 10^{16}$ atm$^{-1}$ (C) 1.1 atm$^{-1}$
(D) 0.94 atm$^{-1}$ (E) $7.4 \times 10^{24}$ atm$^{-1}$

***Solution***

The equilibrium constant obtained will be $K_p$ since thermodynamically gases are defined in terms of partial pressures (for this case $K_c$ would then be $K_p(RT)$, with $R$ = 0.0821 L-atm/mole-K). To find $K_p$, $\Delta G^\circ_{rxn}$ must be evaluated. This can be done in either of two ways: (i) directly from the tabulated $\Delta \overline{G}^\circ_f$ data or

(ii) by first calculating $\Delta H^\circ_{rxn}$ and coupling this with $\Delta S^\circ_{rxn}$.

(i) $\Delta G^\circ_{rxn} = 2(-371.1) - 1(0) - 2(-300.19)$ kJ
$= -141.8$ kJ

(ii) $\Delta H^\circ_{rxn} = 2(-395.72) - 1(0) - 2(-296.83) = -197.78$ kJ
$\Delta G^\circ_{rxn} = \Delta H^\circ_{rxn} - T\Delta S^\circ_{rxn}$
$= -197.78 - (298)(-0.1880) = -141.8$ kJ

[If method (ii) is used, it is important that the units of $\Delta H^\circ$ and $\Delta S^\circ$ are consistent!]

Having $\Delta G^\circ = -141.8$ kJ, $K_p$ can now be calculated from $\Delta G^\circ = -RT\,\ell\text{n}\,K = -2.303RT\log K$ (See log table on p. 347.)

$$K = e^{-\Delta G^\circ/RT} = e^{-(-141{,}800)/8.31(298)} = 7.38 \times 10^{24}\ \text{atm}^{-1} \qquad \text{[E]}$$

or

$$K = 10^{-\Delta G^\circ/2.303RT} = 10^{-(-141.8)/8.31 \times 10^{-3} \cdot 298 \cdot 2.303}$$

$$= 7.31 \times 10^{24}\ \text{atm}^{-1} \qquad \text{[E]}$$

Note that $R = 8.31$ J/mole-K $= 8.31 \times 10^{-3}$ kJ/mole-K, and that in the setups it is emphasized that the units of $\Delta G^\circ$ and $R$ must be consistent. The fact that the two values of $K$ disagree in the third significant figure is common with exponentials—the 2.303 conversion factor is rounded, a more exact value being 2.3025850····.

## Exercises on Free Energy and the Equilibrium Constant

33. Which of the following can be used directly to calculate the value of an equilibrium constant at a given temperature for a reaction?

    (A) the entropy change, $\Delta S$ (B) the standard free energy change, $\Delta G^\circ$
    (C) the value of $\Delta G$ at equilibrium (D) the reaction quotient, $Q$
    (E) the standard enthalpy change, $\Delta H^\circ$

34. Which of the following is TRUE?

    (A) $\Delta G = -RT \log K$ (B) $\Delta G = RT\,\ell\text{n}\,\frac{Q}{K}$ (C) $\Delta G^\circ = RT\,\ell\text{n}\,K$
    (D) If $Q = K$, $\Delta G^\circ = 0$ (E) $\Delta G = \Delta H + T\Delta S$ (at const. $T$ and $P$)

35. $\Delta G^\circ$ for a reaction at constant $T$ and $P$ equals zero if:

    (A) the reaction is exothermic (B) $\Delta S_{syst} > 0$ (C) $Q = 1$
    (D) the system is at equilibrium (E) **both** (C) and (D) are true

36. If at 900 K, $\Delta G^\circ = +25$ kJ for the reaction,

$$A_{(g)} + 2\,B_{(g)} \rightleftharpoons AB_{2(g)}$$

## THERMODYNAMICS AND $K_p$

***Student's Name*** ______________________

For each of the given reactions calculate the required thermodynamic quantities and $K_p$. Use the Tables from the back of this book as needed. The numerical subscripts refer to the Kelvin temperature. In the calculation of $\Delta G_{298}$ all reactants have partial pressures of 10. atm and all reactants have partial pressures of 0.010 atm. You may be asked to turn in this sheet.

| Reaction | $\Delta H^\circ_{298}$, kJ | $\Delta S^\circ_{298}$, J | $\Delta G^\circ_{298}$ kJ | $K_{p298}$ | $\Delta G_{298}$, kJ | Estimated Value for | | |
|---|---|---|---|---|---|---|---|---|
| | | | | | | $\Delta G^\circ_{1200}$, kJ | $K_{p1200}$ | T, °C, at which $K_p = 1$ |
| $CO_{2(g)} + H_{2(g)} = CO_{(g)} + H_2O_{(g)}$ | | | | | | | | |
| $\frac{1}{2} N_{2(g)} + \frac{3}{2} H_{2(g)} = NH_{3(g)}$ | | | | | | | | |
| $CH_{4(g)} + \frac{1}{2} C_2H_{6(g)} + \frac{1}{2} H_{2(g)}$ | | | | | | | | |
| ______ | | | | | | | | |

the equilibrium constant is:

(A) $K_c = 3.54 \times 10^{-2}\ \text{atm}^{-2}$ (B) $K_p = 28.2\ \text{atm}^{-2}$
(C) $K_c = 3.54 \times 10^{-2}\ M^{-2}$ (D) $K_c = 8.67 \times 10^{-7}\ M^{-2}$
(E) $K_c = 28.2\ M^{-2}$

37. If $\Delta G^\circ_{500} = 13.2$ kcal for a certain reaction, what is the value of $K$ for this reaction at 500 K?

(A) $5.8 \times 10^{-5}$ (B) $1.7 \times 10^{-6}$ (C) $1.7 \times 10^{4}$ (D) $5.8 \times 10^{5}$
(E) $2.2 \times 10^{-10}$

38. $K_p$ for the reaction $A_{(g)} + B_{(g)} \rightleftharpoons 2C_{(g)}$ is equal to 0.5 at $T = 298$ K. If initially, $p_A = 1$ atm, $p_B = 10$ atm, and $p_C = 5$ atm, then initially:

(A) $\Delta G < 0$ (B) $\Delta G^\circ = 0$ (C) $\Delta H = T\Delta S$ (D) $\Delta G > 0$ (E) $\Delta G = 0$

39. If at 25°C, $\Delta G^\circ = -100.$ kJ for the reaction,

$$A_{(g)} + 2\,B_{(g)} \rightleftharpoons AB_{2\,(g)}$$

the equilibrium constant, $K_p$, is:

(A) $3.38 \times 10^{17}\ \text{atm}^{-2}$ (B) $3.38 \times 10^{17}\ M^{-2}$ (C) $2.96 \times 10^{-18}\ \text{atm}^{-2}$
(D) $2.96 \times 10^{-18}\ M^{-2}$ (E) $1.71 \times 10^{73}\ \text{atm}^{-2}$

40. For the reaction,

$$A_{(aq)} + 2\,B_{(aq)} \rightarrow 3\,C_{(aq)}, \qquad \Delta S^\circ_{298} = +37.2\ \text{cal/K}, \quad \Delta H^\circ_{298} = +17.1\ \text{kcal}$$

At what temperature would this system be at equilibrium with $[A]_{eq} = [B]_{eq} = [C]_{eq} = 1.00\,M$?

(A) 187°C (B) 0.460°C (C) -272.5°C (D) 460°C (E) -271°C

41. Calculate the equilibrium vapor pressure of water at 25°C using $\Delta G^\circ_f$ data. (Compare your answer to that tabulated in most any lab manual.) (‡)

(A) 23.8 torr (B) 22.4 torr (C) 19.7 torr (D) 30.1 torr (E) 26.9 torr

42. For $H_2O_{(\ell)} \rightarrow H_2O_{(g)}$, $\Delta H^\circ_{298} = 10.519$ kcal and $\Delta S^\circ_{298} = 28.394$ cal/K. What is the value of $K_p$ for this process **at 309K**?

(A) 0.0585 atm (B) 0.0983 atm (C) 0.117 atm (D) 0.230 atm
(E) 0.445 atm

43. Estimate the normal boiling temperature of $CCl_4$. (‡)

(A) 50°C (B) 350 K (C) 377 K (D) 273 K (E) 100°C

44. Estimate the temperature at which the partial pressure of iodine vapor **in equilibrium with** the pure solid is 1.00 atm. (‡)

(A) 160°C (B) 433°C (C) 416°C (D) 143°C (E) 239°C

45. The vapor pressure of $Br_{2\,(\ell)}$ at 25°C is: (‡)

(A) 0.28 atm (B) 3.55 atm (C) 1.00 atm (D) 100 torr
(E) $3.8 \times 10^{-6}$ atm

46. For a process at 25°C having $\Delta G° = -10.00$ kJ and $Q = 56.61$, $\Delta G$ (in kJ) =:

    (A) 0.00 (B) -10.00 (C) -14.04 (D) +4.04 (E) -5.96

47. What is $\Delta G$ at 25°C for a process having $\Delta G° = +10.0$ kJ and $Q = 1.00 \times 10^{-5}$?

    (A) -11.5 kJ (B) +10.0 kJ (C) +3.85 kJ (D) 0.00 kJ (E) -18.5 kJ

48. The reaction $A_{(g)} + B_{(s)} \rightarrow 2\ C_{(g)}$ is **spontaneous** and **endothermic.** Which of the following is TRUE for this reaction?

    (A) $\Delta H<0; \Delta S>0$ (B) $\Delta H<0; \Delta S<0$ (C) $\Delta H>0; \Delta S<0$
    (D) $\Delta H>0; \Delta S>0$ (E) $\Delta H = \Delta E + 2RT$

49. For boiling HCl, $\Delta H°_{vap} = 16.15$ kJ/mole and $\Delta S°_{vap} = 85.77$ J/mole-K. Estimate the normal boiling temperature in °C for HCl.

    (A) 5.3° (B) 188° (C) -85° (D) -110° (E) -34°

50. Estimate the partial pressure of iodine vapor **in equilibrium with** the pure solid at 25°C. (‡)

    (A) $2 \times 10^{-4}$ atm (B) $4 \times 10^{-4}$ atm (C) $6 \times 10^{-3}$ atm
    (D) $8 \times 10^{-3}$ atm (E) $4 \times 10^{-2}$ atm

51. Calculate $K_p$ at 25°C for the reaction: (‡)

    $$H_{2\,(g)} + Br_{2\,(\ell)} \rightleftharpoons 2\ HBr_{(g)}$$

    (A) $2.2 \times 10^{9}$ (B) $5.9 \times 10^{12}$ (C) $5.1 \times 10^{18}$ (D) $3.5 \times 10^{25}$
    (E) $6.3 \times 10^{8}$

---

## GENERAL EXERCISES ON CHEMICAL EQUILIBRIUM

52. An equilibrium mixture for the reaction:

    $$2\ H_2S_{(g)} \rightleftharpoons 2\ H_{2\,(g)} + S_{2\,(g)}$$

    had 1.0 mole of $H_2S$, 0.20 mole of $H_2$, and 0.80 mole of $S_2$ in a 2.0-L vessel. At this temperature $K_c$ equals:

    (A) $4.0 \times 10^{-3}$ $M$ (B) $8.0 \times 10^{-2}$ $M$ (C) 0.16 $M$
    (D) $1.6 \times 10^{-2}$ $M$ (E) $3.2 \times 10^{-2}$ $M$

53. Which of the following equilibria is **not** affected by a pressure change (resulting from a volume change at constant $T$)?

(A) $2\ NaCl_{(s)} \rightleftharpoons 2\ Na_{(s)} + Cl_{2\,(g)}$ (B) $2\ NO_{2\,(g)} \rightleftharpoons N_2O_{4\,(g)}$
(C) $PCl_{5\,(g)} \rightleftharpoons PCl_{3\,(g)} + Cl_{2\,(g)}$ (D) $H_{2\,(g)} + I_{2\,(g)} \rightleftharpoons 2\ HI_{(g)}$
(E) $2\ O_{3\,(g)} \rightleftharpoons 3\ O_{2\,(g)}$

54. Given: $2\ A_{(g)} + B_{(g)} \rightleftharpoons 3\ C_{(g)} + D_{(g)}$. If equal numbers of moles of A and B are added to an empty flask, then the following **must** be true when equilibrium is attained:

(A) [D] = [B] (B) [B] = [A] (C) [B] < [A]
(D) [A] < [B] (E) [A] + [B] > [C] + [D]

55. $K_p$ for the equilibrium, $NH_{3\,(g)} + H_2S_{(g)} \rightleftharpoons NH_4HS_{(s)}$, is equal to:

(A) $\frac{[NH_4HS]}{[NH_3][H_2S]}$ (B) $p_{NH_3} \cdot p_{H_2S}$ (C) $\frac{p_{NH_4HS}}{p_{NH_3} \cdot p_{H_2S}}$
(D) $p_{NH_4HS} - p_{NH_3} - p_{H_2S}$ (E) none of these

56. At a given $T$ the system $NH_{3\,(g)} + H_2S_{(g)} \rightleftharpoons NH_4HS_{(s)}$ has $K_c$ = 400. $L^2/mole^2$. If in a 10.0-L vessel at this temperature, one places 1.00 mole **each** of $NH_3$ and $H_2S$, how many **moles** of $NH_4HS$ will be present at equilibrium?

(A) 0.500 (B) 0.100 (C) 0.0500 (D) 0.0200 (E) 0.400

57.*Suppose that 0.125 mole of $N_2O_4$ is initially placed into a container whose volume can be varied while holding $T$ at 303 $K$:

$$N_2O_{4\,(g)} \rightleftharpoons 2\ NO_{2\,(g)},\ K_{303} = 5.00 \times 10^{-3} \text{ mole/liter}$$

At 30°C, what container volume would be necessary in order that 80.0% of the original $N_2O_4$ will be dissociated into $NO_2$ at equilibrium?

(A) 5.0 L (B) 32 L (C) $1.2 \times 10^2$ L (D) $3.2 \times 10^2$ L
(E) $8.0 \times 10^2$ L

58. At a given temperature the equilibrium constant is 5.0 for

$$CO_{(g)} + H_2O_{(g)} \rightleftharpoons CO_{2\,(g)} + H_{2\,(g)}$$

Analysis showed that an equilibrium mixture at this temperature contained 0.90 mole of CO, 0.25 mole of $H_2O$, and 0.50 mole of $H_2$ in 5.0 L. How many **moles** of $CO_2$ were there in the equilibrium mixture?

(A) 2.3 (B) 0.010 (C) 0.45 (D) 5.0 (E) 0.90

59. Under the same conditions as in Question 58, 0.40 mole of $H_2$ and 0.40 mole of carbon dioxide were placed in a 1.0-L vessel. The equilibrium **concentration** of water vapor will be:

(A) 0.28 $M$ (B) 0.40 $M$ (C) 0.80 $M$ (D) 0.12 $M$ (E) none of these

60. At a different $T$, the **reverse** of the Question 58 reaction has $K_c$ equal to 4.0. If in a 10-L vessel 0.40 mole of $H_2O$ and of CO along with 0.20 mole **each** of $H_2$ and $CO_2$ are introduced, the [CO] at equilibrium will be:

(A) 0.23 $M$ (B) 0.17 $M$ (C) 0.33 $M$ (D) 0.27 $M$ (E) 0.040 $M$

61. If to the equilibrium system in Question 60, 0.40 mole of $CO_2$ and 0.20 mole of $H_2O$ are added, the equilibrium would "shift"

(A) so as to increase $K_c$ (B) to the right (C) to the left
(D) so as to decrease $K_c$ (E) not at all

62. The equilibrium constant is 85.0 for the extraction of $I_2$ from water by $CCl_4$.

$$I_{2\,(aq)} \rightleftharpoons I_{2\,(CCl_4)}$$

Initially 0.0340 g of iodine was dissolved in 100. mL of water. How much $I_2$ is left in the aqueous phase after one extraction with 10.0 mL of $CCl_4$?

(A) 0.0004 g (B) 0.0036 g (C) 0.030 g (D) 0.0040 g (E) 0.0019 g

63. The equilibrium constant for $I_{2\,(aq)} \rightleftharpoons I_{2\,(CCl_4)}$ is 85.0. If 0.0340 g of iodine are dissolved in 100. mL of water, how much $I_2$ is left in the water after extraction with 20. mL of $CCl_4$? See Question 62.

(A) 0.0019 g (B) 0.007 g (C) 0.0004 g (D) 0.032 g (E) 0.0001 g

64.*How much iodine is left in the water after two extractions, but this time using 10. mL of $CCl_4$ per extraction? See Question 63.

(A) 0.0019 g (B) 0.0007 g (C) 0.0004 g (D) 0.032 g (E) 0.0001 g

65. Given: $2\ A_{(g)} + B_{(g)} \rightleftharpoons C_{(g)}$. If this equilibrium is established by starting with equal numbers of moles of *B* and of *C*, **no** A, at equilibrium it is always true that:

(A) [A] = [B] (B) [B] = [C] (C) [A] < [C]
(D) [B] > [C] (E) [A] > [C]

66. Which of the following reactions has the least tendency to go to completion?

(A) $2\ H_{2\,(g)} + O_{2\,(g)} \rightleftharpoons 2\ H_2O_{(g)}$ $K = 1.7 \times 10^{+27}$
(B) $N_{2\,(g)} + O_{2\,(g)} \rightleftharpoons 2\ NO_{(g)}$ $K = 5.0 \times 10^{-31}$
(C) $H_{2\,(g)} + Cl_{2\,(g)} \rightleftharpoons 2\ HCl_{(g)}$ $K = 3.2 \times 10^{+16}$
(D) $2\ HF_{(g)} \rightleftharpoons H_{2\,(g)} + F_{2\,(g)}$ $K = 1.0 \times 10^{-13}$
(E) $2\ NOCl_{(g)} \rightleftharpoons 2\ NO_{(g)} + Cl_{2\,(g)}$ $K = 4.7 \times 10^{-4}$

67. If $K_c$ for the reaction $A + B \rightleftharpoons C$ is 4.0 L/mole, and $K_c$ for the reaction $2\ A + D \rightleftharpoons C$ is 6.0 $L^2/mole^2$, what is the value of $K_c$ for the reaction $C + D \rightleftharpoons 2\ B$?

(A) 0.38 (B) 0.67 (C) 1.5 (D) 2.7 (E) 9.0

68. Given: $N_{2\,(g)} + 3\ H_{2\,(g)} \rightleftharpoons 2\ NH_{3\,(g)}$. At a given temperature, 0.045 mole of $N_2$, 0.985 mole of $H_2$, and 0.020 mole of $NH_3$ are added to an evacuated 1.00-L vessel. Later it is found that 1.000 mole of $H_2$ is present at equilibrium. The value of $K_c$ at this temperature is:

(A) 5.0 (B) $2.0 \times 10^{-3}$ (C) $4.0 \times 10^{-2}$ (D) $8.9 \times 10^{-3}$
(E) $5.0 \times 10^2$

69. At 990°C, $K_c$ for $H_{2\,(g)} + CO_{2\,(g)} \rightleftharpoons H_2O_{(g)} + CO_{(g)}$ is 1.6. If we start with 0.50 mole of $H_2$, 0.80 mole of $CO_2$, 1.20 mole of $H_2O$, and 1.00 mole of CO in a 20.0-L vessel, calculate the number of moles of water at equilibrium.

(A) 0.38 (B) 0.49 (C) 0.72 (D) 1.07 (E) 0.94

70. The system $2\ NO_{(g)} + Cl_{2\,(g)} \rightleftharpoons 2\ NOCl_{(g)}$ at a given $T$ contains at equilibrium 1.000 $M$ NO, 0.1000 $M$ $Cl_2$, and 1.000 $M$ NOCl, which yields $K_c$ = 10.00. How many moles **per liter** of $Cl_2$ need be added to increase [NOCl] to 1.100 $M$ at the new equilibrium?

(A) 0.656 (B) 0.099 (C) 0.328 (D) 0.149 (E) 0.050

71. At a particular temperature, $K_c$ = 10.00 for the system:

$$2\ NO_{(g)} + Cl_{2\,(g)} \rightleftharpoons 2\ NOCl_{(g)}$$

If the equilibrium concentrations for this system are [NOCl] = [NO] = 1.000; [$Cl_2$] = 0.1000; how many moles **per liter** of NO need be **added** to **decrease** the [$Cl_2$] to 0.0500 $M$?

(A) 0.656 (B) 1.22 (C) 0.981 (D) 2.45 (E) 0.750

72. At 30°C an equilibrium mixture of $N_2O_{4\,(g)}$ and $NO_{2\,(g)}$ exerted a total pressure of 0.750 atm. The partial pressure of $N_2O_4$ was 0.500 atm. $K_p$ for the reaction $N_2O_{4\,(g)} \rightleftharpoons 2\ NO_{2\,(g)}$ is:

(A) 1.50 (B) 0.667 (C) 0.125 (D) 0.500 (E) 1.00

73. At 30°C a 5.0-L flask of an equilibrium mixture of $N_2O_4$ and $NO_2$ gases was found to contain 0.100 moles of $N_2O_4$. How many **moles** of $NO_2$ were in the flask? See Question 72.

(A) 0.010 (B) 0.050 (C) 0.200 (D) 0.025 (E) 0.100

74. At ~425°C a 1.00-L vessel contained 0.100 moles of $I_2$, 0.100 moles of $H_2$, and 0.700 moles of HI in equilibrium. $K_c$ for the reaction, $H_{2\,(g)} + I_{2\,(g)} = 2\ HI_{(g)}$, is:

(A) 70.0 (B) 49.0 (C) 14.0 (D) 6.90 (E) 2.70

75. If 0.0500 moles of **each** of the three gases were added to the equilibrium mixture in Question 74, what would be the [HI] after reestablishment of equilibrium?

(A) 0.033 (B) 0.037 (C) 0.850 (D) 0.825 (E) 0.817

76. Suppose that the equilibrium,

$$La_2(C_2O_4)_{3\,(s)} \rightleftharpoons La_2O_{3\,(s)} + 3\ CO_{(g)} + 3\ CO_{2\,(g)}$$

is established by placing 0.1000 mole of $La_2(C_2O_4)_3$ in an evacuated 10.0-L container held at temperature $T$. If at equilibrium the total pressure is 0.200 atm, then $K_p$ equals:

(A) $6.40 \times 10^{-5}$ atm$^6$ (B) $1.00 \times 10^{-6}$ atm$^6$ (C) $4.00 \times 10^{-2}$ atm$^2$ (D) $1.00 \times 10^{-2}$ atm$^2$ (E) $1.30 \times 10^{-9}$ atm$^{-6}$

77.*If in Question 76, $T$ equals 298 K, how many moles of lanthanum oxalate remain at equilibrium?

(A) 0.0864 (B) 0.0136 (C) 0.0494 (D) 0.0728 (E) 0.0272

78. At a temperature of about 425°C, the equilibrium mixture (EQ) given below prevails:

$$H_{2\,(g)} + I_{2\,(g)} \rightleftharpoons 2\,HI_{(g)} \quad \Delta H^\circ = -9.46\text{ kJ}$$

EQ: 0.1000 $M$ 0.1000 $M$ 0.7000 $M$

If suddenly 0.0500 mole/liter of $H_2$, 0.0500 mole/liter of $I_2$, and 0.3500 mole/liter of HI were added, the result would be that:

(A) more HI would form (B) more $H_2$ and $I_2$ would form
(C) the equilibrium would be shifted to the right
(D) heat would be absorbed by the system
(E) no net chemical reaction would occur

79.*Starting with the initial conditions described in Q-78, how many mole/L of $I_2$ would need be added in order to give $[I_2]$ = 0.2000 $M$ at the new equilibrium composition?

(A) 0.1000 mole/liter (B) 0.1384 mole/liter (C) 0.0768 mole/liter
(D) 0.0692 mole/liter (E) 0.1500 mole/liter

80. What is $K_p$ at 25°C for the reaction $N_{2\,(g)} + 3\,H_{2\,(g)} \rightleftharpoons 2\,NH_{3\,(g)}$? (‡)

(A) $7.7 \times 10^2$ (B) $1.08 \times 10^3$ (C) $5.9 \times 10^5$ (D) $1.3 \times 10^{-3}$
(E) $1.7 \times 10^{-6}$

81. What is $\Delta G$ in kJ for a process at 25°C having $K = 1.0 \times 10^{-5}$ and $Q = 1.0 \times 10^{-4}$?

(A) +5.70 (B) -10.0 (C) -5.70 (D) +10.0 (E) +8.31

82.*It is experimentally determined that at 25°C a solution having $[NH_4^+]$ = $[NH_3]$ has $[H_3O^+] = 5.56 \times 10^{-10}$. From solution calorimetry, it is determined that $\Delta H^\circ$ = -51.97 kJ for

$$NH_{3\,(aq)} + H^+_{(aq)} \rightarrow NH^+_{4\,(aq)}$$

Using **only** these data, find $\Delta S^\circ$ for this reaction in J/K.

(A) +28.3 (B) -2.71 (C) -0.352 (D) -13.6 (E) +2.79

**Questions 83-87 deal with the following reversible reaction:**

$$N_2O_{5\,(g)} \rightleftharpoons 2\,NO_{2\,(g)} + 1/2\,O_{2\,(g)}$$

| Substance | $\Delta \overline{H}^\circ_f$, kcal/mole | $\Delta \overline{G}^\circ_f$, kcal/mole | $\overline{S}^\circ$, cal/mole-K |
|---|---|---|---|
| (T = 298.15K) | | | |
| $N_2O_{5\,(g)}$ | 3.35 | 28.18 | 85.00 |
| $NO_{2\,(g)}$ | 8.09 | 12.39 | 57.47 |
| $O_{2\,(g)}$ | - | - | 49.00 |

83. What is $\Delta G^\circ_{rxn}$ at 298.15 K?

(A) -15.79 kcal (B) -11.36 kcal (C) -3.40 kcal (D) +16.18 kcal (E) +28.18 kcal

84. What is the approximate value of $K_p$ at 298 K?

(A) ~$3 \times 10^2$ (B) ~$2 \times 10^5$ (C) ~$1 \times 10^{-6}$ (D) ~10 (E) ~1

85. According to LeChatelier's principle, an increase in $T$ would result in all of the following EXCEPT:

(A) an increase in the value of $K$ (B) an increase in the value of $k$
(C) an increase in the equilibrium partial pressure of oxygen
(D) an increase in the % decomposition of $N_2O_5$
(E) an increase in the value of $E_a$

86. What is the approximate value of $K_p$ at 373 K?

(A) ~$2 \times 10^4$ (B) ~$3 \times 10^5$ (C) ~$4 \times 10^{-6}$ (D) ~0.1 (E) ~1

87. What are the units of $K_p$?

(A) atm (B) $atm^2$ (C) $atm^{3/2}$ (D) $atm^{-1/2}$ (E) $atm^{-1}$

88. Suppose that the equilibrium, $A_{(aq)} + 2\,B_{(aq)} \rightleftharpoons C_{(aq)}$, is established from the *original* concentrations: $[A]_0 = [B]_0 = [C]_0 = 0.100\ M$. At equilibrium it is found that $[C] = 0.040\ M$. What is the value of the equilibrium constant, $K_c$?

(A) 1.1 $M^{-2}$ (B) 5.2 $M^{-2}$ (C) 0.19 $M^{-2}$ (D) 0.88 $M^{-2}$ (E) 46 $M^{-2}$

89. If $K_c = 1.69$ for $H_{2\,(g)} + CO_{2\,(g)} \rightleftharpoons H_2O_{(g)} + CO_{(g)}$ at 990°C, then the concentration of CO in equilibrium with 1.00 $M$ $H_2$ and $CO_2$ and 0.65 $M$ $H_2O$ is:

(A) 2.6 $M$ (B) 0.65 $M$ (C) 1.3 $M$ (D) 1.7 $M$ (E) 0.38 $M$

90. For the reaction in Question 89, $\Delta H = +10$ kcal. Which of the following factors would **increase** the number of moles of CO at equilibrium?

(A) increasing the container volume at constant $T$
(B) increasing the pressure by halving the volume at constant $T$
(C) adding $H_2O_{(g)}$
(D) increasing $T$ to 1200°C
(E) removing some of the hydrogen from the mixture

91. The value of $K_c$ is $2.70 \times 10^{-3}$ for the reaction:

$$CaCO_{3\,(s)} \rightleftharpoons CaO_{(s)} + CO_{2\,(g)},\ \text{at } 800^\circ C$$

A 1.00-g sample of calcium carbonate is placed in each of three containers having volumes of 1 liter, 2 liters, and 4 liters and heated to 800°C. Which of the following statements is TRUE?

(A) The $[CO_2]$ is 0.0027 $M$ in all of the containers.

(B) The $[CO_2]$ is 0.0027 $M$ in the two smaller containers and larger than that in the 4-liter container.
(C) The $[CO_2]$ is 0.0027 $M$ in the two smaller containers but less than that in the 4-liter container.
(D) The $[CO_2]$ is 0.0027 $M$ only in the 1-liter container.
(E) The $[CO_2]$ is less than 0.0027 $M$ in all three containers.

92. For the vaporization of liquid water, $\Delta G°$ = +2.053 kcal/mole at 25°C. Find $K_p$ for $H_2O_{(\ell)} \rightleftarrows H_2O_{(g)}$ at this temperature ($R$ = 1.99 cal/mole-K).

(A) 0.0314 atm (B) 0.0384 atm (C) 0.0284 atm (D) 1.00 atm
(E) 0.284 atm

---

## ANSWERS TO CHAPTER 14 PROBLEMS

| | | | | | |
|---|---|---|---|---|---|
| 1. C | 17. A | 33. B | 49. C | 65. D | 81. A |
| 2. E | 18. D | 34. B | 50. B | 66. B | 82. E |
| 3. B | 19. C | 35. E | 51. C | 67. A | 83. C |
| 4. C | 20. B | 36. A | 52. D | 68. B | 84. A |
| 5. A | 21. A | 37. B | 53. D | 69. D | 85. E |
| 6. D | 22. A | 38. D | 54. D | 70. B | 86. A |
| 7. B | 23. E | 39. A | 55. E | 71. A | 87. C |
| 8. D | 24. B | 40. A | 56. A | 72. C | 88. B |
| 9. E | 25. A | 41. A | 57. D | 73. B | 89. A |
| 10. B | 26. B | 42. A | 58. A | 74. B | 90. D |
| 11. E | 27. E | 43. B | 59. D | 75. E | 91. C |
| 12. C | 28. E | 44. A | 60. E | 76. B | 92. A |
| 13. D | 29. B | 45. A | 61. B | 77. A | |
| 14. B | 30. C | 46. A | 62. B | 78. E | |
| 15. D | 31. C | 47. E | 63. A | 79. B | |
| 16. E | 32. A | 48. D | 64. C | 80. C | |

## DRILL ON FREE ENERGY CHANGE, $Q_p$ and $K_p$ *Student's Name* ______

A mixture originally containing $H_2$ and $N_2$, each at 5.00 atm, is introduced into a constant volume container. At constant temperature, T, the following reaction occurs:

$$N_{2(g)} + 3\ H_{2(g)} = 2\ NH_{3(g)}$$

Complete the following table for the reaction.

| Total Pressure at Equilibrium | $K_p$ | $\Delta G°$ | $\Delta G$ for Initially, $P_{N_2} = P_{H_2} = P_{NH_3} = 1.00$ atm | $\Delta G$ for Initially $P_{N_2} = P_{H_2} = 10.00$ atm; $P_{NH_3} = 0.10$ atm |
|---|---|---|---|---|
| 9.78 atm | | | | |
| 9.00 atm | | | | |
| 8.20 atm | | | | |
| ______ | | | | |

# 15
# Acid-Base Reactions and Equilibria

## BRØNSTED-LOWRY ACIDS AND BASES

Although there are a variety of definitions (*Arrhenius, Brønsted-Lowry, solvent system, etc.*) that narrow, or expand, the terms *acid* and *base*, the *Brønsted-Lowry* (B-L) *definition* is the most easily understood and the most useful when dealing with aqueous or other *protonic solvents* (*e.g.*, $NH_{3(\ell)}$, $CH_3COOH_{(\ell)}$, $H_2SO_{4(\ell)}$, $HCl_{(\ell)}$). A Brønsted acid is a species (ion or molecule) that donates a proton (hence, must contain hydrogen); a Brønsted base is a species (ion or molecule) that accepts a proton (hence, must have an unshared electron pair); and the B-L acid-base reaction is the transfer of a proton ($H^+$) from an acid (proton donor) to a base (proton acceptor). The labeling of a substance as "acid" or "base" depends not so much on the inherent nature of the species, but on **how this species behaves in the particular reaction**. Thus, $CH_3COOH$ may accept a proton to form the ion, $CH_3COOH_2^+$, or under different conditions, it may give up a proton to form the acetate ion, $CH_3COO^-$; in the former case $CH_3COOH$ would be termed a base, in the latter case an acid.

An acid, upon losing a proton, always forms a species having more pronounced basicity (*i.e.*, richer in electrons, since an additional unshared pair of electrons is available for combination with a proton); this species is termed the *conjugate base*. A base, upon gaining a proton, always forms a species having more pronounced acidic properties, its *conjugate acid*. The **products of a B-L acid-base reaction are always a new acid and a new base**, the respective conjugates, *e.g.*,

$$CH_3COOH + OH^- \rightarrow CH_3COO^- + H_2O$$

| $CH_3COOH$ | $OH^-$ | $CH_3COO^-$ | $H_2O$ |
|---|---|---|---|
| Acid | Base | Conjugate base of $CH_3COOH$ | Conjugate acid of $OH^-$ |

In the above example $CH_3COOH$ and $CH_3COO^-$ constitute a *conjugate acid-base pair,* as do $H_2O$ and $OH^-$. The species making up a conjugate pair always differ by **one** proton.

Many species can act either as an acid or as a base, depending on the particular environment. Thus, water behaves as a base in accepting a proton to form $H_3O^+$ but, under other conditions, water can act as an acid by losing a proton to form $OH^-$. Such behavior is referred to as *amphiprotic* or *amphoteric,* and is exhibited by numerous species, *e.g.*, $HCO_3^-$, $HS^-$, $NH_3$.

Protons are more easily removed from some acids than others (*strong vs. weak*); protons are more readily accepted by some bases than others (again, *strong vs. weak*). Ordering of *strengths* of acids (or bases) is accomplished by measuring the extent of reaction with respect to a reference; this most frequently is water. An acid which when added to water completely loses its proton is said to be a *strong acid.* Common examples, which should be remembered, are $HClO_4$, HI, BHr, $H_2SO_4$ (first proton only), HCl, and $HNO_3$. For other acids, *weak acids,* the reaction

$$HA + H_2O \rightarrow A^- + H_3O^+$$

is not complete and, hence, the equilibrium constant,

$$K_a = \frac{[H_3O^+][A^-]}{[HA]}$$

can be written, which indicates the strength of the acid; the larger the value of $K_a$, the greater is the strength of the acid. Similarly, for a base, if the reaction with water is complete, the base is said to be a *strong base.* Some common examples, which also should be remembered, are $CH_3O^-$, $O^{2-}$, $NH_2^-$, $N^{3-}$, and $H^-$. For other bases, *weak bases,* the reaction,

$$B + H_2O \rightarrow BH^+ + OH^-$$

or for an anionic weak base,

$$C^- + H_2O \rightarrow HC + OH^-$$

is not complete, and again an equilibrium constant of the type,

$$K_b = \frac{[BH^+][OH^-]}{[B]}$$

can be written; the larger the value of $K_b$, the greater the strength of the base. It is axiomatic that for a conjugate acid-base pair, the stronger one of the components, the weaker the other. The reaction with water does not differentiate between various strong acids (or strong bases), since they all react completely to give the strongest acid $H_3O^+$ (or base, $OH^-$) that can exist in water. The strong acids (or bases) are not of equal strength but, because of their complete reaction with water in aqueous solutions, have been *leveled* to the strength of $H_3O^+$ (or $OH^-$). By using a different solvent (*e.g.*, acetic acid for acids, liquid ammonia for bases), ordering of strong acids (or bases) may be obtained. The accompanying acid-base table is a list of common acids in order of decreasing strengths (note $K_a$ values for aqueous solutions) along with their conjugate bases in order of increasing base strengths. The extent of an acid-base reaction is governed by the relative strengths of the acid and base formed; thus

$$H_2S_{(aq)} + CN^- \rightleftharpoons HCN_{(aq)} + HS^- \qquad \text{(large extent, } K_c > 1\text{)}$$

but

$$HS^- + CN^- \rightleftharpoons HCN_{(aq)} + S^{2-} \qquad \text{(small extent, } K_c < 1\text{)}$$

In the first case, the product acid, HCN ($K_a = 4.0 \times 10^{-10}$), is weaker than the reactant acid, $H_2S$ ($K_a = 1.0 \times 10^{-7}$), as is the product base, $HS^-$ ($K_b = 1.0 \times 10^{-7}$) when compared to the starting base, $CN^-$ ($K_b = 2.5 \times 10^{-5}$). Such is not the case in the second example: $K_a$ for HCN > $K_a$ for $HS^-$, and $K_b$ for $S^{2-}$ > $K_b$ for $CN^-$. A general rule for the use of the acid-base table is that $K_c$ will be greater than one if the position of the reactant acid is above the position of the reactant base on the table. Quantitatively, $K_c$ for the first example equals $K_{a(H_2S)}/K_{a(HCN)} = 2.5 \times 10^2$, a reaction that occurs to a large extent; for the second example, $K_c = 3.2 \times 10^{-4}$, a reaction that proceeds only to a small extent. The acid-base table is, thus, a convenient tool in helping to write acid-base equations and determining the extent to which a given reaction will proceed.

The following steps need be considered in writing the equation for an acid-base reaction, *e.g.*, the reaction of a sodium sulfate solution with hydrochloric acid:

1. Write down the species that are present to an appreciable extent and that have acid-base properties. Of our potential reactants $Na^+$ and $Cl^-$ can be dismissed, since as acids or bases they are weaker than the solvent, here water. In the sodium sulfate solution, we need consider $SO_4^{2-}$ and $H_2O$; in hydrochloric acid, $H_3O^+$ and $H_2O$.

2. Allow the best acid(s) to react with the best base(s) [*i.e.*, transfer a proton(s)] without allowing the formation of acids (or bases) too strong to exist in the solvent. [That is, in water the product acid(s) cannot be stronger than $H_3O^+$; the product base(s) cannot be stronger than $OH^-$.] In our reaction, the best acid is $H_3O^+$, second best is $H_2O$; the best base is $SO_4^{2-}$, second best is $H_2O$. Hence, upon transfer of one proton, one can write

$$H_3O^+ + SO_4^{2-} \rightarrow HSO_4^- + H_2O$$
$$H_3O^+ + H_2O \rightarrow H_3O^+ + H_2O \qquad \text{(no net reaction)}$$

## RELATIVE STRENGTHS OF ACIDS AND BASES

Acid Strength

| Conjugate Acid | | | Conjugate Base | |
|---|---|---|---|---|
| Name | Formula | $K_a$ ¶ | Name | Formula |
| Perchloric acid$_{(\ell)}$ | $HClO_4$ | $\sim 10^{10}$ | Perchlorate ion | $ClO_4^-$ |
| Hydrogen iodide$_{(g)}$ | $HI$ | $\sim 10^{10}$ | Iodide ion | $I^-$ |
| Hydrogen bromide$_{(g)}$ | $HBr$ | $\sim 10^{9}$ | Bromide ion | $Br^-$ |
| Hydrogen chloride$_{(g)}$ | $HCl$ | $\sim 10^{7}$ | Chloride ion | $Cl^-$ |
| Sulfuric acid$_{(\ell)}$ | $H_2SO_4$ | $\sim 10^{3}$ | Bisulfate ion* | $HSO_4^-$ |
| Nitric acid$_{(\ell)}$ | $HNO_3$ | $\sim 10^{2}$ | Nitrate ion | $NO_3^-$ |
| Hydronium ion | $H_3^+O$ or $(H^+_{(aq)})$ | 56 § | Water | $H_2O$ |
| Bisulfate ion* | $HSO_4^-$ | $1.2 \times 10^{-2}$ | Sulfate ion | $SO_4^{2-}$ |
| Sulfurous acid | $H_2SO_3$ | $1.2 \times 10^{-2}$ | Bisulfite ion* | $HSO_3^-$ |
| Phosphoric acid | $H_3PO_4$ | $7.5 \times 10^{-3}$ | Dihydrogenphosphate ion | $H_2PO_4^-$ |
| Hexaquoiron(III) ion | $Fe(H_2O)_6^{3+}$ | $9 \times 10^{-4}$ | | $Fe(H_2O)_5OH^{2+}$ |
| Hydrofluoric acid | $HF$ | $7.2 \times 10^{-4}$ | Fluoride ion | $F^-$ |
| Nitrous acid | $HNO_2$ | $4.5 \times 10^{-4}$ | Nitrite ion | $NO_2^-$ |
| Acetic acid | $HC_2H_3O_2$ | $1.8 \times 10^{-5}$ | Acetate ion | $C_2H_3O_2^-$ |
| Hexaquoaluminum ion | $Al(H_2O)_6^{3+}$ | $7 \times 10^{-6}$ | | $Al(H_2O)_5OH^{2+}$ |
| Carbonic acid | $H_2CO_3$ | $4.3 \times 10^{-7}$ $(2.6 \times 10^{-4})$† | Bicarbonate ion* | $HCO_3^-$ |
| Bisulfite ion* | $HSO_3^-$ | $2.8 \times 10^{-7}$ | Sulfite ion | $SO_3^{2-}$ |
| Hydrogen sulfide | $H_2S$ | $1.0 \times 10^{-7}$ | Hydrosulfide ion | $HS^-$ |
| Dihydrogenphosphate ion | $H_2PO_4^-$ | $6.2 \times 10^{-8}$ | Monohydrogenphosphate ion | $HPO_4^{2-}$ |

Increasing

| Decreasing | | | | |
|---|---|---|---|---|
| Phenolphthalein | HPhth | $3 \times 10^{-9}$ | Phenolphthalein anion | $Phth^-$ |
| Ammonium ion | $NH_4^+$ | $5.6 \times 10^{-10}$ | Ammonia | $NH_3$ |
| Hydrocyanic acid | HCN | $4.0 \times 10^{-10}$ | Cyanide ion | $CN^-$ |
| Hexaquoiron(II) ion | $Fe(H_2O)_6^{2+}$ | $3 \times 10^{-10}$ | | $Fe(H_2O)_5OH^+$ |
| Bicarbonate ion* | $HCO_3^-$ | $4.7 \times 10^{-11}$ | Carbonate ion | $CO_3^{2-}$ |
| Hydrogen peroxide | $H_2O_2$ | $2.4 \times 10^{-12}$ | Hydroperoxide ion | $HO_2^-$ |
| Monohydrogenphosphate ion | $HPO_4^{2-}$ | $1.0 \times 10^{-12}$ | Phosphate ion | $PO_4^{3-}$ |
| Hydrosulfide ion | $HS^-$ | $1.3 \times 10^{-13}$ | Sulfide ion | $S^{2-}$ |
| Water | $H_2O$ | $1.8 \times 10^{-16}$ § | Hydroxide ion | $OH^-$ |
| Methanol | $CH_3OH$ | $\sim 10^{-16}$ | Methoxide ion | $CH_3O^-$ |
| Ammonia | $NH_3$ | $\sim 10^{-35}$ | Amide ion | $NH_2^-$ |
| Hydroxide ion | $OH^-$ | $\sim 10^{-36}$ | Oxide ion | $O^{2-}$ |
| Hydrogen | $H_2$ | $\sim 10^{-36}$ | Hydride ion | $H^-$ |
| Amide | $NH_2^-$ | very weak | Imide ion | $NH^{2-}$ |
| Imide ion | $NH^{2-}$ | very weak | Nitride ion | $N^{3-}$ |
| Methane | $CH_4$ | $\sim 10^{-58}$ | Methide | $CH_3^-$ |

Base Strength

*Accepted systematic nomenclature of ions derived from diprotic acids upon removal of a single proton calls for attaching the word "hydrogen" directly to the name of the parent ion, *e.g.*, $HCO_3^-$ would be the hydrogencarbonate ion. Commonly, however, the prefix "bi" is used to denote this single hydrogen; hence $HCO_3^-$ more usually is called the bicarbonate ion.

¶ The $K_a$ values given here for strong acids and strong bases are in most instances gross estimates and may be in error by several orders of magnitude. Indeed, values for a number of these species are the subject of considerable debate.

†The $K_a$ value of $4.3 \times 10^{-7}$ is based on the total $CO_2$ in the solution. Most of this is present as dissolved $CO_2$ rather than $H_2CO_3$ molecules. $H_2CO_3$ is a much stronger acid than indicated by this value. The value in parentheses, $2.6 \times 10^{-4}$, is the $K_a$ calculated on the basis of the actual concentration of molecular $H_2CO_3$.

§ The equilibrium, $2\,H_2O \rightleftharpoons H_3O^+ + OH^-$, exists in all aqueous solutions and its equilibrium constant, usually called the *ion product* ($K_w$), is $1.0 \times 10^{-14}$ at 25°C. The product of these two numbers is $K_w$; they were obtained by taking into account that $[H_2O]$ is 55.5 moles/liter.

$H_2O + SO_4^{2-} \rightarrow HSO_4^- + OH^-$ (but $OH^-$ cannot exist in appreciable quantity in a solution containing $H_3O^+$)

or transferring two protons, one might try

$2\ H_3O^+ + SO_4^{2-} \rightarrow H_2SO_4 + 2\ H_2O$ (but an acid is formed, $H_2SO_4$, which is too strong to exist in water)

Hence, the only important reaction is the first one, which would proceed to a large extent since the product acid and base, respectively, are weaker than the reactant acid and base. It would be well worthwhile to review equation writing for acid-base reactions in Chapter 7 before proceeding with this section. The following Table, "Relative Strengths of Acids and Bases", is similar to that in Chapter 7. It now includes $K_a$ values that will be used in the problems in this chapter.

## AQUEOUS SOLUTIONS OF STRONG ACIDS AND BASES–THE pH SCALE

Water is essentially a molecular substance, but even in pure water ionization occurs to a very small extent, yielding $H_3O^+$ (sometimes abbreviated, $H^+$) and $OH^-$ ions, *i.e.*,

$$2\ H_2O \rightleftharpoons H_3O^+ + OH^- \quad \text{or} \quad H_2O \rightleftharpoons H^+ + OH^-$$

The usual expression,

$$K_{eq} = \frac{[H_3O^+][OH^-]}{[H_2O]^2} \quad \text{or} \quad K_{eq} = \frac{[H^+][OH^-]}{[H_2O]}$$

(where the brackets [] are read as the molar concentration of the species in the brackets) may be set up for the equilibrium, but since in dilute solution the concentration of water is essentially constant, 1000 g/L ÷ 18.0 g/mole = 55.5 mole/L, the denominator may be incorporated with $K_{eq}$, resulting in the more commonly used *ion product,* $K_w$:

$$K_w = [H_3O^+][OH^-] \qquad \textbf{[Eqn. 1]}$$

At 25°C, $K_w = 1.0 \times 10^{-14}$ and, hence, the $[H_3O^+] = [OH^-] = 1.0 \times 10^{-7}$ *M*. Both $K_{eq}$ and $K_w$, like other equilibrium constants, are temperature dependent (at 100°C, $K_w = 1.0 \times 10^{-13}$ and $[H_3O^+] = [OH^-] = 3.2 \times 10^{-7}$ *M*). Pure water, or an aqueous solution in which $[H_3O^+] = [OH^-]$, is said to be *neutral.*

Addition of a soluble acid or base to water disturbs the $K_w$ equilibrium; acids cause an increase in $[H_3O^+]$, bases cause an increase in $[OH^-]$. In either case neutrality is destroyed, and the resulting solution is said to be either acidic or basic. Very small concentrations of $H_3O^+$ and/or $OH^-$ at times play important roles in chemistry, which means that it is necessary to express very wide concentration ranges of these ions. This is commonly done by defining the logarithmic (to the base 10) functions pH and pOH:†

$$pH = -\log[H_3O^+] = \log\frac{1}{[H_3O^+]}, \text{ or } [H_3O^+] = 10^{-pH} \qquad \text{[Eqn. 2]}$$

$$pOH = -\log[OH^-] = \log\frac{1}{[OH^-]}, \text{ or } [OH^-] = 10^{-pOH} \qquad \text{[Eqn. 3]}$$

The above functions are related to each other through Eqn. 1:

$$1.0 \times 10^{-14} = [H_3O^+][OH^-] \text{ at } 25°C,$$

Taking negative logs of both sides gives

$$14 = -\log[H_3O^+] - \log[OH^-] = pH + pOH \qquad \text{[Eqn. 4]}$$

If any one of the quantities, $[H_3O^+]$, $[OH^-]$, pH, or pOH, is known for any aqueous solution at room temperature (25°C), the remaining three values may readily be calculated by using the relationships given in Eqns. 1 to 4. Although pH and pOH values commonly range between 0 and 14, negative values occur in very acidic solutions, values greater than 14 in very basic solutions.

A few of the common acids and bases are strong (Ch. 14) and the $[H_3O^+]$, $[OH^-]$, pH, and pOH can be directly related to their molar concentrations. (For example, for 0.10 *M* $Ba(OH)_2$, $[OH^-] = 0.20$ *M*; pOH = $-\log 0.20$ = 0.70; pH = 14.00 − 0.70 = 13.30, and $[H_3O^+] = K_w/[OH^-] = 5.0 \times 10^{-14}$ *M*.) Most acids and bases are weak (*i.e.*, less than 100% dissociated) and the $[H_3O^+]$ and $[OH^-]$ must be calculated using the equilibrium constants for the acid ($K_a$) or base ($K_b$) ionizations rather than directly from the "makeup" concentration (*C*) of the acid or base solute. The acid-base table (p. 254) gives the extent of acid ionization in water by the tabulated $K_a$ value. For the conjugate base of a given acid, $K_b$ is readily calculated from the expression $K_aK_b = K_w$.

---

†See pp. 444 for a log table and brief description of the evaluation of logarithms.

## Example Problems Involving Aqueous Solutions of Strong Acids and Bases—The pH Scale

### 15-A. Hydronium ion Concentration, pH and pOH of a Solution of a Strong Acid

In a hydrochloric acid solution with a pH = 4.00 at 25°C, the concentration of hydronium ion is ______ times greater than the hydroxide ion concentration; the pOH is ______ times the pH. Fill the blanks, respectively.

(A) ten thousand; 2.00 (B) one thousand; 3.00 (C) one million; 2.5
(D) ten million; 1.43 (E) one billion; 1.75

#### *Solution*

If pH = 4.00, $[H_3O^+] = 10^{-4.00}$ molar ($1.0 \times 10^{-4}$ *M*); then by way of $K_w$,

$$[OH^-] = \frac{K_w}{[OH^-]} = 10^{-10.00}\ M.$$

Thus, pOH = $-\log[OH^-]$ = 10.00, and $[H^+]/[OH^-] = 1.00 \times 10^{-4}/1.00 \times 10^{-10} = 1.00 \times 10^6$ (one million); likewise, pOH/pH = 10.00/4.00 = 2.50–the pOH is 2.5 times greater numerically than the pH. **[C]**

### 15-B. Reaction of Strong Acid with Strong Base

If **one liter** of nitric acid solution (pH = 1.000) is mixed with **nine liters** of barium hydroxide solution (pH = 11.000), what will be the pH of the resulting solution?

(A) 2.04 (B) 3.55 (C) 7.30 (D) 8.36 (E) 8.98

#### *Solution*

Nitric acid is a strong acid, and barium hydroxide is a strong base. If the pH of the acid is 1.000, this means that the solution contains $[H_3O^+]$ at $10^{-1.000}$ *M* (0.100 *M* $HNO_3$). Likewise, if the barium hydroxide solution has the pH = 11.000, this means that pOH = 3.000 and $[OH^-] = 10^{-3.000}$ *M* ($5.00 \times 10^{-4}$ *M* $Ba(OH)_2$).

The reaction of strong acid ($H_3O^+$) with strong base ($OH^-$) in solution is essentially quantitative:

$$H_3O^+ + OH^- = 2\,H_2O \quad \text{or equivalently,} \quad H^+ + OH^- = H_2O$$

(Note that the equilibrium constant for this reaction is $10^{+14}$, the inverse of the $K_w$ equilibrium constant.) Thus we can say:

| | |
|---|---|
| One liter of $HNO_3$ solution provides: | 0.100 moles of $H^+$ |
| Nine liters of $Ba(OH)_2$ solution provide: | 0.00900 moles of $OH^-$ |

The result is that ten liters of solution are formed having 0.091 moles of excess $H^+$. Thus,

$$[H^+] = \frac{0.091 \text{ moles}}{10 \text{ liters}} = 9.1 \times 10^{-3}\ M$$

The pH is $-\log(9.1 \times 10^{-3}) = 3 - \log 9.1 = 3 - 0.9590414 = 2.0409586$. Since the moles of excess $H^+$ is good only to two significant figures, the pH must be rounded off accordingly, *i.e.*, pH = 2.04. One further point: students often worry about the water formed by the neutralization reaction, and how it affects the final volume of solution. In this case, since $OH^-$ is the limiting reagent, 0.00900 moles of water have been formed. This amounts to a volume of:

$$(0.00900 \text{ moles})(18.0 \text{ g/mole})(1 \text{ mL/g}) = 0.162 \text{ mL of water formed}$$

This is entirely insignificant in comparison to the ten liters of solution—such is the case whenever dilute solutions of acids and bases react.

## Exercises on Aqueous Solutions on Strong Acids and Bases—The pH Scale

1. The $[H_3O^+]$ in a 0.050 *M* solution of $Ba(OH)_2$ is:

   (A) $1.0 \times 10^{-5}\ M$ (B) $5.0 \times 10^{-2}\ M$ (C) $1.0 \times 10^{-13}\ M$
   (D) $5.0 \times 10^{-10}\ M$ (E) $2.0 \times 10^{-5}\ M$

2. The $[OH^-]$ in a solution of pH = 7.80 is:

   (A) $1.6 \times 10^{-8}\ M$ (B) $7.0 \times 10^{-8}\ M$ (C) $2.5 \times 10^{-6}\ M$
   (D) $9.4 \times 10^{-6}\ M$ (E) $6.3 \times 10^{-7}\ M$

3. A typical fresh egg white will have a pH of 7.80. This corresponds to a:

   (A) $[H_3O^+]$ of $8.0 \times 10^{-7}$ and $[OH^-]$ of $1.3 \times 10^{-8}$
   (B) $[H_3O^+]$ of $7.0 \times 10^{-8}$ and $[OH^-]$ of $1.4 \times 10^{-7}$
   (C) $[H_3O^+]$ of $8.5 \times 10^{-7}$ and $[OH^-]$ of $5.5 \times 10^{-7}$
   (D) $[H_3O^+]$ of $3.0 \times 10^{-8}$ and $[OH^-]$ of $3.3 \times 10^{-7}$
   (E) $[H_3O^+]$ of $1.6 \times 10^{-8}$ and $[OH^-]$ of $6.3 \times 10^{-7}$

4. Calculate the $[H_3O^+]$, $[OH^-]$, pH, and pOH of 0.01 *M* HCl:

| | $[H_3O^+]$ | $[OH^-]$ | pH | pOH |
|---|---|---|---|---|
| (A) | 0.01 | 0.01 | 1 | 1 |
| (B) | 0.02 | 0.1 | 2 | 12 |
| (C) | $10^{-2}$ | $10^{-12}$ | 2 | 12 |
| (D) | $10^{-1}$ | $10^{-13}$ | 1 | 13 |
| (E) | 0.01 | 0.1 | 0.01 | 0.12 |

5. Mixing 50. mL of 0.010 *M* $Ba(OH)_2$ and 150. mL of 0.010 *M* $HNO_3$ will result in a solution with a pH of ______.

   (A) 2.30 (B) 11.70 (C) 13.70 (D) 0.30 (E) 2.60

6. What is the pH of 6.0 *M* HCl?

   (A) -0.78 (B) -0.22 (C) 0.00 (D) 0.22 (E) 0.78

7. To what volume must 20.0 mL of 1.00 *M* HCl be diluted with water to give a solution that has a pH of 1.30?

   (A) 0.20 L (B) 0.25 L (C) 0.40 L (D) 0.50 L (E) none of these

8. What is the pH of the solution obtained by mixing 10. mL of 1.0 *M* HCl with 75 mL of water and 15 mL of 1.0 *M* NaOH?

   (A) 11.3 (B) 1.7 (C) 2.0 (D) 12.7 (E) 7.0

---

## WEAK MONOPROTIC ACIDS AND BASES

It frequently is desired to calculate for a weak acid (or base) solution the pH, pOH, and the actual concentrations of the various species in solution from the makeup concentration (C) and $K_a$ (or $K_b$). One **must begin with the chemical equation to which $K_a$ (or $K_b$) refers.** Concentrations are then determined in a two-step approach: (1) the weak acid (or base) is taken as being undissociated and at its makeup concentration; and then (2) the change, Δ, is considered in which water reacts with "$x$" moles/L of the weak acid (or base), leaving the equilibrium (EQ) concentrations of the weak acid (or base) and of the corresponding conjugate species. This is conveniently done by setting up a table under the equation for the most important equilibrium occurring in a given solution. Thus, for an aqueous solution of a weak monoprotic acid, HA, the equilibrium reaction is:

| | | | |
|---|---|---|---|
| | $HA + H_2O$ ⇌ | $H_3O^+$ | $+ A^-$ |
| or equivalently, | $HA$ ⇌ | $H^+_{(aq)}$ | $+ A^-$ |
| Makeup ($C$), molar: | $C$ | 0 | 0 |
| Change (Δ), molar: | $-x$ | $x$ | $x$ |
| Equilibrium (EQ), molar: | $C-x$ | $x$ | $x$ |

Then: $K_a = \dfrac{x^2}{C-x} \cong \dfrac{x^2}{C}$ (if $x << C$).

Thus, given $K_a$ and $C$, $[H_3O^+] = [A^-] = x$, $[HA] = C-x$, $[OH^-] = K_w/[H_3O^+]$, and pH and pOH may be calculated by simple substitution.

For a solution containing only a weak base (B) the principal equilibrium is

$$B + H_2O \rightleftharpoons BH^+ + OH^-$$

for which $K_b = [BH^+][OH^-]/[B]$. As in the previous case, the equilibrium concentrations of $BH^+$ and $OH^-$ may be defined as "$x$" and that of B as $(C–x)$, with $C$ being the formal, "makeup", concentration of B. Then $K_b = x^2/(C–x)$, which again may be approximated as $K_b \cong x^2/C$, if $x << C$.

In both of the above setups, it has been assumed that the $[H^+]$ (or $[OH^-]$) in solution has its origin in HA (or B), *i.e.*, the contribution from the ionization of water has been neglected. Since the acids (or bases) normally encountered are stronger acids (or bases) than water itself, Le Chatelier's principle predicts that the ionization of water will be repressed (*common-ion effect*); this assumption is valid when $K_a$ or $K_b$ is several orders of magnitude greater than $K_w$ and when $C >> 10^{-7}$. Also, the above calculations involved finding a real solution to a quadratic equation; however, in many cases the calculation may be simplified by using the indicated approximation, but each time the validity of the approximation **must** be checked. There are two ways of doing this, *i.e.*, finding if $(C–x) \cong C$: (a) make the approximation and check the value of "$x$" to determine if "$x$" is indeed negligible when subtracted from $C$ (to the correct number of significant figures); or (b) first evaluate the ratio, $K/C$–if this ratio is $\leqslant 10^{-3}$, "$x$" can be neglected for almost every $C$ normally encountered.

## Hydrolysis

The term hydrolysis is sometimes used to refer to the reaction of an ionic weak acid (or base) with water and the equilibrium constant for such a reaction is designated as $K_h$. $K_h$ is exactly the same as $K_a$ for a cationic weak acid (*e.g.*, $NH_4^+$) or $K_b$ for an anionic weak base (*e.g.*, $C_2H_3O_2^-$, $SO_4^{2-}$, $F^-$).

## Example Problems Involving Weak Monoprotic Acids and Bases

### 15-C. The pH of a Solution of a Weak Acid–Use of the Quadratic Equation

The pH of a 0.100 $M$ aqueous solution of sodium bisulfate is:

(A) 1.92 (B) 1.15 (C) 1.54 (D) 1.00 (E) 1.46

*Solution*

The solution contains $H_2O, Na^+$, and $HSO_4^-$ along with relatively small amounts of $H_3O^+$ and $SO_4^{2-}$, and a **minute** amount of $OH^-$. The $SO_4^{2-}$ and essentially all of the $H_3O^+$ arise from the dissociation of the weak acid $HSO_4^-$ ($K_a = 1.2 \times 10^{-2}$, see acid-base table). This, then, is the predominant equilibrium and the $K_a$ equation and table of concentrations should be set up:

| | $HSO_4^-$ | $+ H_2O \rightleftharpoons$ | $H_3O^+$ | $+ SO_4^{2-}$ |
|---|---|---|---|---|
| or alternately | $HSO_4^-$ | $\rightleftharpoons$ | $H^+_{(aq)}$ | $+ SO_4^{2-}$ |
| Makeup, (*C*), *M*: | 0.100 | | 0 | 0 |
| $\Delta$, *M*: | $-x$ | | $x$ | $x$ |
| EQ, *M*: | $0.100 - x$ | | $x$ | $x$ |

$$K_a = 1.2 \times 10^{-2} = \frac{[H_3O^+][SO_4^{2-}]}{[HSO_4^-]} = \frac{x^2}{(0.100-x)}$$

To find pH, the $[H_3O^+]$ (*i.e.*, $x$) needs to be found. This involves a quadratic equation, the solution of which can be avoided if $x \ll 0.100$. To test if the approximation is valid, we can go ahead and assume that the right-hand term is $\cong x^2/0.100$. This yields (to **two** significant figures):

$$x = [H_3O^+] = [SO_4^{2-}] = 0.035 \quad \text{and} \quad \text{pH} = 1.46 \qquad \textbf{[E]}$$

However, we must inspect to see if the approximation was justified, *i.e.*, $[HSO_4^-] = 0.100 - x \cong 0.100$. $x$ is 0.035 and **certainly** is **not** negligible when compared to 0.100; thus the quadratic must be solved. (Alternatively, the test $K_a/C = 0.12 \gg 10^{-3}$ would have led to the same conclusion.) Then, from the original setup,

$$x^2 + (1.2 \times 10^{-2})x - (1.2 \times 10^{-3}) = 0$$

for which

$$x = \frac{-(1.2 \times 10^{-2}) \pm \sqrt{(1.2 \times 10^{-2})^2 + 4(1.2 \times 10^{-3})}}{2}$$

$$x = -0.041 \quad \text{and} \quad x = +0.029$$

Since the first of these roots has no chemical significance ($[H_3O^+]$ cannot be negative), $x = 0.029$ and (again to two significant figures),

$$[H_3O^+] = [SO_4^{2-}] = 0.029\,M; \text{pH} = 1.54$$
$$[HSO_4^-] = 0.100 - x = 0.071\,M$$

(Note that, by indiscriminately making the approximation, the $[H_3O^+]$ would be too high by 21%. Note also that conservation of sulfur atoms requires that the total concentration of sulfur-containing species be conserved; *i.e.*, $[HSO_4^-] + [SO_4^{2-}] = 0.100$.)

## 15-D. The pH of a Solution Containing a Weak Anionic Base

The pH of a 0.10 *M* solution of barium acetate is:

(A) 4.96 (B) 10.25 (C) 8.87 (D) 9.04 (E) 9.25

### *Solution*

The principal species in solution are $H_2O$, $Ba^{2+}$, and $C_2H_3O_2^-$. Of these, $Ba^{2+}$ has no acid-base properties; $H_2O$ has only **extremely** weak acid-base properties; and $C_2H_3O_2^-$, although categorized as weak, is a base of considerable repute (see acid-base table, $K_b = K_w/1.8 \times 10^{-5}$). The acetate equilibrium is the predominant equilibrium and hence determines the $H_3O^+$ and $OH^-$ concentrations in the solution. So, the $K_b$ reaction for acetate ion is written along with the concentrations of involved species:

$$C_2H_3O_2^- + H_2O \rightleftharpoons HC_2H_3O_2 + OH^-$$

| | $C_2H_3O_2^-$ | $HC_2H_3O_2$ | $OH^-$ |
|---|---|---|---|
| Makeup ($C$), $M$: | 0.20 | 0 | 0 |
| $\Delta$, $M$: | $-x$ | $x$ | $x$ |
| EQ, $M$: | $0.20 - x$ | $x$ | $x$ |

$$K_b = \frac{1.0 \times 10^{-14}}{1.8 \times 10^{-5}} = 5.6 \times 10^{-10} = \frac{[HC_2H_3O_2][OH^-]}{[C_2H_3O_2^-]}$$

$$= \frac{x^2}{(0.20-x)}$$

(Note that 0.10 mole/L $Ba(C_2H_3O_2)_2$ provides a makeup concentration of 0.20 $M$ acetate ion.) If we make the approximation that

$x << 0.20$–note $K_b/C \cong 10^{-9} << 10^{-3}$, $K_b \cong x^2/0.20$

$$x \cong [OH^-] = [HC_2H_3O_2] = 1.1 \times 10^{-5}\ M$$

$(0.20-x) = [C_2H_3O_2^-] = (0.20 - 0.000011) = 0.20\ M$

Since $[H_3O^+] = K_w/[OH^-] = 9.1 \times 10^{-10}$; pH = 9.04 **[D]**

Alternatively, pOH = $-\log(1.1 \times 10^{-5}) = 4.96$

and then pH = 14.00 - 4.96 = 9.04. **[D]**

## Exercises on Weak Monoprotic Acids and Bases

9. The term "$K_a$ for the ammonium ion" refers **directly** to:

   (A) $NH_3 + H_2O \rightleftharpoons NH_4^+ + OH^-$ (B) $NH_4^+ + H_2O \rightleftharpoons NH_3 + H_3O^+$
   (C) $NH_3 + H_3O^+ \rightleftharpoons NH_4^+ + H_2O$ (D) $NH_4^+ + OH^- \rightleftharpoons NH_3 + H_2O$
   (E) none of these

10. If enough of each of the following compounds were dissolved in water to give a 1.00 $M$ solution of each, which would have the **highest** pH?

    (A) HCl (B) NaCl (C) $NaC_2H_3O_2$ (D) NaF (E) $NH_4Cl$

11. Which of the following solutions has the smallest $[H_3O^+]$?

    (A) 0.10 $M$ HCl (B) 0.10 $M$ $H_2SO_4$ (C) 0.010 $M$ $HNO_3$
    (D) 0.10 $M$ $HC_2H_3O_2$ (E) 0.10 $M$ $NH_4Cl$

12. The pH of 0.50 $M$ HCN is:

    (A) between 4.5 and 5.0 (B) between 3.5 and 4.5
    (C) between 5.0 and 5.5 (D) between 5.5 and 6.0
    (E) between 9.0 and 9.5

13. Which one of the following solutions will have the **lowest** hydronium ion concentration?

    (A) 0.15 $M$ NaF (B) 0.15 $M$ $HC_2H_3O_2$ (C) 0.001 $M$ HCl
    (D) 0.10 $M$ $HNO_2$ (E) 0.22 $M$ NaCl

14. Which of the following, when added to water, will **not** change the pH?

    (A) $NaHCO_3$ (B) $NH_4Cl$ (C) KCN (D) KCl (E) any of these

15. A weak acid, HA, is 1.0% ionized in a 0.010 $M$ solution. $K_a$ is:

    (A) $1.0 \times 10^{-6}$ (B) $1.0 \times 10^{-4}$ (C) $1.0 \times 10^{-3}$
    (D) $1.0 \times 10^{-2}$ (E) 0.010

16. What is the pH of a 0.10 $M$ $Ca(C_2H_3O_2)_2$ solution?

    (A) 8.87 (B) 5.13 (C) 4.98 (D) 10.25 (E) 9.02

17. How many moles of $NaHSO_4$ are required to prepare 1.0 liter of a solution with the pH = 1.30?

    (A) 0.26 (B) 0.050 (C) 0.025 (D) 0.42 (E) 0.21

18. An 0.50 $M$ solution of acetic acid has a pH of 2.52, corresponding to an acetate/acetic acid molecule mole ratio of 6:1000. If solid NaOH were added to 0.50 $M$ $HC_2H_3O_2$ until the pH became 5.00, this acetate/acetic acid mole ratio would be increased **by a factor** of:

    (A) 1000 (B) 500 (C) 300 (D) 50 (E) 20,000

**For Questions 19–21 consider the following five solutions sitting on a lab bench at 25°C:**

A. 0.10 $M$ ammonium fluoride

B. 0.10 $M$ ammonium cyanide

C. 0.10 $M$ ammonium chloride

D. 0.10 $M$ ammonium bisulfate

E. 0.10 $M$ ammonium acetate

19. Which solution has the HIGHEST pH?

    (A) A (B) B (C) C (D) D (E) E

20. Which solution(s) are **acidic** solutions?

    (A) all of them (B) only D (C) all **EXCEPT** E (D) only E
    (E) all **EXCEPT B and E**

## DRILL ON pH, pOH, ACIDS AND BASES

***Student's Name*** ______________________

Complete the following table. You may be asked to turn in this sheet.

| Concentration of Solute | Solute Formula | $[H^+]$ | $[OH^-]$ | pH | pOH |
|---|---|---|---|---|---|
| 0.15 *M* | NaOH | | | | |
| | $Ba(OH)_2$ | | | 12.7 | |
| 0.15 *M* | $HNO_3$ | | | | |
| 0.15 *M* | HF | | | | |
| 0.15 *M* | $H_2SO_4$ | | | | |
| $1.0 \times 10^{-4}$ *M* | HF | | | | |
| | | | | | |

21. Which solution has the highest $[H_3O^+]$?

 (A) A (B) B (C) C (D) D (E) E

22. What is the percent ionization of 1.2 $M$ HF?

 (A) 2.4% (B) 4.2% (C) 0.84% (D) 0.082% (E) 0.22%

---

## REACTIONS OF WEAK ACIDS OR BASES–BUFFERS

*Buffers* are solutions containing simultaneously **appreciable** and **comparable** concentrations of **both** a weak acid **and** its conjugate base. Since neither a solution of a weak acid **alone** nor one of a weak base alone can meet these conditions, the typical buffer consists of **two** solutes, a weak acid, HA, (or a weak base, B) and a "salt" (strong electrolyte) containing the conjugate base, $A^-$ (or conjugate acid, $HB^+$). Common examples of buffers include $NH_3$ plus $NH_4Cl$, acetic acid plus sodium acetate, $H_3PO_4$ plus $KH_2PO_4$, *etc.* Buffer solutions are, within stoichiometric limits, resistant to large pH changes ($\Delta$pH) upon addition of $H_3O^+$ or $OH^-$; the buffer solution contains a **compatible** acid-base pair that (essentially quantitatively) consumes these added species. Thus, upon adding hydrochloric acid to a $HC_2H_3O_2/C_2H_3O_2^-$ buffer, the $H_3O^+$ is consumed by the reaction:

$$C_2H_3O_2^- + H_3O^+ \rightleftharpoons HC_2H_3O_2 + H_2O$$

Upon addition of added base, $OH^-$ is consumed by the reaction:

$$HC_2H_3O_2 + OH^- \rightleftharpoons C_2H_3O_2^- + H_2O$$

Buffer calculations can be based both on the $K_a$ and the $K_b$ equilibria, but usually if the buffer is going to have a pH $< 7$, calculations are based on the $K_a$ equilibrium; if the buffer pH $> 7$, it is more convenient to set up calculations based on $K_b$. In any event the calculations require careful bookkeeping on species from both solutes. Consider a buffer solution $C$ molar in HA and $C'$ molar in NaA (chief source of $A^-$, the common ion). Basing our calculations on the $K_a$ equilibrium, we set up the following table:

| | HA | + $H_2O$ ⇌ | $H_3O^+$ | + $A^-$ | |
|---|---|---|---|---|---|
| Makeup, $M$ | $C$ | | 0 | 0 | (from source HA) |
| | 0 | | 0 | $C'$ | (from source NaA) |
| $\Delta$, $M$ | $-x$ | | $+x$ | $+x$ | |
| EQ, $M$ | $C-x$ | | $x$ | $C'+x$ | |

Then, $K_a = [H_3O^+][A^-]/[HA] = (x)(C'+x)/(C-x)$. Now because of the common ion, $[A^-]$ from the source HA has been diminished, and $(C' + x) \cong C'$ (the chief source of the $A^-$ in solution is from the strong electrolyte, NaA) and $(C-x) \cong C$. Then, $K_a \cong xC'/C$. By taking logarithms (to the base 10) of both sides and rearranging, one obtains the commonly cited equation,

$$pH = pK_a + \log \frac{n_{conj.\ base}}{n_{conj.\ acid}}$$

where $n$ is the number of moles (or millimoles).

Buffers are prepared in two ways: (a) by taking both solutes and dissolving them in a given solution; or (b) by **partial** "neutralization" of a weak acid by a strong base (which quantitatively **forms** the conjugate base and **leaves** some weak acid in solution) or by **partial** "neutralization" of a weak base by a strong acid (which quantitatively forms the conjugate acid and leaves some weak base in solution). **It should be clearly noted, that in situations where chemical reactions take place, the gross reaction must be taken into account first; one may then attempt calculations involving reactions that proceed only to a small extent.** See Examples 15-E and 15-F.

## Acid-Base Titrations

A plot of pH versus volume of added acid (or base) is quite illustrative of what occurs during an acid-base reaction. The curve drawn below is the titration curve for the addition of increments of strong base to a sample of a weak acid.

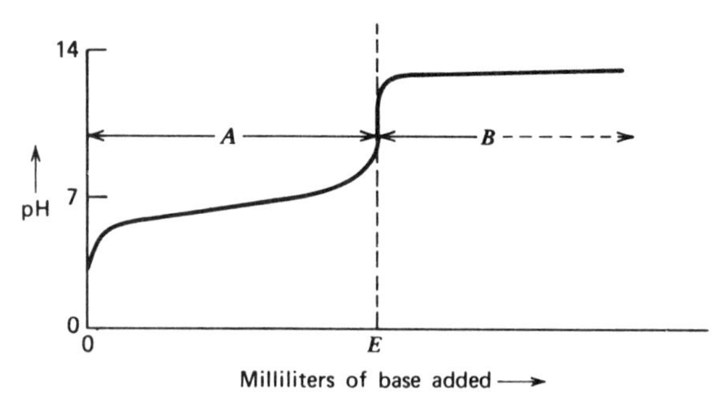


### Calculation of Points on the Curve:

Point $O$ (origin): Contains only a weak acid. See 15-C.

Point $E$ (equivalence point): Contains only the salt of a weak acid, *i.e.*, the weak base, $A^-$. See 15-D.

Any point in Region *A* (buffer region): Mixtures of a weak acid and its conjugate base (salt). See 15-F.

Any point in region *B* (excess base): Mixtures of the conjugate base and excess strong base. Other than cases immediately beyond the equivalence point, the strong base predominates to such an extent that calculations of pH may be made on the basis of the strong base content alone.

## Example Problems Involving Reactions of Weak Acids or Bases—Buffers

### 15-E. Reaction of a Strong Acid with a Weak Base in a Buffer

If 0.35 mole of ammonia, 0.10 mole of ammonium nitrate, and 0.20 mole of nitric acid are added to one liter of water, the final solution will have an ammonia/ammonium ion **mole ratio** of:

(A) 1/2 (B) 3/2 (C) 2/3 (D) 4/3 (E) 3/4

#### *Solution*

Before adding the nitric acid (a strong acid, hence $H_3O^+$) the solution contains a compatible, non-reacting, acid-base mixture of $NH_3$ (conjugate base) and $NH_4^+$ (conjugate acid). When the strong acid is added, it reacts with the base member of this conjugate pair:

$$NH_3 + H_3O^+ \rightarrow NH_4^+ + H_2O$$

Note that this can be regarded as a quantitative reaction, since the equilibrium constant equals

$$\frac{1}{K_{a(NH_4^+)}} = 1.8 \times 10^9$$

Thus, addition of 0.20 mole of $H_3O^+$ (the limiting reactant) will **use up** 0.20 mole of $NH_3$ and convert it into 0.20 mole of **additional** $NH_4^+$. For "bookkeeping" purposes this can be conveniently set up as:

| | $NH_3$ | + $H^+$ | → $NH_4^+$ |
|---|---|---|---|
| Before adding $H_3O^+$, ($H^+$): | 0.35 mole | | 0.10 mole |
| $H_3O^+$ added: | | 0.20 mole | |
| Left in solution: | 0.15 mole | ~0.0 mole | 0.30 mole |

Therefore the mole ratio, $\frac{NH_3}{NH_4^+} = \frac{0.15}{0.30} = \frac{1}{2}$. [A]

Reactions of a strong acid with a weak base or a strong base with a weak acid are gross (extensive) reactions–they **must** be handled prior to an equilibrium calculation dealing with species left in solution after reaction. See also Example 15-F.

## 15-F. pH Changes in Buffered Solutions

20. mL of 1.0 $M$ NaOH is added to 200. mL of a solution that is 0.40 $M$ in $NH_3$ and also 0.50 $M$ in $NH_4NO_3$. By this addition the pH should change from ______ to ______.

(A) 4.77 to 11.42 (B) 9.56 to 10.02 (C) 9.16 to 9.36 (D) 11.42 to 12.96
(E) 9.35 to 9.64

### *Solution*

The $NH_3/NH_4NO_3$ solution is a basic buffer since it contains a compatible conjugate acid-base pair ($NH_4^+/NH_3$) at appreciable and comparable concentrations. The pH of the original solution can be calculated by considering the $K_b$ reaction:

| | $NH_3$ + $H_2O$ ⇌ | $NH_4^+$ + | $OH^-$ |
|---|---|---|---|
| Makeup ($C$), $M$: | 0.40 | 0 | 0 (from $NH_3$) |
| | 0 | 0.50 | 0 (from $NH_4NO_3$) |
| $\Delta, M$: | $-x$ | $x$ | $x$ |
| $EQ, M$: | $0.40-x$ | $0.50+x$ | $x$ |

$$K_b = 1.8 \times 10^{-5} = \frac{[NH_4^+][OH^-]}{[NH_3]} = \frac{(0.50+x)(x)}{(0.40-x)} \cong \left(\frac{0.50}{0.40}\right)x$$

$$x = [OH^-] = 1.4 \times 10^{-5};\ [H_3O^+] = 7.1 \times 10^{-10};\ pH = 9.16 \qquad \textbf{[C]}$$

(Note that the $K_b$ equilibrium was used, and furthermore, it was assumed that $NH_3$ rather than $NH_4^+$ would react. This assumption was made on the basis that $NH_3$ is stronger as a base than $NH_4^+$ is as an acid; if the assumption had been erroneous, $x$ would have shown up negative, indicating that the $K_a$ equilibrium was the more important one under that given set of conditions. Also note that by the $K_b/C$ test $x \ll 0.50$ or 0.40. In this case, the approximation is even better than in the earlier examples because of the two sources of the ammonium ion.)

After we add $OH^-$, the reaction $OH^- + NH_4^+ \rightarrow NH_3 + H_2O$ ($K_{rxn} = 5.6 \times 10^4$) occurs essentially quantitatively. Thus **additional** $NH_3$ is produced and **some** $NH_4^+$ is consumed. In the final (220 mL) volume, the makeup concentrations are

$$C_{NH_3} = \frac{(80 + 20)\text{ mmole}}{220\text{ mL}} = 0.455\ M$$

$$C_{NH_4^+} = \frac{(100 - 20)\text{ mmole}}{220\text{ mL}} = 0.36\ M$$

Then, substituting these into the $K_b$ expression above (since $x$ will be small)

$$1.8 \times 10^{-5} = \frac{0.36}{0.455}[OH^-] = 2.3 \times 10^{-5}$$

$$[H_3O^+] = 4.3 \times 10^{-10}; pH = 9.36 \qquad [C]$$

Alternatively, the initial and final pH values could have been calculated using the relationship:

$$pH = pK_a + \log \frac{n_{conj.\ base}}{n_{conj.\ acid}} \quad \text{with } pK_a = 9.25$$

This would have had the advantage in that $n$ (moles or mmoles) could have been used directly, and the dilution calculation would have been avoided.

## 15-G. $K_a$ from Titration Data

100. mL of a solution of a weak, monoprotic acid required 32.00 mL of 0.1500 $M$ KOH for titration to the equivalence point. After adding 24.00 mL of KOH solution, the pH was observed to be 7.97. $K_a$ for the acid must be:

(A) $3.2 \times 10^{-8}$ (B) $1.8 \times 10^{-5}$ (C) $4.8 \times 10^{-4}$ (D) $7.0 \times 10^{-6}$
(E) $3.0 \times 10^{-9}$

### *Solution*

The fact that a titration can be carried out and the equivalence point determined means that the reaction between the weak acid and $OH^-$ is essentially quantitative, *i.e.*, $HA + OH^- \rightarrow A^- + H_2O$. Therefore, (32.00)(0.1500) = 4.800 mmoles of the acid must originally have been present and, after adding 24.00 mL of KOH solution, 3.600 mmoles of HA would have reacted, leaving in the solution (pH = 7.97) 3.600 mmoles of $A^-$ and 1.200 mmoles of unreacted HA. Now that the reaction has been disposed of, we need to consider the principal equilibrium, which is

$$HA + H_2O \rightleftharpoons H_3O^+ + A^-$$

$$K_a = \frac{[H_3O^+][A^-]}{[HA]} = (10^{-7.97}) \frac{(3.600 \text{ mmoles } A^-/124 \text{ mL})}{(1.200 \text{ mmoles HA}/124 \text{ mL})}$$

$$K_a = 1.07 \times 10^{-8}(3) = 3.2 \times 10^{-8}$$

## Exercises on Reactions of Weak Acids or Bases—Buffers

23. If you mix 0.1 mole of NaOH, 0.1 mole of $HC_2H_3O_2$, and 1.0 liter of water, you will have a solution that is:

    (A) red (B) blue (C) acidic (D) basic (E) neutral

24. A solution prepared by mixing 0.10 mole of $NH_4Cl$ and 0.10 mole of NaOH in enough water to make 1.0 L of solution has a pH between:

(A) 4 and 6 (B) 6.5 and 7.5 (C) 8.5 and 9.5 (D) 10 and 11
(E) 11 and 12

25. What is the pH of a solution which is 0.0100 $M$ in HA and also 0.0020 $M$ in NaA ($K_a = 9.0 \times 10^{-6}$)?

    (A) 4.35 (B) 5.65 (C) 6.65 (D) 7.15 (E) 2.40

26. Calculate the approximate pH of an aqueous solution made by adding 0.0150 mole of HCl to 1.00 liter of 0.0010 $M$ HA ($K_a$ for HA is $9.0 \times 10^{-6}$).

    (A) 0.82 (B) 1.18 (C) 1.82 (D) 2.18 (E) $10^{-3}$

27. Calculate the pH of a solution prepared by adding 0.0100 mole of HCl to 1.00 L of a solution that is 0.0100 $M$ HA and also 0.0200 $M$ NaA ($K_a$ for HA is $9.0 \times 10^{-6}$).

    (A) 3.68 (B) 4.74 (C) 5.26 (D) 5.72 (E) 6.20

**Questions 28 through 32 refer to the following titration curve obtained upon titration of 20.0 mL of an acetic acid solution with 0.050 $M$ sodium hydroxide:**

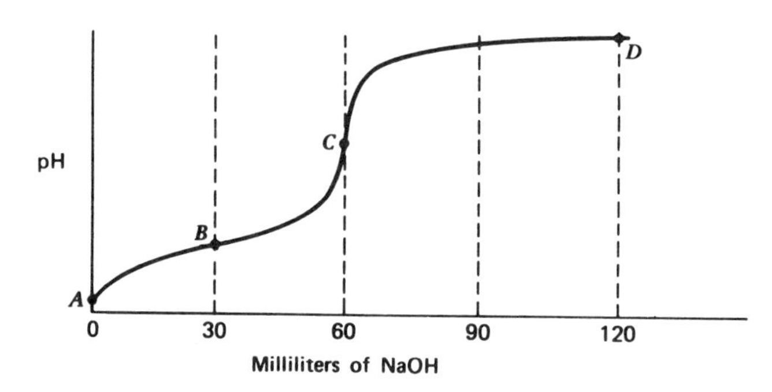


28. What is the molarity of the original acetic acid solution?

    (A) 0.15 $M$ (B) 0.038 $M$ (C) 0.10 $M$ (D) 0.050 $M$ (E) $1.8 \times 10^{-5}$ $M$

29. What is the acetate ion concentration at point $A$?

    (A) $1.6 \times 10^{-3}$ $M$ (B) $1.3 \times 10^{-3}$ $M$ (C) 0.10 $M$ (D) 0.038 $M$
    (E) none of these

30. At which point on the curve is the pH of the solution 7.00?

    (A) $A$ (B) $B$ (C) $C$ (D) $D$ (E) none of these

31. What is the pH of the solution at point $B$?

    (A) 1.5 (B) 4.7 (C) 4.9 (D) 3.5 (E) 7.0

Titration ______; Student's Name ________________________________

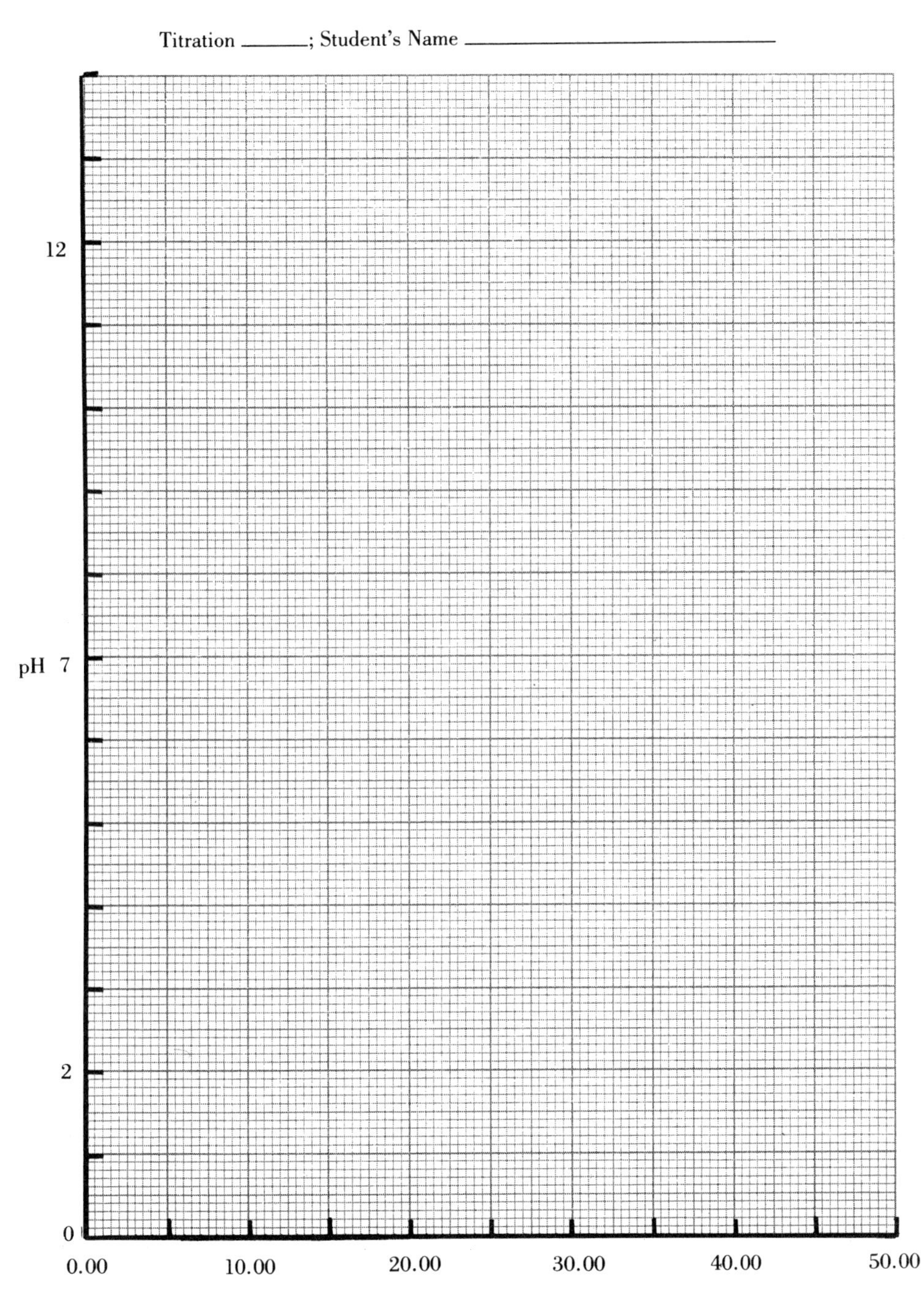

Titration ______; Student's Name ______________________________

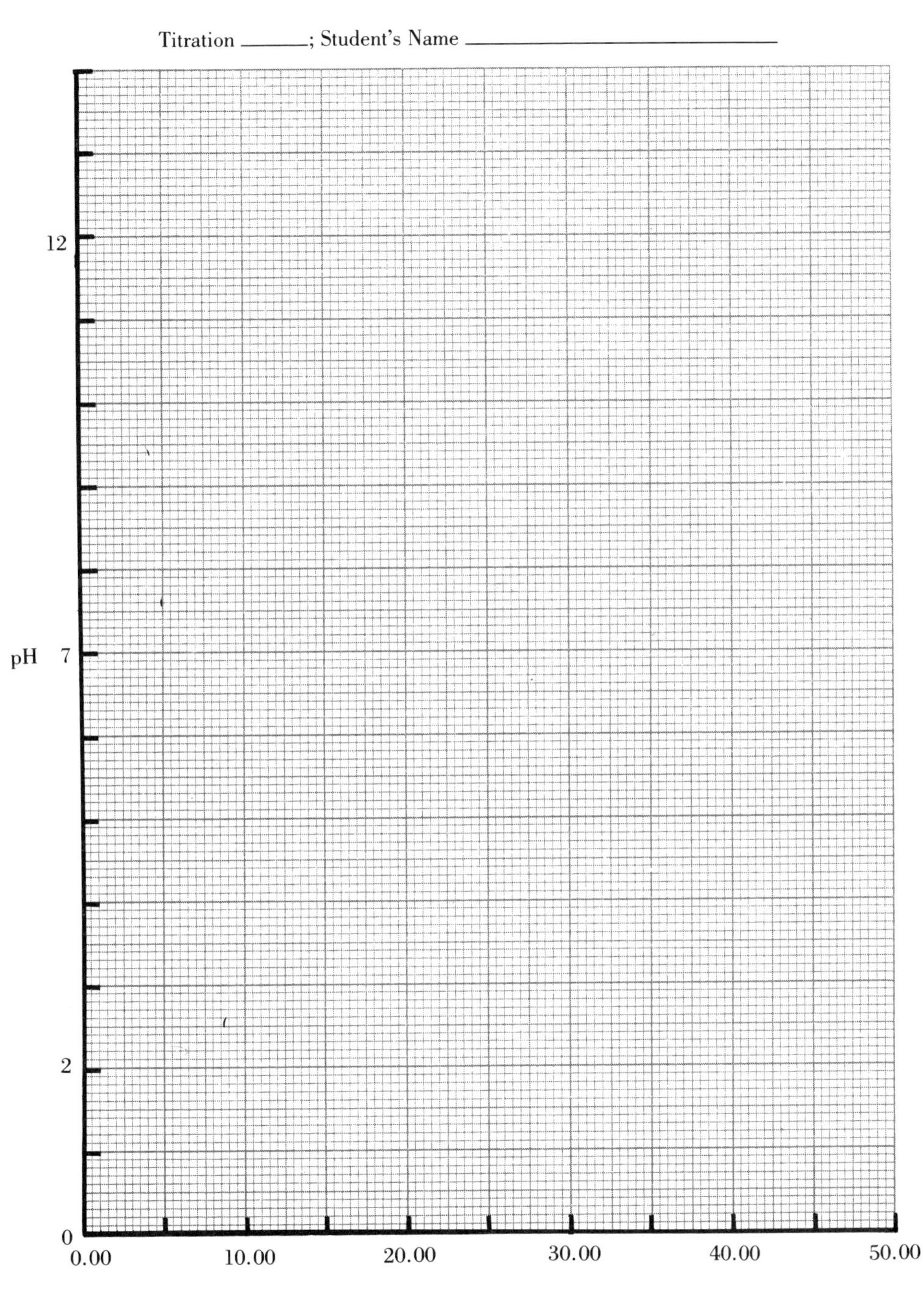

## DRILL ON ACID-BASE TITRATION

***Student's Name*** ______________________

Complete the following table by adding appropriate data for the titration of 50.0 mL of HA (a weak monoprotic acid) solution with 0.100 *M* NaOH. You may be asked to turn in this sheet.

| Acid No. | Volume of Base Required to Reach the End Point | pH at equivalence point | pH at 40% titration | $K_a$ for HA | Initial pH of the HA Solution |
|---|---|---|---|---|---|
| 1 | 25.00 mL | 10.00 | | | |
| 2 | 35.00 mL | | 2.96 | | |
| 3 | 45.00 mL | | | $5.6 \times 10^{-10}$ | |
| 4 | | | | | |

## TITRATION CURVES

***Student's Name*** ____________________

Calculate the pH of the solutions that would result during the titration of:

(A) 1.0 mmole of $HC_2H_3O_2$ in 50.0 mL of solution with 0.0500 $M$ NaOH

(B) 2.5 mmole of $NH_3$ in 75.0 mL of solution with 0.100 $M$ HCl

(C) 50.0 mL of 0.300 $M$ $HNO_3$ with 0.500 $M$ KOH

Plot the data on the graph paper provided.

| mL Titrant Added | pH | | | mL Titrant Added | pH | | |
|---|---|---|---|---|---|---|---|
| | (A) | (B) | (C) | | (A) | (B) | (C) |
| 0.00 | | | | 25.00 | | | |
| 5.00 | | | | 27.50 | | | |
| 10.00 | | | | 30.00 | | | |
| 15.00 | | | | 32.50 | | | |
| 17.50 | | | | 35.00 | | | |
| 20.00 | | | | 40.00 | | | |
| 22.50 | | | | 45.00 | | | |
| 23.25 | | | | 50.00 | | | |

32. At which point is the solution buffered?

    (A) *A* (B) *B* (C) *C* (D) *D* (E) none of these

33. What is the pH of a solution made by mixing equal volumes of 0.10 *M* acetic acid and 0.050 *M* NaOH?

    (A) 5.04 (B) 4.74 (C) 4.04 (D) 8.36 (E) 2.87

34. Which one of the following mixtures will be a buffer when dissolved in a liter of water?

    (A) 0.2 mole of NaOH and 0.2 mole of HBr
    (B) 0.2 mole of NaCl and 0.3 mole of HCl
    (C) 0.4 mole of $HC_2H_3O_2$ and 0.2 mole of NaOH
    (D) 0.5 mole of $NH_3$ and 0.5 mole of HCl
    (E) 0.3 mole of KOH and 0.2 mole of HBr

---

## POLYPROTIC ACIDS

These are compounds that have two or more acidic hydrogens and dissociate in a stepwise manner, *e.g.*, in the case of a diprotic acid, $H_2A$,

$$H_2A + H_2O \rightleftharpoons H_3O^+ + HA^- \quad K_{a_1}$$

$$HA^- + H_2O \rightleftharpoons H_3O^+ + A^{2-} \quad K_{a_2}$$

The calculation of the concentrations of the various species in solution falls into three different categories depending on the relative magnitudes of the $K_a$'s. If all dissociations are weak and the $K_a$ values for successive dissociations differ by at least $10^3$, *i.e.*, $1 >> K_{a_1} >> K_{a_2}$, only the first equilibrium needs to be considered in the calculation of $[H_3O^+] \cong [HA^-]$ and of $[H_2A]$. The second equilibrium needs to be invoked only when the $[A^{2-}]$ is desired: $[A^{2-}] \cong K_{a_2}$ (see Example 15-H). In the second type of situation some of the $K$'s are strong and some are weak (see Example 15-I), *i.e.*, for a diprotic acid, $K_{a_1} >> 1 >> K_{a_2}$. $H_2A$ completely dissociates, yielding $H_3O^+$ and $HA^-$, and a significant amount of $H_3O^+$ may be formed by dissociation of some of the $HA^-$. The $[H_3O^+]$ is the sum of that arising from the two steps; the $[HA^-]$ and $[A^{2-}]$ are calculated by considering the second step. The third case is one where successive $K$'s are about the same order of magnitude. These calculations involve the solution of higher-ordered algebraic equations and are not taken up in the beginning chemistry course.

## Example Problems Involving Polyprotic Acids

### 15-H. Concentrations of Species Present in a Solution of a Weak Diprotic Acid

Find the sulfide ion concentration in an 0.080 *M* solution of hydrogen sulfide.

(A) 0.080 *M* (B) $8.9 \times 10^{-5}$ *M* (C) 0.040 *M* (D) $1.3 \times 10^{-13}$ *M*
(E) $6.4 \times 10^{-8}$ *M*

#### *Solution*

$H_2S$ is a diprotic acid of the type $1 >> K_{a_1} >> K_{a_2}$:

$$H_2S + H_2O \rightleftharpoons H_3O^+ + HS^- \quad K_{a_1} = 1.0 \times 10^{-7}$$

$$HS^- + H_2O \rightleftharpoons H_3O^+ + S^{2-} \quad K_{a_2} = 1.3 \times 10^{-13}$$

Thus, in the 0.080 *M* solution of $H_2S$, the conservation of sulfur-containing species requires that $0.080 = [H_2S] + [HS^-] + [S^{2-}]$. Since the $K_{a_1}$ dissociation is so predominant, the first stage of ionization (which supplies $H_3O^+$ and $HS^-$ as species common to the second stage) may be handled separately first; and then these results are applied to the second stage (since this is the source of $S^{2-}$):

| *Step1:* | $H_2S + H_2O \rightleftharpoons$ | $H_3O^+$ | $+ HS^-$ |
|---|---|---|---|
| Makeup (*C*), *M*: | 0.080 | 0 | 0 |
| $\Delta_1, M$: | $-x$ | $x$ | $x$ |
| Allowing only 1st step: | $0.080-x$ | $x$ | $x$ |

| Then, *Step 2:* | $HS^- + H_2O \rightleftharpoons$ | $H_3O^+$ | $+ S^{2-}$ |
|---|---|---|---|
| From Step 1, *M*: | $x$ | $x$ | 0 |
| $\Delta_2, M$: | $-y$ | $+y$ | $+y$ |
| EQ, *M*: | $x-y$ | $x+y$ | $y$ |

After considering Step 1, we see that (since $K_{a_1}/C \cong 10^{-6} << 10^{-3}$) $x$ is small compared to 0.080, so that $[H_2S] \cong 0.080$ molar and $x = [H_3O^+] = [HS^-] \cong 8.9 \times 10^{-5}$ (neglecting the second stage). As we start to consider Step 2, we note that very little of the $HS^-$ dissociates (since $K_{a_2}/8.9 \times 10^{-5} \cong 10^{-9}$ and since the common ion, $H_3O^+$, is already present); thus, $y << x$. Then,

$$K_{a_2} = 1.3 \times 10^{-13} = \frac{[H_3O^+][S^{2-}]}{[HS^-]} = \frac{(x+y)y}{(x-y)} \cong y$$

and

$$[H_3O^+] = (x+y) \cong x \cong 8.9 \times 10^{-5}\ M$$
$$[H_2S] = (0.080 - x) \cong 0.080\ M$$
$$[HS^-] = (x-y) \cong x \cong 8.9 \times 10^{-5}\ M$$
$$[S^{2-}] = y \cong K_{a_2} \cong 1.3 \times 10^{-13}\ M$$

**[D]**

Checking by mass balance (conservation of sulfur species):

$$0.080 = [H_2S] + [HS^-] + [S^{2-}]$$
$$= 0.080 + 8.9 \times 10^{-5} + 1.3 \times 10^{-13}$$
$$= 0.080 \text{ to 2 significant figures}$$

## 15-I. The Concentrations of Ions Present in a Sulfuric Acid Solution

Find the sulfate ion concentration in a solution prepared by diluting with water 13.3 mL of 3.0 *M* sulfuric acid to a final volume of 0.50 liter.

(A) 0.012 *M* (B) 0.0040 *M* (C) 0.0095 *M* (D) 0.070 *M* (E) 0.080 *M*

### *Solution*

The dilution requires that the makeup concentration of the final solution will be (13.3)(3.0)/500 = 0.080 *M* $H_2SO_4$. Sulfuric acid is a diprotic acid of the type $K_{a_1} >> 1 >> K_{a_2}$ with $K_{a_1} \cong \infty$ and $K_{a_2} = 1.2 \times 10^{-2}$. This means that the first-stage ionization is quantitative,

$$H_2SO_4 + H_2O \xrightarrow{\sim 100\%} H_3O^+ + HSO_4^-$$

yielding, at this point, $[H_3O^+] = [HSO_4^-]$, *i.e.*, not taking the second step into account. The second-stage ionization, the source of the $SO_4^{2-}$ ion, may now be treated, remembering that $H_3O^+$ and $HSO_4^-$ from the first stage are already present:

$$HSO_4^- + H_2O \rightleftharpoons H_3O^+ + SO_4^{2-}$$

| | $HSO_4^-$ | $H_3O^+$ | $SO_4^{2-}$ |
|---|---|---|---|
| Makeup (*C*), *M*: | 0.080 | 0.080 | 0 |
| Δ, *M*: | $-x$ | $+x$ | $x$ |
| EQ, *M*: | $0.080-x$ | $0.080+x$ | $x$ |

$$K_{a_2} = 0.012 = \frac{[H_3O^+][SO_4^{2-}]}{[HSO_4^-]} = \frac{(0.080 + x)(x)}{(0.080 - x)}$$

which, on neglecting $x$ in the additive terms (assuming $x << 0.080$) yields $x = [SO_4^{2-}] = 0.012$ *M*. This value, however, is 15% of 0.080 and is hardly negligible. (Note that, **despite the repression** by the common ion, $H_3O^+$, produced in the first step, $K_{a_2}/C = 0.15 >> 10^{-3}$.) Thus, the quadratic must be solved to find $x$, the $[SO_4^{2-}]$; $x^2 + 0.092x - 0.00096 = 0$.

$$x = -0.10 \quad \text{and} \quad x = 0.0095$$

Since $[SO_4^{2-}]$ cannot be negative, the valid root is 0.0095 and the composition of the solution is:

$$[H_2SO_4] = 0$$
$$[H_3O^+] = 0.090$$
$$[HSO_4^-] = 0.070$$
$$[SO_4^{2-}] = 0.0095$$

[C]

Cycling these values back into the $K_{a_2}$ expression gives a check on the answers, yielding $K_{a_2} = 0.012$ to 2 significant figures.

## Exercises on Polyprotic Acids

35. The $[H_3O^+]$ in 0.0050 *M* $H_2S$ is:

(A) $1.0 \times 10^{-7}$ *M* (B) $2.2 \times 10^{-5}$ *M* (C) $1.0 \times 10^{-4}$ *M*
(D) $5.0 \times 10^{-4}$ *M* (E) none of these

36. The $[H_3O^+]$ in 0.100 *M* $H_2SO_4$ is:

(A) 0.100 *M* (B) 0.200 *M* (C) between 0.180 *M* and 0.200 *M*
(D) between 0.100 *M* and 0.120 *M* (E) between 0.145 *M* and 0.155 *M*

37. The $[CO_3^{2-}]$ in 0.050 *M* $H_2CO_3$ solution (*e.g.*, "Pepsi") is:

(A) $1.5 \times 10^{-4}$ *M* (B) $4.3 \times 10^{-7}$ *M* (C) $1.0 \times 10^{-7}$ *M*
(D) $4.7 \times 10^{-11}$ (E) $1.3 \times 10^{-13}$ *M*

38. The pH of the solution described in Question 37 is:

(A) less than 3.0 (B) more than 4.5 (C) between 3.0 and 3.5
(D) between 4.0 and 4.5 (E) between 3.5 and 4.0

39. The pH of a 0.10 *M* $Na_2CO_3$ solution is:

(A) 2.43 (B) 3.68 (C) 11.66 (D) 8.00 (E) 10.32

40. What percentage of the total sulfur-containing species exists in the form of bisulfate ions in a solution that is nominally 0.0500 *M* sulfuric acid?

(A) 55% (B) 72% (C) 15% (D) 83% (E) 48%

41. Approximately what percentage of the total carbon-containing species exists in the form of bicarbonate ion in a solution that is nominally 0.0500 *M* "carbonic acid"?

(A) 15% (B) 0.3% (C) 1% (D) 4% (E) 0.02%

---

## GENERAL EXERCISES ON CHAPTER 15—ACID-BASE EQUILIBRIA IN AQUEOUS SOLUTION—pH, BUFFERS, AND TITRATION CURVES

42. Calculate $[H_3O^+]$ for a 0.0500 *M* solution of HF.

(A) $5.0 \times 10^{-3}$ *M* (B) $5.4 \times 10^{-3}$ *M* (C) $5.7 \times 10^{-3}$ *M*
(D) $6.0 \times 10^{-3}$ *M* (E) $6.4 \times 10^{-3}$ *M*

43. Calculate the pH of a solution that is 0.050 *M* in NaF.

(A) 7.79 (B) 6.08 (C) 8.21 (D) 7.92 (E) 9.67

44. The pH of a solution 0.050 *M* in HF and also 0.050 *M* in NaF is:

(A) 3.14 (B) 4.13 (C) 9.87 (D) 7.00 (E) 10.86

**For Questions 45–47, consider a solution in a bottle labeled:**

| HF, 2.22 mole/liter |
|---|

45. What is the concentration of hydronium ions in this solution?

(A) $7.2 \times 10^{-4}$ *M* (B) 0.040 *M* (C) 2.2 *M* (D) $1.6 \times 10^{-4}$ *M*
(E) 1.5 *M*

46. What is the approximate % ionization of HF in this solution?

(A) ~0.5% (B) ~1% (C) ~0.1% (D) ~5% (E) ~2%

47. What should be the approximate pH of the solution resulting from the addition of 0.40 mole of solid KOH to 1.00 liter of 2.22 *M* HF?

(A) ~1.5 (B) ~2.0 (C) ~2.5 (D) ~3.0 (E) ~3.5

48.*How many milliliters of a standard 0.1000 *M* KOH solution should be added to 50.00 mL of a nitric acid solution with pH = 0.000, in order that the final mixture will have pH = 11.96? (Assume that volumes are additive.)

(A) 105 mL (B) 175 mL (C) 305 mL (D) 475 mL (E) 555 mL

49. By using data from the Acid-Base Table, pp. 348-9, calculate $pK_a$ for acetic acid.

(A) 4.74 (B) -4.74 (C) 0.26 (D) 5.26 (E) -0.26

50. Given the following hypothetical acids, with $pK_a$ values in parentheses: $HP^-$ (1.92); HQ (10.33); $H_2R$ (7.21); $H_4Z^+$ (9.25) and $H_2X$ (11.62). Which of these is the strongest acid?

(A) $HP^-$ (B) HQ (C) $H_2R$ (D) $H_4Z^+$ (E) $H_2X$

51. If $K_b = 1.5 \times 10^{-16}$ for $X^-$, $pK_a$ for HX is:

(A) 15.82 (B) 1.18 (C) -1.82 (D) 16.18 (E) none of these

**Questions 52 through 54 refer to the following data. A 0.500-g sample of an impure monoprotic acid is known to contain, in addition to the acid, only inert ingredients. The sample was dissolved in ~10 mL of water and titrated with 0.100 *M* NaOH. After 25.00 mL of the NaOH solution was added the pH was 5.20; the endpoint of the titration was reached upon addition of 40.00 mL of NaOH solution. The pure acid was known to have a molecular weight of 100.**

52. The percent acid in the original sample is:

(A) 20.0% (B) 40.0% (C) 60.0% (D) 80.0% (E) 55.0%

53. The dissociation constant, $K_a$, for the acid is:

(A) $1.0 \times 10^{-6}$ (B) $1.0 \times 10^{-5}$ (C) $1.0 \times 10^{-4}$ (D) $1.8 \times 10^{-5}$
(E) $1.8 \times 10^{-4}$

54. Which of the following indicators would be suitable for detection of the equivalence point in this titration?

(A) methyl orange (red → yellow; pH 3.1-4.4)
(B) methyl red (red → yellow; pH 4.8-6.0)

(C) bromthymol blue (yellow → blue; pH 6.0-7.6)
(D) thymol blue (yellow → blue; pH 8.0-9.6)
(E) thymolphthalein (colorless → blue; pH 9.3-10.5)

55. The pH of 0.10 *M* HA is 4.70. What is the value of $K_a$ for HA?

(A) $4.0 \times 10^{-8}$ (B) $2.5 \times 10^{-7}$ (C) $4.0 \times 10^{-5}$ (D) $2.5 \times 10^{-6}$
(E) none of these

56. A 0.10 *M* solution of aqueous ammonia also is 0.10 *M* in ammonium chloride. Calculate $[OH^-]$.

(A) $1.8 \times 10^{-5}$ *M* (B) $1.0 \times 10^{-14}$ *M* (C) 0.10 *M* (D) $5.5 \times 10^{-10}$ *M*
(E) It is too small to be worth calculating.

57. What is the pH of the solution obtained when 150.0 mL of 0.100 *M* $NH_3$ is mixed with 50.0 mL of 0.200 *M* HCl?

(A) 9.26 (B) 10.98 (C) 8.96 (D) 10.26 (E) 9.56

58. Calculate the $[H_3O^+]$ of a solution made by mixing 40. mL of 2.0 *M* $HC_2H_3O_2$ with 20. mL of 1.0 *M* NaOH.

(A) $5.4 \times 10^{-5}$ *M* (B) $6.9 \times 10^{-3}$ *M* (C) $4.2 \times 10^{-3}$ *M*
(D) $6.0 \times 10^{-6}$ *M* (E) $1.8 \times 10^{-5}$ *M*

59. Calculate the $[H_3O^+]$ in the solution in Question 58 after 2.0 mL of 1.0 *M* HCl has been added.

(A) $4.8 \times 10^{-5}$ *M* (B) $2.1 \times 10^{-5}$ *M* (C) $5.0 \times 10^{-3}$ *M*
(D) $8.0 \times 10^{-3}$ *M* (E) $6.2 \times 10^{-5}$ *M*

60. How many moles of pure $NaHSO_4$ would be required to prepare 1.00 liter of a solution having a pH = 2.00?

(A) 0.018 (B) 0.0096 (C) 0.192 (D) 0.0083 (E) 0.0048

61. What is the concentration of $SO_4^{2-}$ in 0.100 *M* $H_2SO_4$?

(A) 0.012 *M* (B) 0.035 *M* (C) 0.017 *M* (D) 0.010 *M* (E) 0.029 *M*

62. What is the pH of a solution made by mixing 0.10 mole of HCl and 0.10 mole of lithium acetate in enough water to make 1.0 liter of solution?

(A) 4.74 (B) 2.87 (C) 2.44 (D) 3.17 (E) 1.00

63. What is $[H_3O^+]$ in a 0.010 *M* solution of $NaHSO_4$?

(A) 0.0052 *M* (B) 0.010 *M* (C) 0.0065 *M* (D) 0.0013 *M*
(E) 0.0032 *M*

64. 2.0 liters of a hydrochloric acid solution with pH = 3.22 is mixed with 1.0 liter of a NaOH solution of pH = 10.78. The final 3.0 liters should have a pH of:

(A) 2.74 (B) 3.22 (C) 3.70 (D) 5.74 (E) 7.00

65. What is the pH of a 0.00010 $M$ HA solution ($K_a$ = 9.0 × $10^{-6}$)?

    (A) 4.58 (B) 7.05 (C) 5.52 (D) 4.52 (E) 5.48

66. A solution is prepared by dissolving 115 g of 85.0% (by weight) phosphoric acid in sufficient water to make 1.00 liter of solution. The resulting solution has a pH = 1.08; [$H_3PO_4$] = 0.92 $M$; [$H_2PO_4^-$] = 0.083 $M$; and [$HPO_4^{2-}$] = 6.2 × $10^{-8}$ $M$. What is the phosphate ion concentration in the solution?

    (A) 4.7 × $10^{-22}$ $M$ (B) 1.0 × $10^{-12}$ $M$ (C) 5.6 × $10^{-16}$ $M$
    (D) 7.5 × $10^{-19}$ $M$ (E) 6.2 × $10^{-20}$ $M$

67. 0.0100 mole of NaA is dissolved in water to make 1.00 liter of solution. The solution pH is 11.95. What is $K_b$ for $A^-$?

    (A) 7.2 × $10^{-2}$ (B) 1.4 × $10^{-4}$ (C) 5.5 × $10^{-3}$ (D) 1.0 × $10^{-3}$
    (E) 6.8 × $10^{-3}$

68. What mole ratio of sodium acetate to acetic acid should be used in preparing a buffer having a pH of 4.35?

    (A) 1.0/1.0 (B) 0.59/1.0 (C) 0.41/1.0 (D) 2.0/1.0 (E) 0.75/1.0

69. What is the pH of a solution prepared by mixing 30.0 mL of 6.0 $M$ acetic acid, 0.250 **moles** of sodium acetate, 40.0 mL of 0.500 $M$ HCl and enough water to give a total volume of 1200 mL?

    (A) 4.80 (B) 3.60 (C) 2.10 (D) 6.35 (E) 4.74

70. What is the pH of 0.10 $M$ hydrogen sulfide?

    (A) 2.87 (B) 3.50 (C) 4.70 (D) 4.00 (E) 5.94

71. What is the [$S^{2-}$] in 0.10 $M$ hydrogen sulfide?

    (A) 1.3 × $10^{-18}$ $M$ (B) 1.3 × $10^{-8}$ $M$ (C) 1.3 × $10^{-13}$ $M$
    (D) 1.3 × $10^{-15}$ $M$ (E) 1.3 × $10^{-21}$ $M$

72.*$NH_{3\,(g)}$ is added to 0.10 $M$ $H_2S$ until the pH = 10.00. What is the [$S^{2-}$] in this solution?

    (A) 0.13 $M$ (B) 1.3 × $10^{-8}$ $M$ (C) 1.3 × $10^{-4}$ $M$ (D) 1.3 × $10^{-2}$ $M$
    (E) 1.3 × $10^{-3}$ $M$

73. In human blood carbonic acid and the bicarbonate ion act as one of the buffer systems that maintains the pH of blood at 7.40. What is the molar ratio of $HCO_3^-$ to $H_2CO_3$ needed?

    (A) 8:1 (B) 20:1 (C) 1:2 (D) 3:10 (E) 11:1

74. What is the pH of a solution made by mixing 1.0 L of 0.10 $M$ acetic acid with 1.0 L of 0.50 $M$ sodium acetate?

    (A) 4.04 (B) 4.26 (C) 5.44 (D) 4.96 (E) 4.74

75. How many grams of NaF (FW = 42.0) would have to be added to 2.00 L of 0.100 *M* HF to yield a solution with a pH = 4.00?

(A) 300 g (B) 36 g (C) 0.84 g (D) 6.9 g (E) 61 g

76. The pH of a solution prepared by addition of 0.200 mole of NaF to 1.0 liter of 2.00 *M* HF is:

(A) 3.00 (B) 2.14 (C) 10.8 (D) 7.00 (E) 6.70

77. The concentration of $SO_4^{2-}$ in 2.0 *M* $H_2SO_4$ is:

(A) $2.0 \times 10^{-1}$ *M* (B) $1.2 \times 10^{-2}$ *M* (C) $4.0 \times 10^{-1}$ *M*
(D) $1.0 \times 10^{-7}$ *M* (E) $4.0 \times 10^{-2}$ *M*

78. What is the pH of a solution made by mixing equal volumes of 0.10 *M* $HC_2H_3O_2$, 0.10 *M* NaOH, and 0.050 *M* HCl?

(A) 5.04 (B) 4.74 (C) 8.36 (D) 11.30 (E) 4.04

79. The value of $K_w$ at 37°C (98.6°F) is $2.42 \times 10^{-14}$. What is the pOH for a neutral solution at this normal temperature of the human body?

(A) 13.62 (B) 7.00 (C) 7.19 (D) 6.81 (E) 6.19

80.*What is the $[PO_4^{3-}]$ in a solution made by adding NaOH to a 0.20 *M* $H_3PO_4$ solution until the pH = 11.00 (assume the volume remains constant)?

(A) 0.20 *M* (B) 0.15 *M* (C) 0.072 *M* (D) 0.018 *M* (E) 0.0034 *M*

81.*What percentage of the total phosphorus in a phosphoric acid solution adjusted to a pH of 12.00 would be in the form of $PO_4^{3-}$?

(A) 10 (B) 30 (C) 50 (D) 70 (E) 100

82. Calculate the pH of a $1 \times 10^{-8}$ *M* solution of $HNO_3$.

(A) 8.0 (B) 6.0 (C) 7.0 (D) 9.0 (E) 5.0

83. What is the pH of a 0.020 *M* phosphoric acid solution?

(A) 1.91 (B) 4.45 (C) 1.70 (D) 1.46 (E) 2.04

84. What is the $[PO_4^{3-}]$ in the solution in Question 83?

(A) $5.3 \times 10^{-14}$ *M* (B) $1.2 \times 10^{-17}$ *M* (C) $3.5 \times 10^{-16}$ *M*
(D) $2.8 \times 10^{-20}$ *M* (E) $6.7 \times 10^{-18}$ *M*

85. Equal volumes of an HCl solution (pH = 3.000) and a NaOH solution (pH = 10.00) are mixed. If we assume that the volumes are additive, the final solution should have a pH of:

(A) 7.00 (B) 3.01 (C) 6.50 (D) 8.53 (E) 3.35

86.*Equal volumes of 0.300 *M* acetic acid and 0.300 *M* sodium nitrite are mixed. What percentage of the total $NO_2^-$ would be converted to nitrous acid, $HNO_2$? (Assume additive volumes.)

(A) 40.% (B)17% (C) 33% (D) 25% (E) 60.%

87. What is the pH of the solution obtained by mixing 50.0 mL of 0.100 $M$ HCl and 150.0 mL of 0.100 $M$ $NH_3$?

    (A) 9.26 (B) 10.98 (C) 8.96 (D) 10.26 (E) 9.56

88. The pH of the solution prepared by adding 50.0 mL of 0.100 $M$ NaOH to 20.0 mL of 1.00 $M$ acetic acid is:

    (A) 5.22 (B) 4.14 (C) 2.57 (D) 5.34 (E) 4.26

89. Methylamine ($CH_3NH_2$) is a weak base with $K_b = 4.0 \times 10^{-4}$. Calculate the hydroxide ion concentration in 0.010 $M$ $CH_3NH_2$.

    (A) $4.0 \times 10^{-6}$ $M$ (B) 0.010 $M$ (C) $1.8 \times 10^{-3}$ $M$ (D) $2.0 \times 10^{-3}$ $M$ (E) 0.20 $M$

90. The pH of a 0.100 $M$ solution of the weak acid HX is 1.340. What is $K_a$ for HX?

    (A) $2.09 \times 10^{-2}$ (B) $3.85 \times 10^{-2}$ (C) $4.57 \times 10^{-1}$ (D) $1.80 \times 10^{-5}$ (E) $3.60 \times 10^{-5}$

91. The pH of a 0.012 $M$ base, methylaminopyrazine, was found to be 11.30. What is the value of $K_b$?

    (A) $3.3 \times 10^{-4}$ (B) $4.0 \times 10^{-4}$ (C) $2.1 \times 10^{-21}$ (D) $1.8 \times 10^{-5}$ (E) $3.6 \times 10^{-6}$

**The following 4 exercises involve calculation of acid-base equilibrium constants from data taken from the table in the Appendix entitled "Thermodynamic Data for Species in Aqueous Solution."**

92. From the thermodynamic properties of aqueous species, calculate $K_a$ for acetic acid. (‡)

    (A) $5.45 \times 10^{-10}$ (B) $1.77 \times 10^{-5}$ (C) $1.83 \times 10^{-5}$ (D) $1.70 \times 10^{-5}$ (E) $1.86 \times 10^{-5}$

93. From the thermodynamic properties of aqueous species, calculate $K_b$ for ammonia. (‡)

    (A) $5.60 \times 10^{-10}$ (B) $5.50 \times 10^{-10}$ (C) $1.88 \times 10^{-5}$ (D) $1.77 \times 10^{-5}$ (E) $1.70 \times 10^{-5}$

94. From the value of $\Delta G°$ for $H_2O_{(\ell)} \rightleftharpoons H^+_{(aq)} + OH^-_{(aq)}$ at 25°C, find $K_w$ using $R$ = 1.99 cal/mole-K and T = 298 K. (‡)

    (A) $1.04 \times 10^{-14}$ (B) $1.00 \times 10^{-14}$ (C) $9.96 \times 10^{-15}$ (D) $1.10 \times 10^{-14}$ (E) $9.89 \times 10^{-15}$

95. Repeat the above calculation (Question 94) using $R$ and $T$ to five signifi-

96. Oxalic acid, $H_2C_2O_4$, is a diprotic acid with $K_{a_1} = 5.36 \times 10^{-2}$ and $K_{a_2} = 5.3 \times 10^{-5}$. Addition of a solution of potassium hydrogenoxalate to a solution of ammonium dihydrogenphosphate would result in formation of: (*cf.*, Acid-Base Table, p. 254).

    (A) ammonia (B) hydrogenphosphate ion (C) phosphoric acid
    (D) oxalic acid (E) ammonium ion

97. Addition of excess acetic acid solution to a solution of sodium oxalate (see question 96 would result in formation of largest amounts of: (*cf.*, Acid-Base Table, p. 254).

    (A) oxalic acid (B) water (C) hydrogenoxalate ion
    (D) carbon dioxide (E) oxalate ion

98. If $K_a$ for HZ is $7.2 \times 10^{-4}$, then it follows that:

    (A) HZ ionizes very extensively in water solution
    (B) $H_2O$ is a stronger base than $Z^-$
    (C) HZ is a stronger acid than $H_3O^+$
    (D) a solution of NaZ should be neutral (pH = 7)
    (E) HZ should be formed upon mixing aqueous solutions of NaZ and HCl

99. The **net ionic** equation for the reaction occuring when sodium acetate and hydrochloric acid solutions are mixed is:

    (A) $H_3O^+ + C_2H_3O_2^- \rightarrow HC_2H_3O_2 + H_2O$
    (B) $H_2O + C_2H_3O_2^- \rightarrow HC_2H_3O_2 + OH^-$
    (C) $C_2H_3O_2^- \rightarrow H^+ + C_2H_2O_2^{2-}$
    (D) $OH^- + HC_2H_3O_2 \rightarrow H_2O + C_2H_3O_2^-$
    (E) $HC_2H_3O_2 \rightarrow H^+ + C_2H_3O_2^-$

100. A sample of maple syrup has a pH of 6.85. This corresponds to a hydroxide ion concentration of:

    (A) $1.4 \times 10^{-7}$ *M* (B) 7.15 *M* (C) $8.5 \times 10^{-6}$ *M* (D) $1.5 \times 10^{-7}$ *M*
    (E) $7.1 \times 10^{-8}$ *M*

101. A 50.0-mL sample of 6.0 *M* HCl is added to 100. mL of 1.20 *M* NaOH. What is the pH of the resulting solution?

    (A) 1.20 (B) 0.18 (C) 0.08 (D) -0.08 (E) -1.20

102. A 0.010 *M* solution of a weak acid, HA, has a pH of 4.20. What is $K_a$ for the acid?

    (A) $4.0 \times 10^{-7}$ (B) $2.5 \times 10^{-6}$ (C) $6.3 \times 10^{-6}$ (D) $1.6 \times 10^{-4}$
    (E) $6.3 \times 10^{-5}$

103. What is the pOH of the solution described in Question 102?

    (A) 4.20 (B) 14.0 (C) 9.80 (D) 7.0 (E) none of these

104. If the pH of the solution described in Question 102 were adjusted to 3.0, what would be the $[A^-]$?

(A) $1.0 \times 10^{-3}$ *M* (B) $2.5 \times 10^{-5}$ *M* (C) $6.0 \times 10^{-4}$ *M*
(D) $4.0 \times 10^{-6}$ *M* (E) $1.6 \times 10^{-9}$ *M*

105. A solution is 0.010 *M* in KA, the potassium salt of the weak acid described in Question 102. The pH of this solution would be:

(A) 4.80 (B) 4.20 (C) 10.20 (D) 8.80 (E) 9.20

106. For a 0.0100 molar solution of HY, the pH is 2.40. What is $K_a$ for HY?

(A) $4.0 \times 10^{-3}$ (B) $2.4 \times 10^{-3}$ (C) $1.6 \times 10^{-3}$ (D) $1.6 \times 10^{-5}$
(E) $2.7 \times 10^{-3}$

107. Calculate the $[H_3O^+]$ in a solution prepared by mixing **equal volumes** of 0.25-molar ammonium chloride **and** 0.40-molar ammonia.

(A) $1.1 \times 10^{-8}$ *M* (B) $1.8 \times 10^{-9}$ *M* (C) $3.5 \times 10^{-10}$ *M*
(D) $1.4 \times 10^{-10}$ *M* (E) $7.1 \times 10^{-11}$ *M*

108. Calculate the pH of a solution prepared by mixing equal volumes of 0.15-molar nitric acid and 0.40-molar ammonia.

(A) 7.95 (B) 8.57 (C) 8.98 (D) 9.21 (E) 9.47

109. Calculate $[OH^-]$ in a solution prepared by adding 100. mL of 0.10 *M* KOH to 200 mL of 0.25 *M* ammonium nitrate.

(A) $1.3 \times 10^{-5}$ (B) $4.5 \times 10^{-6}$ (C) $5.1 \times 10^{-7}$ (D) $6.3 \times 10^{-8}$
(E) $2.6 \times 10^{-9}$

110. A solution contains 2.0 moles of $NH_3$ and 3.0 moles of $NH_4Cl$ and shows pH = 9.07. If 100 mL of 6.00 *M* HCl were added to this solution, the pH would decrease by ______ unit. Fill in the blank.

(A) 0.16 (B) 0.23 (C) 0.38 (D) 0.08 (E) 0.01

111. A solution contains 1.00 mole of acetic acid and has pH = 2.50. How many **grams** of $NaC_2H_3O_2 \cdot 3H_2O$ (FW = 136) should be dissolved in this solution to raise the pH to 4.00?

(A) 10 g (B) 15 g (C) 20 g (D) 25 g (E) 35 g

112. A bottle labeled 0.150 *M* sulfuric acid contains a solution which is ______ *M* in sulfate ions? The number needed to fill the blank is:

(A) 0.011 (B) 0.150 (C) 0.075 (D) 0.139 (E) 0.161

113. Starting with a saturated solution of $H_2S$ (0.10 *M*) the pH is slowly raised by adding base. At what pH would $[S^{2-}]$ reach $1.5 \times 10^{-15}$ *M*?

(A) -0.94 (B) 2.30 (C) 3.03 (D) 7.07 (E) 10.97

114. A phosphoric acid-solution is neutralized until the $[HPO_4^{2-}]$ is one-thousand times greater than $[PO_4^{3-}]$. What is the pH of this solution?

(A) 5.0 (B) 7.0 (C) 9.0 (D) 11.0 (E) 13.0

115.*For the solution in Q-114, approximately what percent of the total phosphorous present is in the form of $H_2PO_4^-$?

(A) 0.5% (B) 1.5% (C) 2.5% (D) 5.5% (E) 9.5%

---

## ANSWERS TO CHAPTER 15 PROBLEMS

| | | | | | | |
|---|---|---|---|---|---|---|
| 1. C | 18. C | 35. B | 52. D | 69. A | 86. B | 103. C |
| 2. E | 19. B | 36. D | 53. B | 70. D | 87. E | 104. D |
| 3. E | 20. E | 37. D | 54. D | 71. C | 88. E | 105. E |
| 4. C | 21. D | 38. E | 55. E | 72. C | 89. C | 106. E |
| 5. E | 22. A | 39. C | 56. A | 73. E | 90. B | 107. C |
| 6. A | 23. D | 40. D | 57. C | 74. C | 91. B | 108. E |
| 7. C | 24. E | 41. B | 58. A | 75. E | 92. B | 109. B |
| 8. D | 25. A | 42. C | 59. E | 76. B | 93. C | 110. B |
| 9. B | 26. C | 43. D | 60. A | 77. B | 94. A | 111. D |
| 10. C | 27. B | 44. A | 61. D | 78. B | 95. A | 112. A |
| 11. E | 28. A | 45. B | 62. B | 79. D | 96. C | 113. C |
| 12. A | 29. A | 46. E | 63. C | 80. D | 97. C | 114. C |
| 13. A | 30. E | 47. C | 64. C | 81. C | 98. E | 115. B |
| 14. D | 31. B | 48. E | 65. A | 82. C | 99. A | |
| 15. A | 32. B | 49. A | 66. D | 83. E | 100. E | |
| 16. E | 33. B | 50. A | 67. A | 84. E | 101. D | |
| 17. A | 34. C | 51. C | 68. C | 85. E | 102. A | |

# 16
# *Solubility and Complex Ion Equilibria*

## SOLUBILITY AND PRECIPITATION EQUILIBRIA

The solubility product constant, $K_{sp}$ is the equilibrium constant for the aqueous *dissolution reaction* involving (most frequently) a slightly soluble strong electrolyte. The equilibrium referred to is between the ions in the *saturated solution* and the undissolved solid. The reverse of dissolution is *precipitation* (or *crystallization*). Since $K_{sp}$ is a measure of the extent of a dissolution (dissolving) process, it is of course related to solubility. Although solubility may be defined in terms of a variety of units (*e.g.*, g/100 mL, %, parts per million, molality), the molar solubility, *S*, the number of moles of a compound that may be dissolved per liter of aqueous solution, has the greatest utility in calculations dealing with $K_{sp}$. The relationship between *S* and $K_{sp}$ may be seen by using the slightly soluble strong electrolyte, $Ag_2CrO_4$, as an example ($K_{sp}$ for $Ag_2CrO_4$ = $9.0 \times 10^{-12}$):

$$Ag_2CrO_{4\,(s)} \rightleftharpoons 2\,Ag^+_{(aq)} + CrO_4^{2-}{}_{(aq)}$$

Dissolving: $S$ molar

EQ (*ss*): $2S$ molar $\quad S$ molar

$$K_{sp} = [Ag^+]^2_{ss}[CrO_4^{2-}]_{ss} = (2S)^2(S) = 4S^3 = 9.0 \times 10^{-12}$$

$$S = \text{solubility of } Ag_2CrO_4 = \sqrt[3]{K_{sp}/4} = 1.3 \times 10^{-4}\ M$$

(*ss* is used to designate a saturated solution)

In addition to its use in calculating solubilities in water, the $K_{sp}$ can be used (1) to predict whether a precipitate will form upon mixing of two solutions, and (2) to calculate the solubility of an ionic compound in the presence of a common ion (*e.g.*, AgCl in 0.1 *M* $AgNO_3$). The solubility of a salt in the presence of a common ion (and in the absence of other equilibria) is less than its solubility in water alone (in accord with Le Chatelier's principle).

To determine whether under given conditions a precipitate will occur, the ion product, $Q_{ip}$, frequently is calculated for initial conditions. The ion product is nothing more than a general reaction quotient for a dissolution reaction. Recalling (from Ch. 14) that $\Delta G_{rxn} = RT \ln Q/K$ and, for this case letting $Q = Q_{ip}$ and $K = K_{sp}$, we have:

| Case | Dissolution Reaction | Precipitation Reaction | Conclusion |
|---|---|---|---|
| $Q_{ip} < K_{sp}$ | $\Delta G < 0$ | $\Delta G > 0$ | **Precipitation will not occur; the solution is** unsaturated, **and more solid may be dissolved until** $Q_{ip} = K_{sp}$. |
| $Q_{ip} > K_{sp}$ | $\Delta G > 0$ | $\Delta G < 0$ | **Precipitation will occur; the solution is** supersaturated, **and solid will crystallize until** $Q_{ip} = K_{sp}$. |
| $Q_{ip} = K_{sp}$ | $\Delta G = 0$ | $\Delta G = 0$ | **The solution is saturated and the system is at equilibrium with constant composition.** |

From the above it is seen that, **if at the instant of mixing,** the $Q_{ip} > K_{sp}$, **precipitation will occur** (at some rate), reducing the ionic concentrations in solution **until the ion product reaches its equilibrium value.** In cases where precipitation occurs upon mixing two solutions, the final composition is best calculated by the following procedure: (a) tentatively assume that the reaction proceeds quantitatively to form the precipitate; and then (b) "back up" the system to equilibrium, using the appropriate $K_{sp}$, taking into consideration sources of common ions, if any.

The $K_{sp}$, like all equilibrium constants, is temperature dependent. If the dissolution reaction is endothermic, $\Delta H_{soln} > 0$, $K_{sp}$ increases with increasing temperature; the reverse is true if $\Delta H_{soln} < 0$.

## Example Problems Involving Solubility and Complex Ion Equilibria

### 16-A. Molar Concentration for a Saturated Solution from $K_{sp}$

Silver sulfate has a $K_{sp}$ of $1.2 \times 10^{-5}$ at 25°C. What is the maximum $[Ag^+]$ that can be attained at 25°C by saturating water with this solid?

(A) $1.4 \times 10^{-2}$ *M* (B) $3.5 \times 10^{-3}$ *M* (C) $3.0 \times 10^{-6}$ *M* (D) 0.028 *M*
(E) 0.0070 *M*

### *Solution*

The dissolution equation for the slightly soluble strong electrolyte $Ag_2SO_4$ is:

$$Ag_2SO_{4\,(s)} \rightleftharpoons 2\,Ag^+_{(aq)} + SO_{4\,(aq)}^{2-}$$

Careful attention must be given to the **stoichiometry** of the dissolution equation in working $K_{sp}$ problems. Let $S$ be the number of moles/liter of $Ag_2SO_4$ that dissolve to give the saturated solution. This gives $[Ag^+] = 2S$, and $[SO_4^{2-}] = S$ molar in the saturated solution. Then,

$$K_{sp} = [Ag^+]^2[SO_4^{2-}] = (2S)^2(S) = 4S^3 = 1.2 \times 10^{-5}$$

$$S = 1.4 \times 10^{-2}\ M$$

This, however, is the **molar solubility** of the compound, $Ag_2SO_4$, at 25°C–it is **not** the $[Ag^+]$. The $[Ag^+]$ has been defined as $2S$ and thus $[Ag^+]_{max} = 0.028\ M$. **[D]**

## 16-B. $K_{sp}$ from the pH of a Saturated Solution of a Slightly Soluble Hydroxide

The pH of an oxygen-free saturated solution of iron(II) hydroxide is 9.40. What is the $K_{sp}$ of $Fe(OH)_2$?

(A) $1.6 \times 10^{-14}$ (B) $1.3 \times 10^{-5}$ (C) $3.0 \times 10^{-10}$ (D) $3.2 \times 10^{-12}$
(E) $7.9 \times 10^{-15}$

### *Solution*

The dissolution equation is:

$$Fe(OH)_{2\,(s)} \rightleftharpoons Fe^{2+}_{(aq)} + 2\,OH^-_{(aq)},\ K_{sp} = [Fe^{2+}][OH^-]^2$$

From the pH, pOH = 4.60 and $[OH^-] = 10^{-4.60}$ or $2.5 \times 10^{-5}$ *M*. From the reaction stoichiometry we note that $[Fe^{2+}] = 1/2[OH^-]$, so that $[Fe^{2+}] = 1.25 \times 10^{-5}$ *M*. Then, for the saturated solution,

$$K_{sp} = [Fe^{2+}][OH^-]^2 = 7.8 \times 10^{-15}$$

Alternatively, had we left $[OH^-] = 10^{-4.60}$ *M* and $[Fe^{2+}] = 1/2 \cdot 10^{-4.60}$ molar, then

$$K_{sp} = [Fe^{2+}][OH^-]^2 = 1/2 \cdot 10^{-4.60}(10^{-4.60})^2 = 1/2 \cdot 10^{-13.80} = 7.9 \times 10^{-15}$$

The fact that the first answer does not exactly agree with (E) is a common problem when dealing with rounded values, particularly when one is involved with exponentials. Compare with 16-A and note that here $S = [Fe^{2+}]$. **[E]**

## 16-C. Common Ion Effect on Molar Solubility

What is the molar solubility of $Ag_2CrO_4$ ($K_{sp} = 9.0 \times 10^{-12}$) in a solution that is 0.10 *M* in $AgNO_3$?

(A) $3.0 \times 10^{-6}$ *M* (B) $2.1 \times 10^{-4}$ *M* (C) $1.3 \times 10^{-10}$ *M*
(D) $9.0 \times 10^{-10}$ *M* (E) $9.0 \times 10^{-11}$ *M*

### *Solution*

This problem illustrates the common ion effect as applied to solubility problems. As in Ch. 15, the common ion must be considered from all sources, *viz.*,

| Dissolution Equation: | $Ag_2CrO_{4\,(s)} =$ | $2\,Ag^+_{(aq)}$ | $+\ CrO_4^{2-}{}_{(aq)}$ |
|---|---|---|---|
| Init.† *M* (from $Ag_2CrO_4$) | | 0 | 0 |
| Init.† *M* (from $AgNO_3$) | | 0.10 | 0 |
| $\Delta$, † *M*: | | $+\ 2S$ | $+S$ |
| EQ† (*ss*), *M*: | | $0.10 + 2S$ | $S$ |

$$K_{sp} = [Ag^+]^2[CrO_4^{2-}] = (0.10 + 2S)^2(S)$$

An "exact" answer would require solution of the above cubic equation. However, even in the absence of the common ion, $S$ must be small, $\sim 10^{-4}$ *M;* because of the common $Ag^+$ already present, it must be even less (Le Chatelier's principle). The net result is that $2S \ll 0.10$, so that the cubic equation may be avoided by the approximation,

$$K_{sp} = [Ag^+]^2[CrO_4^{2-}] = (2S + 0.10)^2(S) \cong (0.10)^2 S$$

giving $S = 9.0 \times 10^{-10}$ *M*. Note that, indeed, $S \ll 0.10$. **[D]**

This approximation is valid almost in every case where there is an appreciable concentration of a common ion present.

## 16-D. Determination of the Hydroxides that Precipitate from a Buffered Solution

A solution is $2.0 \times 10^{-3}$ *M* in $Co^{2+}$, $Fe^{2+}$, and $La^{3+}$. The solution volume is exactly doubled by adding an $NH_3/NH_4Cl$ buffer that maintains the pH of the solution at 8.00. The $K_{sp}$'s for the hydroxides are: $Co(OH)_2$, $2.0 \times 10^{-16}$; $Fe(OH)_2$, $7.9 \times 10^{-15}$; $La(OH)_3$, $1.0 \times 10^{-19}$. Which hydroxides should precipitate?

(A) only $Co(OH)_2$ (B) only $La(OH)_3$ (C) only $La(OH)_3$ and $Co(OH)_2$
(D) $Co(OH)_2$, $Fe(OH)_2$, and $La(OH)_3$ (E) none

### *Solution*

Since the volume is doubled, all metal ions are initially (before allowing for any possible reaction) at a concentration of $1.0 \times 10^{-3}$ *M*. Since the pH = 8.00,

---

†"Init." refers to the time-zero situation, when no $Ag_2CrO_4$ has yet gone into solution; here $\Delta = S$, the number of moles per liter of $Ag_2CrO_4$ that must dissolve to create the saturated solution; EQ refers to the equilibrium condition, *i.e.*, the saturated solution (*ss*).

the pOH = 6.00 and the $[OH^-] = 1.0 \times 10^{-6}$ *M*. From the dissolution equations:

$$Co(OH)_{2\,(s)} = Co^{2+}_{(aq)} + 2\,OH^-_{(aq)} \quad K_{sp} = 2.0 \times 10^{-16}$$
$$Fe(OH)_{2\,(s)} = Fe^{2+}_{(aq)} + 2\,OH^-_{(aq)} \quad K_{sp} = 7.9 \times 10^{-15}$$
$$La(OH)_{3\,(s)} = La^{3+}_{(aq)} + 3\,OH^-_{(aq)} \quad K_{sp} = 1.0 \times 10^{-19}$$

The ion product, $Q_{ip}$, is calculated for each and compared with the appropriate $K_{sp}$.

For $Co(OH)_2$: $Q_{ip} = (1.0 \times 10^{-3})(1.0 \times 10^{-6})^2 = 1.0 \times 10^{-15} > K_{sp}$

For $Fe(OH)_2$: $Q_{ip} = (1.0 \times 10^{-3})(1.0 \times 10^{-6})^2 = 1.0 \times 10^{-15} < K_{sp}$

For $La(OH)_3$: $Q_{ip} = (1.0 \times 10^{-3})(1.0 \times 10^{-6})^3 = 1.0 \times 10^{-21} < K_{sp}$

Since only the $K_{sp}$ of $Co(OH)_2$ is exceeded, this will be the only precipitate to form, the other ions remaining in solution. [A]

## Exercises on Solubility and Precipitation Equilibria

1. The expression for the solubility product of $SrF_2$ is $K_{sp}$ =

   (A) $[Sr^{2+}][2F^-]^2$ (B) $[Sr^+][F^-]$ (C) $[Sr^{2+}][F^-]$ (D) $[Sr^{2+}][F^-]^2$
   (E) $[Sr^{2+}][2F^-]$

2. A number of inorganic salts exhibit brilliant colors, which along with their low solubilities, makes them valuable commercial pigments for paints. Cobalt violet is $Co_3(PO_4)_2$; the expression for its solubility product is $K_{sp}$ equals:

   (A) $3[Co^{2+}] + [PO_4^{3-}]$ (B) $Co_3(PO_4)_2$ (C) $[Co^{2+}]^3[PO_4^{3-}]^2$
   (D) $[Co^{2+}]^3[P^{5+}]^2[O^{2-}]^8$ (E) $[3Co^{2+}]^3[2PO_4^{3-}]^2$

3. The molar solubility, *S*, of $Bi_2S_3$, in terms of its $K_{sp}$, is:

   (A) $S = [K_{sp}]^{1/2}$ (B) $S = [K_{sp}/6]^{1/5}$ (C) $S = [K_{sp}/25]^{1/5}$
   (D) $S = [K_{sp}/108]^{1/5}$ (E) none of these

4. The $K_{sp}$ for the hypothetical electrolyte AB is $1.0 \times 10^{-12}$. Its molar solubility is:

   (A) $1.0 \times 10^{-12}$ *M* (B) $5.6 \times 10^{-2}$ *M* (C) 0.010 *M*
   (D) $1.0 \times 10^{-6}$ *M* (E) 0.0010 *M*

5. The hypothetical salt, $AD_2$, has a molar solubility of $1.0 \times 10^{-7}$ *M*. Calculate the value of its $K_{sp}$.

   (A) $4.0 \times 10^{-21}$ (B) $1.0 \times 10^{-21}$ (C) $1.0 \times 10^{-14}$ (D) $1.0 \times 10^{-7}$
   (E) $2.0 \times 10^{-21}$

6. The $K_{sp}$ of $X(OH)_2$ is $8.0 \times 10^{-12}$. What is the pH of a saturated solution of $X(OH)_2$?

   (A) 10.40 (B) 11.10 (C) 9.70 (D) 10.10 (E) 10.70

7. The solubility of silver sulfide is $2.00 \times 10^{-5}$ g/100 mL. Calculate the $K_{sp}$ for silver sulfide (FW = 248 g/mole).

   (A) $1.28 \times 10^{-12}$ (B) $3.16 \times 10^{-16}$ (C) $4.00 \times 10^{-21}$
   (D) $2.09 \times 10^{-18}$ (E) $5.12 \times 10^{-19}$

8. A saturated solution of the strong base $M(OH)_2$ has a pH of 10.00. Calculate $K_{sp}$ for $M(OH)_2$.

   (A) $5.0 \times 10^{-13}$ (B) $1.0 \times 10^{-12}$ (C) $2.0 \times 10^{-12}$ (D) $5.0 \times 10^{-8}$
   (E) $1.0 \times 10^{-30}$

9. $MF_2$ (FW = 100 g/mole) has a $K_{sp}$ of $4.0 \times 10^{-12}$. What is the aqueous solubility of $MF_2$ in **grams per liter**?

   (A) 0.016 (B) 0.00020 (C) 0.010 (D) 0.040 (E) 0.0040

**Questions 10 through 15 deal with $PbCl_2$ for which $K_{sp} = 1.7 \times 10^{-5}$. In any problems involving mixing of solutions, assume additive volumes.**

10. What is the molar solubility of $PbCl_2$ in 1.0 *M* NaCl?

    (A) $1.6 \times 10^{-2}$ *M* (B) $4.1 \times 10^{-3}$ *M* (C) $1.7 \times 10^{-4}$ *M*
    (D) $2.6 \times 10^{-2}$ *M* (E) $1.7 \times 10^{-5}$ *M*

11. What % by weight of a $2.78 \times 10^{-2}$-g sample of $PbCl_2$ could dissolve when the sample is washed with 1.0 mL of 0.10 *M* HCl?

    (A) 16% (B) 0.041% (C) 1.7% (D) 0.16% (E) 0.38%

12. What is the $[Cl^-]$ in a solution made by mixing 20.0 mL of 1.0 *M* $Pb(NO_3)_2$ with 20.0 mL of 1.0 *M* KCl?

    (A) $4.1 \times 10^{-3}$ *M* (B) $3.4 \times 10^{-4}$ *M* (C) $8.2 \times 10^{-3}$ *M*
    (D) $6.8 \times 10^{-4}$ *M* (E) $1.7 \times 10^{-4}$ *M*

13. Calculate the final $[Pb^{2+}]$ in solution after mixing 50.0 mL of 0.200 *M* $Pb(NO_3)_2$ with 150. mL of 0.100 *M* $AlCl_3$.

    (A) $1.1 \times 10^{-3}$ *M* (B) $2.7 \times 10^{-2}$ *M* (C) $3.4 \times 10^{-4}$ *M*
    (D) $4.0 \times 10^{-5}$ *M* (E) $5.6 \times 10^{-4}$ *M*

14. Calculate the final $[Pb^{2+}]$ in solution after mixing 20 drops of 0.30 *M* lead nitrate with 30 drops of 0.40 *M* $FeCl_3$.

    (A) $7.4 \times 10^{-5}$ *M* (B) $1.6 \times 10^{-2}$ *M* (C) $9.5 \times 10^{-4}$ *M*
    (D) $3.5 \times 10^{-5}$ *M* (E) $2.6 \times 10^{-2}$ *M*

15. The molar solubility of $PbCl_2$ in 0.20 *M* $Pb(NO_3)_2$ solution is:

(A) $1.7 \times 10^{-4}\ M$ (B) $9.2 \times 10^{-3}\ M$ (C) $1.7 \times 10^{-5}\ M$
(D) $4.6 \times 10^{-3}\ M$ (E) $8.5 \times 10^{-5}\ M$

---

## SIMULTANEOUS EQUILIBRIA

There are many situations where two or more equilibria may be important simultaneously in the same solution. The concentrations of the species involved then have to adjust themselves so that all equilibrium conditions are met. We have already considered simultaneous equilibria in the situation where more than one weak acid-base pair was present in a solution, *e.g.*,

$$NH_4^+ + F^- \rightleftharpoons NH_3 + HF$$

and found that $K_c$ for the reaction equals $K_{a(NH_4^+)}/K_{a(HF)}$. Now we consider two additional types of simultaneous equilibria: (a) situations in which one equilibrium involves a *complex ion* and one involves a slightly soluble substance; and (b) situations in which one equilibrium involves weak acids or bases and the other involves a slightly soluble substance.

### Dissolution of Precipitates by Complex Ion Formation

A complex ion usually consists of a metal ion to which one or more molecules or ions (*ligands*) are bound to the central metal ion. Common ligands are $H_2O$, $NH_3$, $CN^-$, $Cl^-$, *etc.*, which may form complex ions such as $Fe(OH_2)_6^{2+}$ (which is $\equiv Fe^{2+}{}_{(aq)}$), $Cu(NH_3)_4^{2+}$, $Ag(NH_3)_2^+$, $Ag(CN)_2^-$, and $Zn(OH)_4^{2-}$. Often complex ions are written enclosed in brackets, *e.g.*, $[Fe(CN)_6]^{4-}$ or $K_4[Fe(CN)_6]$, which would be regarded as 6 $CN^-$ ligands attached to a central $Fe^{2+}$; in one instance it has been shown simply as the ion, in the other as part of the compound. Care must be used in distinguishing this notation from the use of square brackets to designate molar concentrations.

For every complex ion there is a *dissociation reaction,* the extent of which is indicated by an equilibrium constant designated $K_{inst}$ (or also $K_{diss}$), the *instability* (or *dissociation*) *constant.* Thus, for $Ag(NH_3)_2^+$, the reaction to which $K_{inst}$ refers is

$$Ag(NH_3)_2^+{}_{(aq)} \rightleftharpoons Ag^+{}_{(aq)} + 2\,NH_3{}_{(aq)}; \quad K_{inst} = \frac{[Ag^+][NH_3]^2}{[Ag(NH_3)_2^+]}$$

## DRILL ON THE USE OF SOLUBILITY PRODUCT *Student's Name* ______

T1OH ($K_{sp} = 6.3 \times 10^{-1}$), $Mg(OH)_2$ ($K_{sp} = 1.5 \times 10^{-11}$), and $La(OH)_3$ ($K_{sp} = 1.0 \times 10^{-19}$) are readily precipitated by mixing sodium hydroxide solution with solutions of the respective nitrates ($M(NO_3)_n$, where 'M' represents T1, Mg or La and 'n' is the appropriate coefficient). Fill in the following table. You may be asked to turn in this sheet.

| System | A | | B | | C | |
|---|---|---|---|---|---|---|
| | $[T1^+]$ | $[OH^-]$ | $[Mg^{2+}]$ | $[OH^-]$ | $[La^{3+}]$ | $[OH^-]$ |
| Pure water saturated with $M(OH)_n$ | | | | | | |
| 0.10 *M* NaOH saturated with $M(OH)^n$ | | | | | | |
| 0.10 *M* $M(NO_3)_n$ saturated with $M(OH)_n$ | | | | | | |
| 10.0 mL 0.10 *M* $M(NO_3)_n$ + 10.0 mL 0.10 *M* NaOH | | | | | | |
| 20.0 mL 0.010 *M* $M(NO_3)_n$ + 30.0 mL 0.010 *M* NaOH | | | | | | |
| | | | | | | |

and for $Al(OH)_4^-$:

$$Al(OH)_4^-{}_{(aq)} \rightleftharpoons Al^{3+}{}_{(aq)} + 4\,OH^-{}_{(aq)};\quad K_{inst} = \frac{[Al^{3+}][OH^-]^4}{[Al(OH)_4^-]}$$

If we have an equilibrium system consisting of an ammonical solution above a precipitate of AgCl, the species involved must simultaneously meet the conditions governed by the two constants, $K_{inst}$ and $K_{sp}$, *i.e.*,

$$K_{sp} = [Ag^+][Cl^-] \qquad \text{and} \qquad K_{inst} = \frac{[Ag^+][NH_3]^2}{[Ag(NH_3)_2^+]}$$

It must be remembered that as long as a single solution is being considered the symbol $[Ag^+]$ is the silver ion concentration in the solution and, hence, has the same value in both equilibria; the concentrations of other species must adjust accordingly.

## Dissolution of Precipitates by Adjusting Acid-Base Equilibria

The formation or dissolution of a slightly soluble salt depends on the concentrations of the specific component ions in solution. In many instances the anion is readily protonated, thereby forming a weak acid at the expense of the concentration of the anion. The anion concentration then is governed by the pH, which in turn governs precipitate formation or dissolution.

A common example of precipitate formation being determined by pH is the precipitation of sulfides using saturated aqueous solutions of $H_2S$ ($K_{a_1} = 1.0 \times 10^{-7}$; $K_{a_2} = 1.3 \times 10^{-13}$). At 1 atm pressure, saturated aqueous $H_2S$ is 0.10 *M*; the $[HS^-]$ and $[S^{2-}]$ depend upon the $[H_3O^+]$. This dependence on $[H_3O^+]$ may be expressed in the combined equilibrium

$$H_2S_{(aq)} + 2\,H_2O \rightleftharpoons 2\,H_3O^+{}_{(aq)} + S^{2-}{}_{(aq)}$$

$$K_{a_{1,2}} = K_{a_1} \cdot K_{a_2} = \frac{[H_3O^+]^2[S^{2-}]}{[H_2S]} = 1.3 \times 10^{-20}$$

Thus, as the pH is lowered, $[S^{2-}]$ decreases; as the pH is raised, $[S^{2-}]$ increases. The sulfide ion concentration and the specific metal ion(s) that precipitate from a saturated $H_2S$ solution can be regulated by varying the $[H_3O^+]$. This particular procedure is the basis for a number of separations in qualitative analysis.

There are many other examples of simultaneous equilibria but the principles are the same as the complex-ion-precipitation and acid-base-precipitation equilibria discussed here. In some instances the algebraic equations can be quite difficult to solve.

## Example Problems Involving Simultaneous Equilibria

### 16-E. Competitive Equilibria: Complex Ion Formation-Precipitation

The molar solubility of AgBr ($K_{sp}$ = 3.3 X $10^{-13}$) in 3.0 $M$ aqueous ammonia is: ($K_{inst}$ for $Ag(NH_3)_2^+$ is 6.8 X $10^{-8}$)

(A) 5.7 X $10^{-7}$ $M$ (B) 6.6 X $10^{-3}$ $M$ (C) 1.3 X $10^{-3}$ $M$
(D) 4.4 X $10^{-6}$ $M$ (E) 4.4 X $10^{-3}$ $M$

### *Solution*

The dissolution reaction described is:

$$AgBr_{(s)} + 2\,NH_{3\,(aq)} \rightleftharpoons Ag(NH_3)_2^+{}_{(aq)} + Br^-{}_{(aq)}$$

It should be recognized that the silver ion is involved in (or is common to) two equilibria: the $K_{sp}$ equilibrium where it is linked to $AgBr_{(s)}$ and $Br^-$ and the $K_{inst}$ equilibrium where it is linked to ammonia and the diamminesilver(I) complex ion. Then the desired reaction can be expressed as the sum of the $K_{sp}$ equilibrium and the **reverse** of the $K_{inst}$ equilibrium:

$$AgBr_{(s)} \rightleftharpoons Ag^+{}_{(aq)} + Br^-{}_{(aq)} \qquad K_{sp}$$

$$Ag^+{}_{(aq)} + 2\,NH_{3\,(aq)} \rightleftharpoons Ag(NH_3)_2^+{}_{(aq)} \qquad 1/K_{inst}$$

$$\text{Net: } AgBr_{(s)} + 2\,NH_{3\,(aq)} \rightleftharpoons Ag(NH_3)_2^+{}_{(aq)} + Br^-{}_{(aq)} \qquad K_c = \frac{K_{sp}}{K_{inst}}$$

for which $K_c$ = 4.9 X $10^{-6}$. The problem can now be set up by using the notation "*Init.*" (before any reaction occurs), "Δ," and "EQ."

| | $AgBr_{(s)}$ | + $2\,NH_{3\,(aq)}$ | ⇌ $Ag(NH_3)_2^+{}_{(aq)}$ | + $Br^-{}_{(aq)}$ |
|---|---|---|---|---|
| *Init., M:* | | 3.0 | 0 | 0 |
| *Δ, M:* | - S | -2S | + S | + S |
| EQ, *M:* | | 3.0 - 2S | S | S |

$$\text{Then, } K_c = \frac{[Ag(NH_3)_2^+][Br^-]}{[NH_3]^2} = \frac{S^2}{(3.0 - 2S)^2} = 4.9 \times 10^{-6}.$$

From the magnitude of $K_c$ it can be seen that S << 3.0, and the (2$S$) in the denominator may be neglected. Then,

$$4.9 \times 10^{-6} \cong \frac{S^2}{(3.0)^2} \quad \text{or} \quad S = 6.6 \times 10^{-3}\,M \qquad \textbf{[B]}$$

This means that 0.0066 moles of AgBr can be dissolved per liter of 3.0 $M$ ammonia—this should be compared with S in pure $H_2O$, 5.7 X $10^{-7}$ $M$, which is much less. Note also that the ionization of $NH_3$ ($K_b$ = 1.8 X $10^{-5}$; $K_b/C$ = 6 X $10^{-6}$) has been ignored and that the "neglect 2S" approximation was valid; this can be verified by substitution of S back into the original equation—this yields $K_c$ = 4.9 X $10^{-6}$ to 2 significant figures.

## 16-F. Control of Precipitation of Sulfides by Adjusting pH

Given $K_{sp}$'s: FeS, $1.0 \times 10^{-19}$; NiS, $3.0 \times 10^{-21}$. A solution that is 0.020 *M* in both $Ni^{2+}$ and $Fe^{2+}$ is to be saturated with $H_2S$ at 25°C and 1 atm pressure. Over which of the following pH ranges would the NiS be precipitated while leaving the $Fe^{2+}$ in solution? $K_{a_{1,2}}$ for $H_2S = 1.3 \times 10^{-20}$.

(A) 0.00–9.90 (B) 1.10–1.70 (C) 1.90–2.50 (D) 2.70–3.30 (E) 3.50–4.10

### *Solution*

From the $K_{sp}$ expression, $K_{sp} = [M^{2+}][S^{2-}]$ (where $M^{2+} = Fe^{2+}$ or $Ni^{2+}$); precipitation will begin when $Q_{ip} = K_{sp}$; for 0.020 *M* $Ni^{2+}$ this would be at $[S^{2-}] = 1.5 \times 10^{-19}$; for 0.020 *M* $Fe^{2+}$, $[S^{2-}] = 5.0 \times 10^{-18}$. These represent the minimum sulfide ion concentrations needed, so that $Q_{ip}$ will equal $K_{sp}$.

Substitution of the $H_2S$ concentration, 0.10 *M* at 25°C and 1 atm pressure, into the $K_{a_{1,2}}$ equation gives the relationship, $[H_3O^+]^2[S^{2-}] = 1.3 \times 10^{-21}$ for saturated $H_2S$ solution. If we substitute in the minimum sulfide ion concentrations required to start precipitation for each of the metal ions, the required $[H_3O^+]$ and pH can be calculated. Initiation of NiS precipitation requires:

$$[H_3O^+] = \left(\frac{1.3 \times 10^{-21}}{1.5 \times 10^{-19}}\right)^{1/2} = 9.3 \times 10^{-2}\ M \quad \text{or} \quad pH = 1.03$$

and FeS requires:

$$[H_3O^+] = \left(\frac{1.3 \times 10^{-21}}{5.0 \times 10^{-18}}\right)^{1/2} = 1.6 \times 10^{-2}\ M \quad \text{or} \quad pH = 1.79$$

At $[H_3O^+] > 0.093$ (pH < 1.03), **neither** NiS nor FeS will be precipitated (*e.g.*, pH = 1.00). If the $[H_3O^+]$ is lowered (as by gradual addition of a base, such as $NaC_2H_3O_2$) **below** 0.016 *M* (pH > 1.79), **both** NiS and FeS will be precipitated, and no separation of the metals will be achieved. Thus, a good range to shoot for to achieve separation would be **between** pH = 1.03 and pH = 1.79; thus answer (B). The **maximum** degree of separation would be when the pH is raised as high as possible, without risking initiation of FeS precipitation. For example at pH = 1.70, $[H_3O^+] = 0.020\ M$ and $[S^{2-}] = 3.2 \times 10^{-18}$; under these conditions:

$$Q_{ip} = 6.4 \times 10^{-20} \ll K_{sp} \text{ for FeS}; > K_{sp} \text{ for NiS}$$

Also, if $[S^{2-}] = 3.2 \times 10^{-18}$, the $[Ni^{2+}]$ at equilibrium would be $9.4 \times 10^{-4}$ *M*, meaning that only about 5% of the original 0.020 *M* $Ni^{2+}$ would remain in solution, while 100% of the original $Fe^{2+}$ remains in solution.

## Exercises on Simultaneous Equilibria

16. Silver ion forms a slightly soluble iodide and also a very stable complex

ion with $CN^-$. In terms of $K_{sp}$ and $K_{inst}$ find the equilibrium constant for the reaction:

$$AgI_{(s)} + 2\,CN^-_{(aq)} \rightleftharpoons Ag(CN)_2^-{}_{(aq)} + I^-_{(aq)}$$

(A) $K_{inst}/K_{sp}$ (B) $K_{sp}/(K_{inst})^2$ (C) $K_{sp}\cdot(K_{inst})^2$ (D) $K_{sp}/K_{inst}$
(E) $K_{sp}\cdot K_{inst}$

17. The salt, $K_3[Fe(CN)_6]$, provides $K^+$ and the complex ion, $Fe(CN)_6^{3-}$, in solution. If $K_{inst} = 1.0 \times 10^{-44}$, calculate the free $[CN^-]$ in a 0.10 *M* $K_3[Fe(CN)_6]$ solution.

    (A) 0.60 *M* (B) $8.0 \times 10^{-8}$ *M* (C) $3.7 \times 10^{-7}$ *M* (D) $3.2 \times 10^{-8}$ *M*
    (E) $4.8 \times 10^{-7}$ *M*

18. If you add ammonia to pure liquid water, the concentration of a chemical species already present will decrease. This species is:

    (A) $O^{2-}$ (B) $H_3O^+$ (C) $H_2O$ (D) $OH^-$ (E) $H^-$

19. An aqueous solution buffered at pH = 4.00 is saturated with hydrogen sulfide. What is the concentration of $S^{2-}$ in this solution?

    (A) $1.3 \times 10^{-9}$ *M* (B) $1.3 \times 10^{-13}$ *M* (C) 0.000013 *M*
    (D) $1.3 \times 10^{-17}$ *M* (E) $1.3 \times 10^{-7}$ *M*

20. A solution at pH = 1.0 contains $[H_2S] = 1.0$ *M* and $Ni^{2+}$ and $Mn^{2+}$ ions at 0.02 *M* each. The $K_{sp}$ values for the sulfides are NiS, $\sim 10^{-21}$ and MnS, $\sim 10^{-16}$. It can be concluded that

    (A) the $[H_3O^+] = 1$ *M* (B) the $[S^{2-}] = 1.1 \times 10^{-21}$ *M*
    (C) both NiS and MnS will precipitate
    (D) neither NiS nor MnS will precipitate
    (E) NiS should precipitate, MnS should not

21. An aqueous solution that contains 0.10 *M* $Mn^{2+}$ is saturated with $H_2S$ at 1 atm and 25°C. If this solution is buffered at pH = 4.00, how much manganese(II) ion will remain in solution? $K_{sp}$ for MnS is $6.0 \times 10^{-16}$.

    (A) 0.10 *M* (B) 0.50 *M* (C) $4.6 \times 10^{-3}$ *M* (D) $1.0 \times 10^{-4}$ *M*
    (E) $1.0 \times 10^{-5}$ *M*

22. An aqueous solution, saturated with $H_2S$ at 1 atm and 25°C, is 0.10 *M* in $CoCl_2$ and also in $MnCl_2$. Calculate the pH range in which CoS may be precipitated without precipitating MnS. $K_{sp}$ for CoS = $3.0 \times 10^{-21}$; for MnS, $6.0 \times 10^{-16}$.

    (A) 1.0-2.3 (B) 1.5-2.7 (C) 0.0-2.7 (D) 3.2-3.7 (E) 0.7-3.2

23. Calculate the $[Cd^{2+}]$ that will remain in a 0.30 *M* HCl solution saturated with $H_2S$. ($K_{sp}$ for CdS = $3.6 \times 10^{-29}$)

(A) $8.3 \times 10^{-10}$ *M* (B) $6.0 \times 10^{-15}$ *M* (C) 0.30 *M* (D) $2.5 \times 10^{-9}$ *M*
(E) $1.4 \times 10^{-20}$ *M*

**For Questions 24-26 consider the following:**

**Equal volumes of 0.20 *M* $AgNO_3$ and 2.00 *M* ammonia are mixed.**

$$K_{inst} = 6.8 \times 10^{-8} \text{ for } Ag(NH_3)_2^+$$

24. What equilibrium concentration of the complex ion, $Ag(NH_3)_2^+$, should result?

    (A) $3.5 \times 10^{-10}$ *M* (B) 0.10 *M* (C) $1.1 \times 10^{-8}$ *M* (D) 0.0035 *M*
    (E) $1.0 \times 10^{-10}$ *M*

25. What equilibrium concentration of "free" ammonia molecules, $[NH_3]$, should result?

    (A) $2.8 \times 10^{-9}$ *M* (B) 1.0 *M* (C) $2.8 \times 10^{-8}$ *M* (D) $8.0 \times 10^{-10}$ *M*
    (E) 0.80 *M*

26. About what % of the total $Ag^+$ would remain "free" (as $Ag^+_{(aq)}$), as opposed to that bound in the complex ion?

    (A) $\sim 10^{-1}$% (B) $\sim 10^{-2}$% (C) $\sim 10^{-3}$% (D) $\sim 10^{-5}$% (E) $\sim 10^{-6}$%

27. Suppose that *xs* solid AgBr ($K_{sp} = 3.3 \times 10^{-13}$) were added to one liter of 2.0 *M* ammonia:

    $$AgBr_{(s)} + 2\, NH_3 = Ag(NH_3)_2^+ + Br^-$$

    What bromide-ion concentration should result when equilibrium has been established? ($K_{inst} = 6.8 \times 10^{-8}$ for $Ag(NH_3)_2^+$)

    (A) $4.9 \times 10^{-6}$ *M* (B) 1.0 *M* (C) $3.5 \times 10^{-10}$ *M* (D) 0.00033 *M*
    (E) $4.4 \times 10^{-3}$ *M*

28. $K_{sp}$ for $M_2SO_4$ is $1.08 \times 10^{-7}$. The solubility of $M_2SO_4$ is:

    (A) $6.00 \times 10^{-3}$ *M* (B) $1.08 \times 10^{-7}$ *M* (C) $3.00 \times 10^{-3}$ *M*
    (D) $3.30 \times 10^{-4}$ *M* (E) $4.76 \times 10^{-3}$ *M*

29. $K_{sp}$ for $X(OH)_2$ is $6.0 \times 10^{-12}$. What is the molar solubility of $X(OH)_2$ in a 0.0010 *M* NaOH solution?

    (A) $3.0 \times 10^{-9}$ *M* (B) $6.0 \times 10^{-9}$ *M* (C) $1.2 \times 10^{-10}$ *M*
    (D) $6.0 \times 10^{-6}$ *M* (E) $1.5 \times 10^{-6}$ *M*

30. If $Q_{ip} < K_{sp}$ for a solution of a slightly soluble electrolyte, which of the following is **true**?

    (A) The solution is saturated. (B) No precipitate will form.

(C) A solid will form. (D) The solution pH = 7.00.
(E) The solution is supersaturated and a precipitate will form.

31. $K_{sp}$ for $Zn(OH)_2$ is $4.5 \times 10^{-17}$ and $K_{inst}$ for the complex ion $Zn(OH)_4^{2-}$ is $3.6 \times 10^{-16}$. What is $K_c$ for the reaction?

$$Zn(OH)_{2\,(s)} + 2\,OH^-_{(aq)} \rightleftharpoons Zn(OH)_4^{2-}{}_{(aq)}$$

(A) 8.0 (B) $1.6 \times 10^{-32}$ (C) 0.90 (D) 0.016 (E) 0.12

32. Which of the following metal sulfides is the most soluble in water?

(A) CdS ($K_{sp} = 3.6 \times 10^{-29}$) (B) FeS ($K_{sp} = 1 \times 10^{-19}$)
(C) MnS ($K_{sp} = 5.6 \times 10^{-16}$) (D) ZnS ($K_{sp} = 1.1 \times 10^{-21}$)
(E) HgS ($K_{sp} = 3 \times 10^{-53}$)

33. $K_{sp}$ for AgCl is $1.2 \times 10^{-10}$. If you mix 10 mL of 0.01 *M* $AgNO_3$ with 10 mL of 0.01 *M* NaCl, then:

(A) the solution will remain clear (no precipitation)
(B) silver chloride (AgCl) will precipitate
(C) the concentration of $Ag^+$ will be 0.05 *M*
(D) silver nitrate will precipitate
(E) sodium chloride will precipitate

34. The $K_{sp}$ for thallium(I) chloride, TlCl, is $1.0 \times 10^{-4}$. What is the maximum number of milligrams of TlCl that can be dissolved in 1.0 mL of water?

(A) 10. (B) 1.0 (C) 2.4 (D) 24 (E) 240

**For Questions 35 through 40 use $K_{sp}$ for AgBr = $3.3 \times 10^{-13}$, $K_{sp}$ for AgCl = $1.2 \times 10^{-10}$, and $K_{inst}$ for $Ag(NH_3)_2^+$ = $6.8 \times 10^{-8}$.**

35. Calculate the equilibrium constant for the dissolution of silver bromide in aqueous ammonia.

(A) $3.3 \times 10^{-13}$ (B) $6.8 \times 10^{-8}$ (C) $2.2 \times 10^{-20}$ (D) $4.9 \times 10^{-6}$
(E) $2.1 \times 10^5$

36. The molar solubility of AgBr in 2.0 *M* aqueous ammonia is:

(A) $5.7 \times 10^{-7}$ *M* (B) 2.0 *M* (C) $3.3 \times 10^{-13}$ *M* (D) $6.8 \times 10^{-8}$ *M*
(E) $4.4 \times 10^{-3}$ *M*

37. What is the $[Ag^+]$ remaining in solution when equal volumes of 2.0 *M* $NH_3$ and 0.010 *M* $AgNO_3$ are mixed?

(A) $3.6 \times 10^{-5}$ *M* (B) $3.4 \times 10^{-10}$ *M* (C) $1.7 \times 10^{-8}$ *M*
(D) $1.8 \times 10^{-5}$ *M* (E) $1.7 \times 10^{-10}$ *M*

38. What is the solubility of AgCl in 0.10 *M* NaCl?

(A) $1.1 \times 10^{-5}$ *M* (B) $1.2 \times 10^{-10}$ *M* (C) $1.2 \times 10^{-9}$ *M*
(D) $4.2 \times 10^{-5}$ *M* (E) none of these

39. What is the equilibrium constant for the dissolution of AgCl in 0.50 *M* $NH_3$?

    (A) $1.2 \times 10^{-10}$ (B) $6.8 \times 10^{-8}$ (C) $8.2 \times 10^{-18}$ (D) $1.8 \times 10^{-3}$ (E) $5.7 \times 10^{-2}$

40. What is the molar solubility of AgCl in 0.50 *M* $NH_3$?

    (A) 0.042 *M* (B) 0.036 *M* (C) 0.019 *M* (D) 0.40 *M* (E) 0.25 *M*

41. Which of the following compounds has the **largest** molar solubility in water?

    (A) $PbC_2O_4$, $K_{sp} = 2.74 \times 10^{-11}$ (B) $Ag_3PO_4$, $K_{sp} = 1.8 \times 10^{-18}$ (C) $PbCrO_4$, $K_{sp} = 1.8 \times 10^{-14}$ (D) $CuCo_3$, $K_{sp} = 1.37 \times 10^{-10}$ (E) $Cd(OH)_2$, $K_{sp} = 1.2 \times 10^{-14}$

42. Which of the following compounds has the **largest** molar solubility in water?

    (A) AgCl, $K_{sp} = 1.2 \times 10^{-10}$ (B) $BaCrO_4$, $K_{sp} = 2 \times 10^{-10}$ (C) $CdCO_3$, $K_{sp} = 2.5 \times 10^{-14}$ (D) $Mn(OH)_2$, $K_{sp} = 4.5 \times 10^{-14}$ (E) $MgNH_4PO_4$, $K_{sp} = 2.5 \times 10^{-13}$

43. How many moles of $MgF_2$ will dissolve in 1.0 liter of 0.20 *M* NaF ($K_{sp}$ for $MgF_2 = 8.0 \times 10^{-8}$)?

    (A) $5.0 \times 10^{-7}$ mole (B) $2.0 \times 10^{-7}$ mole (C) 0.20 mole (D) $8.0 \times 10^{-8}$ mole (E) $2.0 \times 10^{-6}$ mole

44. $K_{sp} = 1.2 \times 10^{-10}$ for AgCl. Calculate the final $[Ag^+]$ in solution after mixing 50. mL of 0.400 *M* $AgNO_3$ with 50. mL of 0.500 *M* $AlCl_3$.

    (A) $2.4 \times 10^{-9}$ *M* (B) $2.2 \times 10^{-10}$ *M* (C) $8.0 \times 10^{-11}$ *M* (D) $1.1 \times 10^{-10}$ *M* (E) $3.9 \times 10^{-10}$ *M*

45. $K_{sp}$ for $PbSO_4 = 1.8 \times 10^{-8}$. The maximum concentration of $Pb^{2+}$ that can exist in a solution in which the sulfate ion concentration is maintained at 0.0045 *M* is:

    (A) $2.5 \times 10^{-5}$ *M* (B) $6.3 \times 10^{-8}$ *M* (C) $2.7 \times 10^{-8}$ *M* (D) $1.8 \times 10^{-8}$ *M* (E) $4.0 \times 10^{-6}$ *M*

46. $K_{sp} = 7.9 \times 10^{-6}$ for calcium hydroxide. What is the molar solubility of calcium hydroxide in water?

    (A) 0.0079 *M* (B) 0.0028 *M* (C) 0.013 *M* (D) 0.0014 *M* (E) 0.020 *M*

47. The pH of a saturated solution of calcium hydroxide is 12.40. What is $K_{sp}$ for calcium hydroxide?

    (A) $6.3 \times 10^{-4}$ (B) $1.6 \times 10^{-5}$ (C) $3.2 \times 10^{-38}$ (D) $1.6 \times 10^{-25}$ (E) $7.8 \times 10^{-6}$

48. A solution is 0.010 *M* in each of $Pb(NO_3)_2$, $Mn(NO_3)_2$, and $Zn(NO_3)_2$.

Solid NaOH is added until the pH of the solution is 8.00. $K_{sp} = 2.8 \times 10^{-16}$ for $Pb(OH)_2$, $4.5 \times 10^{-14}$ for $Mn(OH)_2$, and $4.5 \times 10^{-17}$ for $Zn(OH)_2$. Which is **true**?

(A) there will be no precipitate formed
(B) only $Zn(OH)_2$ will precipitate
(C) only $Mn(OH)_2$ will precipitate
(D) all three hydroxides will precipitate
(E) only $Zn(OH)_2$ and $Pb(OH)_2$ will precipitate

49. $K_{sp} = 5.6 \times 10^{-20}$ for $Cu(OH)_2$. At what pH will $Cu(OH)_2$ start to precipitate from a solution with $[Cu^{2+}] = 0.20\ M$?

(A) 0.28 (B) 3.76 (C) 6.62 (D) 4.72 (E) 7.38

50. $K_{sp} = 1.9 \times 10^{-33}$ for $Al(OH)_3$. At what pH will $Al(OH)_3$ start to precipitate from a solution with $[Al^{3+}] = 0.010\ M$?

(A) 3.76 (B) 5.94 (C) 5.46 (D) 8.54 (E) 4.24

51. A solution that is $1.0 \times 10^{-3}\ M$ each in $Cr^{3+}$, $Al^{3+}$, and $Fe^{3+}$ is adjusted to, and maintained at, pH = 4.00. What, if any, precipitate is formed? $K_{sp} = 6.7 \times 10^{-31}$ for $Cr(OH)_3$, $1.9 \times 10^{-33}$ for $Al(OH)_3$, and $1.1 \times 10^{-36}$ for $Fe(OH)_3$.

(A) no precipitate forms (B) only $Cr(OH)_3$ precipitates
(C) only $Fe(OH)_3$ precipitates (D) all three precipitate
(E) only $Fe(OH)_3$ and $Al(OH)_3$ precipitate

52. A saturated solution of calcium phosphate is in contact with some of the solid. The addition of a small amount of strong acid to the system will:

(A) increase the concentration of $PO_4^{3-}$ in the solution
(B) decrease the concentration of $Ca^{2+}$ in the solution
(C) increase the amount of solid in contact with the solution
(D) raise the pH of the solution a **very** small amount
(E) produce none of the preceding changes

53. Calculate the molar solubility of $PbBr_2$ in 0.10 $M$ $Pb(NO_3)_2$. $K_{sp}$ for $PbBr_2$ is $6.3 \times 10^{-6}$

(A) 0.019 $M$ (B) 0.0040 $M$ (C) 0.0079 $M$ (D) 0.00012 $M$
(E) 0.012 $M$

54.*A solution is made up to contain $[Cd^{2+}] = 0.40\ M$ and $[H_3O^+] = 0.10\ M$. Then the solution was treated with hydrogen sulfide until it was saturated. If $K_{sp} = 3.6 \times 10^{-29}$ for CdS, what concentration of $Cd^{2+}$ remains at equilibrium?

(A) $3.6 \times 10^{-28}\ M$ (B) 0.30 $M$ (C) $2.8 \times 10^{-10}\ M$ (D) $2.2 \times 10^{-8}\ M$
(E) $2.8 \times 10^{-15}\ M$

**For Questions 55-59, consider the following:**

The solutions below are maintained **saturated** with hydrogen sulfide at 25°C. At this temperature the actual concentration of $H_2S$ molecules is 0.10-molar.

| For $H_2S$: | $K_{a_1} = 1.0 \times 10^{-7}$ | $K_{sp}$'s: | MnS, $6.0 \times 10^{-16}$ |
|---|---|---|---|
| | $K_{a_2} = 1.3 \times 10^{-13}$ | | FeS, $1.0 \times 10^{-19}$ |

I. $H_2S_{(aq)}$

II. $H_2S_{(aq)}$, adjusted to pH = 7.0 by adding NaOH

III. $H_2S_{(aq)}$ + $[Mn^{2+}]$ = 0.0010 $M$, adjusted to pH = 2.0 by adding HCl

IV. $H_2S_{(aq)}$ + $[Fe^{2+}]$ = 0.0010 $M$, adjusted to pH = 2.0 by adding HCl

V. $H_2S_{(aq)}$ + $[Fe^{2+}]$ = $[Mn^{2+}]$ = 0.0010 $M$

55. What is the pH of Solution I?

    (A) 1.00 (B) 2.00 (C) 3.00 (D) 4.00 (E) 5.00

56. What is the concentration of $S^{2-}$ in Solution II?

    (A) $4.2 \times 10^{-5}$ $M$ (B) $1.3 \times 10^{-7}$ $M$ (C) $2.6 \times 10^{-9}$ $M$
    (D) $1.0 \times 10^{-11}$ $M$ (E) $1.3 \times 10^{-13}$ $M$

57. What % of the total $Mn^{2+}$ will **be precipitated** as MnS in Solution III, once equilibrium is established?

    (A) 0% (B) 50% (C) 90% (D) 95% (E) essentially 100%

58. What % of the total $Fe^{2+}$ will **be precipitated** as FeS in Solution IV, once equilibrium is established?

    (A) 0% (B) 50% (C) 90% (D) 95% (E) 99%

59. For Solution **V**, which of the following is the MAXIMUM pH-value at which FeS **would be** precipitated, but MnS **would not**?

    (A) 1.0 (B) 2.0 (C) 3.0 (D) 4.0 (E) 5.0

60.*Magnesium hydroxide ($K_{sp} = 1.5 \times 10^{-11}$) may be dissolved by the acid, $NH_4^+$, yielding $Mg^{2+}$, $NH_3$, and $H_2O$. **Per liter**, how many moles of solid $NH_4NO_3$ are needed to dissolve the magnesium hydroxide in 100 "ant-acid tablets" (290 mg of magnesium hydroxide/tablet)?

    (A) 1.0 (B) 3.3 (C) 11 (D) 0.55 (E) 4.3

61.*Calculate the pH of a $1.00 \times 10^{-8}$ $M$ $HNO_3$ solution. ($K_w$ for water = $1.00 \times 10^{-14}$)

    (A) 7.00 (B) 6.98 (C) 8.00 (D) 6.91 (E) 6.86

62. In calculating $[H_3O^+]$ and $[OH^-]$ of solutions of acids and bases, the contribution of these ions from the dissociation of water has usually been

ignored. For which of the following solutions would we have to include the contribution from water to find the $[H_3O^+]$ to 2 significant figures?

(A) $5.0 \times 10^{-6}$ $M$ $HNO_3$
(B) $1.0 \times 10^{-3}$ $M$ $HC_2H_3O_2$ ($K_a = 1.8 \times 10^{-5}$)
(C) saturated $Mg(OH)_2$ ($K_{sp} = 1.5 \times 10^{-11}$)
(D) $1.0 \times 10^{-3}$ $M$ $NH_3$ ($K_b = 1.8 \times 10^{-5}$)
(E) saturated $Fe(OH)_3$ ($K_{sp} = 1.1 \times 10^{-36}$)

63.*$K_{sp}$ for $Sn(OH)_2$ is $5.0 \times 10^{-26}$. What is the molar solubility of $Sn(OH)_2$ in water?

(A) $2.3 \times 10^{-9}$ $M$ (B) $6.1 \times 10^{-11}$ $M$ (C) $5.0 \times 10^{-12}$ $M$
(D) $3.7 \times 10^{-9}$ $M$ (E) $1.3 \times 10^{-12}$ $M$

64.*$K_{sp}$ for cobalt(III) hydroxide is $2.5 \times 10^{-43}$. What is the molar solubility of cobalt(III) hydroxide in water?

(A) $9.8 \times 10^{-12}$ $M$ (B) $2.5 \times 10^{-22}$ $M$ (C) $5.0 \times 10^{-22}$ $M$
(D) $9.6 \times 10^{-23}$ $M$ (E) $2.2 \times 10^{-11}$ $M$

65. About how many milligrams of silver iodide (FW = 235; $K_{sp} = 8.5 \times 10^{-17}$) could be dissolved in a 500,000-liter swimming pool containing pure water?

(A) ~0.001 mg (B) ~0.01 mg (C) ~0.1 mg (D) ~100 mg
(E) ~1000 mg

**For Questions 66–69, consider the following:**

A 100.-mL mixture is prepared by mixing 20.0 mL of 0.100 $M$ $La(NO_3)_3$ with 80.0 mL of 0.100 $M$ KOH.

$$K_{sp} = 1.00 \times 10^{-19} \text{ for } La(OH)_3$$

66. At the instant of mixing, the ion product ($Q_{ip}$) for lanthanum hydroxide will exceed its $K_{sp}$ by **a factor** of:

(A) $\sim 10^8$ (B) $\sim 10^{10}$ (C) $\sim 10^{12}$ (D) $\sim 10^{14}$ (E) $\sim 10^6$

67. The % of the total $La^{3+}$ that will remain in solution, once equilibrium is established, is:

(A) $>10^{-5}\%$ (B) between $10^{-5}\%$ and $10^{-7}\%$
(C) between $10^{-7}\%$ and $10^{-9}\%$ (D) between $10^{-9}\%$ and $10^{-10}\%$
(E) $<10^{-10}\%$

68. What will be the pH of the eventual solution?

(A) ~8.2 (B) ~9.4 (C) ~10.6 (D) ~11.5 (E) ~12.3

69. Suppose that now (after #67, 68 above) the pH of the solution is lowered to a final equilibrium value of 7.00 by adding a few drops of concentrated

strong acid to the resulting mixture (assume constant volume). It could be concluded that:

(A) the La $(OH)_{3s}$ would **partially** dissolve
(B) the La $(OH)_{3s}$ would **completely** dissolve
(C) the amount of La $(OH)_{3s}$ would remain unchanged
(D) the amount of La $(OH)_{3s}$ would increase
(E) None of the above

70. A saturated solution of lanthanum hydroxide, $La(OH)_3$, has a pH of 9.37. What is the $K_{sp}$ for this compound?

(A) $1.3 \times 10^{-14}$ (B) $1.1 \times 10^{-38}$ (C) $3.3 \times 10^{-38}$ (D) $1.0 \times 10^{-19}$
(E) $3.0 \times 10^{-19}$

71. Find the $[Cl^-]$ in a saturated solution of *calomel,* mercury(I) chloride ($Hg_2Cl_2$), given its $K_{sp} = 1.1 \times 10^{-18}$?

(A) $1.3 \times 10^{-6}$ $M$ (B) $1.0 \times 10^{-9}$ $M$ (C) $6.5 \times 10^{-7}$ $M$
(D) $2.0 \times 10^{-9}$ $M$ (E) $5.0 \times 10^{-10}$ $M$

72. One-quarter of a mole of $La_2(SO_4)_3$ has been dissolved in sufficient water to make 1.0 liter of solution. This solution is now to be saturated with calcium sulfate ($K_{sp} = 2.4 \times 10^{-5}$). Approximately how many milligrams of calcium sulfate can be dissolved in the solution?

(A) ~1 mg (B) ~4 mg (C) ~10 mg (D) ~50 mg (E) ~200 mg

73. A solution is $2.0 \times 10^{-3}$ $M$ in **each** of the following ions: $Co^{2+}$, $Fe^{2+}$, and $La^{3+}$. The solution volume **is exactly doubled** while adding base until the pH reaches 8.50. Which metallic hydroxides should precipitate?

$K_{sp}$'s: $Co(OH)_2$, $2.0 \times 10^{-16}$; $Fe(OH)_2$, $7.9 \times 10^{-15}$;
$La(OH)_3$, $1.0 \times 10^{-19}$

(A) only $Co(OH)_2$ (B) only $Co(OH)_2$ and $Fe(OH)_2$
(C) only $Co(OH)_2$ and $La(OH)_3$ (D) only $La(OH)_3$
(E) all three: $Co(OH)_2$, $Fe(OH)_2$, and $La(OH)_3$

74. For $PbCl_2$, $K_{sp} = 1.7 \times 10^{-5}$. What chloride ion concentration would be left in solution after mixing 20.0 mL of 0.580-molar $Pb(NO_3)_2$ with 20.0 mL of 0.0800-molar KCl? (Assume final volume = 40.0 mL.)

(A) $6.3 \times 10^{-5}$ $M$ (B) 0.0063 $M$ (C) 0.0058 $M$ (D) 0.040 $M$
(E) 0.0079 $M$

75. Equal volumes of 0.20 $M$ $Cu^{2+}$ and 1.20 $M$ $NH_3$ are mixed. $K_{inst}$ for $Cu(NH_3)_4^{2+}$ is $2.1 \times 10^{-13}$. Approximately what percent of the copper would be present in the solution as uncomplexed $Cu^{2+}$?

(A) 10% (B) 1% (C) $10^{-2}$% (D) $10^{-5}$% (E) $10^{-8}$%

76. For the red modification of HgO, $K = 3.0 \times 10^{-26}$ for

$$HgO_{(s, red)} + H_2O_{(\ell)} \rightleftharpoons Hg^{2+} + 2\,OH^-$$

Given that $K_{inst} = 1.1 \times 10^{-16}$ for $HgCl_4^{2-}$, find the pH of a 0.10 molar NaCl solution that is saturated with HgO. This is expressed by the equilibrium:

$$HgO_{(s)} + H_2O_{(\ell)} + 4\,Cl^- \rightleftharpoons HgCl_4^{2-} + 2\,OH^-$$

(A) 5.59 (B) 12.76 (C) 8.48 (D) 9.58 (E) 7.95

77.*Equal volumes of 0.050 *M* magnesium chloride and 0.50 *M* ammonia are to be mixed. **Per liter,** what is the minimum number of grams of solid ammonium chloride that need be added to prevent precipitation of magnesium hydroxide, $K_{sp} = 1.5 \times 10^{-11}$?

(A) 1.0 g (B) 5.0 g (C) 10. g (D) 15 g (E) 20. g

78. A solution at pH = 1.0 contains $[H_2S] = 0.1\ M$ and $[Ni^{2+}] = [Mn^{2+}] = 0.02\ M$. Given the $K_{sp}$ values: NiS, $10^{-21}$; MnS, $10^{-15}$. It can be concluded that:

(A) the $[S^{2-}]$ is $1.3 \times 10^{-21}$
(B) both NiS and MnS would precipitate
(C) neither NiS nor MnS would precipitate
(D) NiS would precipitate; MnS would not precipitate
(E) MnS would precipitate; NiS would not precipitate

---

## ANSWERS TO CHAPTER 16 PROBLEMS

| | | | | | |
|---|---|---|---|---|---|
| 1. D | 14. A | 27. E | 40. C | 53. B | 66. D |
| 2. C | 15. D | 28. C | 41. B | 54. D | 67. E |
| 3. D | 16. D | 29. D | 42. E | 55. D | 68. E |
| 4. D | 17. E | 30. B | 43. E | 56. B | 69. B |
| 5. A | 18. B | 31. E | 44. B | 57. A | 70. D |
| 6. A | 19. B | 32. C | 45. E | 58. A | 71. A |
| 7. D | 20. E | 33. B | 46. C | 59. D | 72. B |
| 8. A | 21. C | 34. C | 47. E | 60. E | 73. B |
| 9. C | 22. E | 35. D | 48. E | 61. B | 74. E |
| 10. E | 23. D | 36. E | 49. D | 62. E | 75. E |
| 11. C | 24. B | 37. B | 50. A | 63. C | 76. D |
| 12. C | 25. E | 38. C | 51. C | 64. B | 77. C |
| 13. A | 26. D | 39. D | 52. E | 65. E | 78. D |

# 17
# *Electrochemistry*

Electrochemistry deals with the relationship between electricity and chemical change, in particular:

1. Nonspontaneous oxidation-reduction reactions that are forced to occur by passing an electric current through a condensed phase (usually a liquid or a solution). In such cases, the system is called an *electrolytic cell* and the reaction is referred to as an *electrolysis.*

2. Spontaneous oxidation-reduction (redox) reactions that are allowed to occur in such a way that the reducing agent (Red.) is isolated from the oxidizing agent (Ox.) and the **spontaneous** electron-transfer (Red. $\xrightarrow{e^-}$ Ox.) is made through an external conductor that connects the *oxidation half-reaction* (Red. $\rightarrow$ Ox. + $ne^-$) to the *reduction half-reaction* (Ox$'$. + $ne^-$ $\rightarrow$ Red$'$.). Such a system **produces** useful electrical energy and is called a *galvanic* (or *voltaic*) *cell, i.e.,* a "battery."

Quantity of electric charge is measured in *coulombs,* C, or in *faradays,* $\mathscr{F}$ (equal to $9.65 \times 10^4$ coulombs). A coulomb is equal to the product, (amperes)(seconds), in which the current in *amperes* (A) is the **rate of flow** of electrical charge, and the time ($t$), in seconds, during which the charge flow occurs. Thus, $C = A \cdot t$ is a relationship fundamental to electrochemistry. The **faraday is of particular importance** because this **represents the total charge of one mole of electrons.** Then $C/\mathscr{F}$ equals the number of moles of electrons involved in an electrochemical process (*i.e.,* the **number** of faradays transferred). This allows faradays (as moles of electrons) to be handled stoichiometrically in chemical equations.

No charge flow will occur, however, if there is no force, or "push," to cause the electric current. The unit analogous to force is the *volt* ($V$), the *potential difference,* or *electromotive force* (*emf,* $\mathscr{E}$). As in work energy (work = force $\times$ distance moved), electrical energy is (force)(charge moved) = $\mathcal{E} \times C = n\mathcal{F}\mathcal{E}$; which is equal to the free energy change, $\Delta G$ (in joules, J) for a chemical reaction, *i.e.*, $\Delta G = -n\mathcal{F}\mathcal{E}$. In a galvanic cell, $\mathcal{E}$ is positive (voltage is **supplied** spontaneously–the force is inherent in the chemical species involved in the reaction); in an electrolytic cell, $\mathcal{E} < 0$, meaning that voltage must be **applied** in order to force the nonspontaneous redox reaction to occur.

Every redox reaction can be expressed as the sum of an oxidation and a reduction half-reaction (*half-cell*). The potential, $\mathcal{E}$, of a redox reaction can be measured; it is the sum of the potentials for the two half-reactions involved,

$$\mathcal{E}_{cell} = \mathcal{E}_{oxid'n\ 1/2\text{-}rxn} + \mathcal{E}_{red'n\ 1/2\text{-}rxn}$$

If the various species involved in the reaction are at standard-state conditions, then a superscript zero is appended to the potentials, $\mathcal{E}^0$, and these then are referred to as standard electrode or cell potentials. By defining $\mathcal{E}^0$ for one half-reaction (the standard hydrogen electrode, S.H.E.), $\mathcal{E}^0$ values for other half-reactions can be readily tabulated. A table of standard potentials for **reduction** half-reactions (at 25°C) relative to $\mathcal{E}^0 = 0.00$ for the S.H.E. is given on p. 313 for reference in solving the exercises.

For conditions other than standard-state conditions, we may substitute $\Delta G = -n\mathcal{F}\mathcal{E}$ and $\Delta G^\circ = -n\mathcal{F}\mathcal{E}^0$ into the equation (*Cf.* Ch. 13),

$$\Delta G = \Delta G^\circ + RT \ln Q$$

to yield the *Nernst equation* (which may be applied to a cell reaction or to a half-cell):

$$\mathcal{E} = \mathcal{E}^0 - \frac{RT}{n\mathcal{F}} \ln Q$$

or at 25°C, with $\ln Q = 2.303 \log Q$, and $R$ in electrical units (8.314 J/mole-K),

$$\mathcal{E} = \mathcal{E}^0 - \frac{0.0592}{n} \log Q$$

where "$n$" is the number of electrons transferred in the balanced redox equation, and $Q$ is the general reaction quotient. When an electrochemical cell comes to equilibrium, $\Delta G = 0$ and $\mathcal{E} = 0$, since no net chemical reaction occurs. At this point the reaction quotient ($Q$) is equal to the equilibrium quotient ($K$, the equilibrium constant for the reaction). Then just as $\Delta G^\circ$ was related to $K$ for any reaction, we have, for a **redox** reaction, $\log K = 16.9n\mathcal{E}^0$.

A table of electrode potentials† is very useful for making predictions concerning the spontaneity and extent of redox reactions. As the table is

†This Table is essentially the same as the one used qualitatively in "The Activity Series" in Chapter 7 (p. 183); however, in this chapter it has been put on a more quantitative basis

set up here (p. 313), the half-reactions are written as reductions, Ox. + $ne^-$ = Red., with oxidizing agents (Ox.) on the left and decreasing in strength (potential, tendency) from top-to-bottom. Reducing agents are on the right–they decrease in strength in the reverse direction. Under typical conditions it can be predicted that any reduction half-reaction in the table is capable of occurring spontaneously, **but** only at the expense of reversing a half-reaction that lies below it in the table. Rigorously this is true only for standard-state conditions at 25°C, but it is usually true for other conditions since the concentration-dependence in the Nernst equation is in the form of a small logarithmic term. Thus, $\mathcal{E}^0$ will be positive, $\Delta G°$ negative, and $K_{rxn} > 1$ for the reactant combination "Left-Upper Ox. + Right-Lower Red." The products will be the conjugate oxidized form of Red. and the conjugate reduced form of Ox., and the balanced equation will be the one in which each half-reaction has been multiplied (**but** NOT the $\mathcal{E}^0$ value!) so as to make the electron loss equal to the electron gain. Thus, for the reaction between $H^+_{(aq)}$ and $Mn_{(s)}$, we have:

Balanced equation: $Mn_{(s)} + 2H^+ \rightarrow Mn^{2+} + H_{2\,(g)}$

Spontaneity: Spontaneous reaction

$\mathcal{E}^0_{cell}$: $\mathcal{E}^0_{oxid'n\ 1/2\text{-}rxn} + \mathcal{E}^0_{red'n\ 1/2\text{-}rxn}$

$+ 1.03\text{ V} \qquad + 0.00\text{ V} = +1.03\text{ V}$

$\Delta G° = -n\mathcal{F}\mathcal{E}^0$: $= -(2\ \cancel{\text{moles}}\ e^-)\left(\dfrac{9.65 \times 10^4\ \text{coul}}{\cancel{\text{mole}}\ e^-}\right)(+1.03\text{ V})$

$= -1.99 \times 10^5$ V-coul or joules

$(1\text{ kcal} = 4.184 \times 10^3\text{ J})$

$= -47.6$ kcal

Equilibrium constant, $K_{rxn}$: $= \dfrac{[Mn^{2+}]p_{H_2}}{[H^+]^2} = 10^{16.9n\mathcal{E}^0} = 6.52 \times 10^{34}$

Thus, "solid manganese can be 'dissolved' by a strong acid," or "$Mn_{(s)}$ displaces hydrogen from a strong acid"; the word "can" implies **nothing** about the **rate** at which the reaction **will** actually occur.

## Example Problems on Electrochemistry

### 17-A. The EMF of a Spontaneous Cell

A galvanic cell can be sketched as follows:

anode|anode compartment||cathode compartment|cathode

by means of standard reduction potentials in volts. A number of these qualitative questions have been repeated here.

where the double vertical lines represent an electrolytic junction (or "salt bridge") connecting two half-cells. Each half-cell consists of an electrode and a solution containing those species involved in the particular half-reaction. The oxidation half-reaction occurs at the anode, the reduction half-reaction (written on the right) at the cathode. For a "zinc-copper cell,"

$$Zn|Zn^{2+}\ (0.0500\ M)||Cu^{2+}\ (1.000\ M)|Cu$$

The spontaneous reaction implied involves the "upper-left Ox." ($Cu^{2+}$) with the "right-lower Red." (Zn):

$$\mathcal{E}^0_{Red.} = +0.34\ \text{V}$$

$$Zn_{(s)} + Cu^{2+} \rightleftharpoons Zn^{2+} + Cu_{(s)}$$

$$\mathcal{E}^0_{Ox.} = -\mathcal{E}^0_{Red.} = +0.76\ \text{V}$$

$\mathcal{E}^0_{cell} = 0.76\ \text{V} + 0.34\ \text{V} = +1.10\ \text{V}$ and $\Delta G^\circ < 0$. It is thus seen that, under standard conditions (*i.e.*, $[Cu^{2+}] = [Zn^{2+}] = 1.00\ M$) at 25°C, the spontaneous reaction can deliver 1.10 V ($\mathcal{E}^0 > 0$). What would be the initial voltage of a cell with the concentrations indicated?

(A) 1.14 V (B) 1.18 V (C) 1.10 V (D) 1.02 V (E) 1.06 V

### *Solution*

Since $[Zn^{2+}] < 1.0\ M$ the emf will not be $\mathcal{E}^0$; in fact, we can tell from Le Chatelier's principle that $\mathcal{E}$ will be greater than $\mathcal{E}^0$ since relative to the standard-state concentration for solutes, the $[Zn^{2+}]$ has been lowered, but not $[Cu^{2+}]$. The equilibrium should be shifted to the right, corresponding to a greater emf. At nonstandard concentrations, $\mathcal{E}$ is related to $\mathcal{E}^0$ and the actual concentrations given by the Nernst equation:

$$\mathcal{E} = \mathcal{E}^0 - \frac{0.0592}{n}\log Q = \mathcal{E}^0 - \frac{0.0592}{2}\log\frac{[Zn^{2+}]}{[Cu^{2+}]}$$

$$= +1.10 - 0.0296\log\frac{0.0500}{1.0000} = 1.14\ \text{V} \qquad \text{[A]}$$

## 17-B. Change in Cell Voltage as Concentration is Altered

Calculate the voltage, $\mathcal{E}$, for the cell given in 17-A when the cell has discharged to the point where $[Cu^{2+}] = 0.010\ M$.

(A) 1.14 V (B) 0.85 V (C) 1.10 V (D) 1.04 V (E) 0.00 V

### *Solution*

As the cell discharges $[Zn^{2+}]$ increases, $[Cu^{2+}]$ decreases, $Q$ becomes larger, and $\mathcal{E}$ decreases. From

## DRILL ON CELL POTENTIALS AND REACTION SPONTANEITY

***Student's Name*** ________________

Given below are some reversible redox systems to be used in making up galvanic cells at 25°C. Fill in the table and decide the direction of the spontaneous cell reaction (← or →) in each case. You may be asked to turn in this sheet.

| System | E°, volt | ΔG°, kJ | $K_{eq}$ | Q | E, volt | ΔG kJ | Dir. ← or → |
|---|---|---|---|---|---|---|---|
| $Fe_{(s)} + Cu^{2+}$ *(1.0 M)* = $Fe^{2+}$ *(1.0 M)* + $Cu_{(s)}$ | | | | | | | |
| $Pb^{2+}$ *(0.00010 M)* + $Sn_{(s)}$ = $Pb_{(s)} + Sn^{2+}$ *(2.0 M)* | | | | | | | |
| $Fe^{2+}$ *(0.0010 M)* + $Ag^{+}$ *(0.0010 M)* = $Fe^{3+}$ *(2.0 M)* + $Ag_{(s)}$ | | | | | | | |
| $Zn_{(s)} + Cd^{2+}$ *(1.0 M)* = $Zn^{2+}$ *(1.0 M)* + $Cd_{(s)}$ | | | | | | | |
| $Cd^{2+}$ + *(0.0010 M)* + $Ni_{(s)}$ = $Ni^{2+}$ *(1.0 M)* + $Cd_{(s)}$ | | | | | | | |
| $2X^{+}$ *(0.10 M)* + $Fe_{(s)}$ = $Fe^{2+}$ *(1.0 M)* + $2\ X_{(s)}$ | | | | | + 0.04 | | |
| | | | | | | | |

## STANDARD REDUCTION POTENTIALS AT 25°C

**The format of this table is Ox. + $ne^-$ ⇌ Red., with all unsubscripted species having the 1 $M$ aqueous solution as the standard state. The bracketed entries [ ] are not standard-state values, but refer to neutral water at 25°C.**

| Half-reaction | $\mathcal{E}^0$, Volts |
|---|---|
| $F_{2(g)} + 2e^- \rightleftharpoons 2F^-$ | +2.87 |
| $S_2O_8^{2-} + 2e^- \rightleftharpoons 2SO_4^{2-}$ | +2.00 |
| $Co^{3+} + e^- \rightleftharpoons Co^{2+}$ | +1.84 |
| $H_2O_2 + 2H^+ + 2e^- \rightleftharpoons 2H_2O_{(\ell)}$ | +1.78 |
| $PbO_{2(s)} + 4H^+ + SO_4^{2-} + 2e^- \rightleftharpoons PbSO_{4(s)} + 2H_2O_{(\ell)}$ | +1.69 |
| $MnO_4^- + 8H^+ + 5e^- \rightleftharpoons Mn^{2+} + 4H_2O_{(\ell)}$ | +1.51 |
| $Cl_{2(g)} + 2e^- \rightleftharpoons 2Cl^-$ | +1.36 |
| $O_{2(g)} + 4H^+ + 4e^- \rightleftharpoons 2H_2O_{(\ell)}$ | +1.23 |
| $Br_{2(\ell)} + 2e^- \rightleftharpoons 2Br^-$ | +1.09 |
| $NO_3^- + 4H^+ + 3e^- \rightleftharpoons NO_{(g)} + 2H_2O_{(\ell)}$ | +0.96 |
| $Hg^{2+} + 2e^- \rightleftharpoons Hg_{(\ell)}$ | +0.85 |
| $[O_{2(g)} + 4H^+ (10^{-7}\ M) + 4e^- \rightleftharpoons 2H_2O_{(\ell)}]$ | [+0.82] |
| $Ag^+ + e^- \rightleftharpoons Ag_{(s)}$ | +0.80 |
| $Fe^{3+} + e^- \rightleftharpoons Fe^{2+}$ | +0.77 |
| $O_{2(g)} + 2H^+ + 2e^- \rightleftharpoons H_2O_2$ | +0.68 |
| $I_{2(s)} + 2e^- \rightleftharpoons 2I^-$ | +0.54 |
| $O_{2(g)} + 2H_2O_{(\ell)} + 4e^- \rightleftharpoons 4OH^-$ | +0.40 |
| $Cu^{2+} + 2e^- \rightleftharpoons Cu_{(s)}$ | +0.34 |
| $AgCl_{(s)} + e^- \rightleftharpoons Ag_{(s)} + Cl^-$ | +0.22 |
| $2H^+ + 2e^- \rightleftharpoons H_{2(g)}$ | Defined as 0.00 |
| $Pb^{2+} + 2e^- \rightleftharpoons Pb_{(s)}$ | -0.13 |
| $Sn^{2+} + 2e^- \rightleftharpoons Sn_{(s)}$ | -0.14 |
| $Ni^{2+} + 2e^- \rightleftharpoons Ni_{(s)}$ | -0.25 |
| $Co^{2+} + 2e^- \rightleftharpoons Co_{(s)}$ | -0.28 |
| $PbSO_{4(s)} + 2e^- \rightleftharpoons Pb_{(s)} + SO_4^{2-}$ | -0.36 |
| $Cd^{2+} + 2e^- \rightleftharpoons Cd_{(s)}$ | -0.40 |
| $[2H_2O_{(\ell)} + 2e^- \rightleftharpoons H_{2(g)} + 2OH^- (10^{-7}\ M)]$ | [-0.41] |
| $Fe^{2+} + 2e^- \rightleftharpoons Fe_{(s)}$ | -0.44 |
| $Zn^{2+} + 2e^- \rightleftharpoons Zn_{(s)}$ | -0.76 |
| $2H_2O_{(\ell)} + 2e^- \rightleftharpoons H_{2(g)} + 2OH^-$ | -0.83 |
| $Mn^{2+} + 2e^- \rightleftharpoons Mn_{(s)}$ | -1.03 |
| $Al^{3+} + 3e^- \rightleftharpoons Al_{(s)}$ | -1.66 |
| $Mg^{2+} + 2e^- \rightleftharpoons Mg_{(s)}$ | -2.37 |
| $Na^+ + e^- \rightleftharpoons Na_{(s)}$ | -2.71 |
| $K^+ + e^- \rightleftharpoons K_{(s)}$ | -2.92 |
| $Li^+ + e^- \rightleftharpoons Li_{(s)}$ | -3.05 |

↑ **Better Oxidizing Agents (More easily reduced)**

↓ **Better Reducing agents (More easily oxidized)**

| | $Cu^{2+} + Zn_{(s)}$ | $\rightleftharpoons Zn^{2+} + Cu_{(s)}$ | |
|---|---|---|---|
| Initial, $M$: | 1.000 | 0.0500 | $\mathcal{E}$ = 1.14 V. |
| $\Delta, M$: | -0.990 | +0.990 | |
| Leaves, $M$: | 0.010 | 1.040 | $\mathcal{E}$ = ? |

Again $\mathcal{E} \neq \mathcal{E}^0$ and, substituting into the Nernst equation,

$$\mathcal{E} = +1.10 - 0.0296 \log (104) = +1.04 \text{ V} \qquad \textbf{[D]}$$

This assumes, of course, that the Zn electrode is of such mass that it has not been all consumed by the reaction. Also, although the concentrations have varied widely from the standard 1.0 $M$, the emf has still remained **near** the $\mathcal{E}^0$ value.

## 17-C. Prediction of Electrolysis Products from Half-Cell Potentials

If a 1.0 $M$ solution of KI is electrolyzed:

(A) the solution should become more basic
(B) potassium should be deposited at the cathode
(C) hydrogen should be evolved at the anode
(D) oxygen should be evolved at the anode
(E) iodine should be formed at the cathode

### *Solution*

The solution contains $K^+$, $I^-$, $H_2O$, and traces of $H_3O^+$ and $OH^-$. Of these, $I^-$, $H_2O$, and $OH^-$ may in principle be oxidized at the anode. $K^+$, $H_3O^+$, and $H_2O$ may be reduced at the cathode. The reaction is best predicted by using the $\mathcal{E}^0$ table. At the anode, **oxidation** half-reaction possibilities must be considered and we must compare $\mathcal{E}^0_{Ox}$, values (the **reverse** of the half-reactions tabulated). These are for $I^-$, -0.54 V, for $H_2O$ (at $[H^+] = [OH^-] = 10^{-7}$ $M$), - 0.82 V. The most positive $\mathcal{E}^0_{Ox.}$ is the process with the greatest driving force, so we write:

$$\textit{Anode:}\ 2\,I^- \rightarrow I_{2\,(s)} + 2\,e^-;\ \mathcal{E}^0_{Ox.} = -0.54 \text{ V}$$

At the cathode, the reduction of $K^+$ has $\mathcal{E}^0_{Red.} = -2.92$ V and that for $H_2O$ (at $[H^+] = 10^{-7}$ $M$) is -0.41 V. Since the reduction of water is by far the easiest,

$$\textit{Cathode:}\ 2\,H_2O_{(\ell)} + 2\,e^- \rightarrow H_{2\,(g)} + 2\,OH^-;\ \mathcal{E}^0_{Red.} = -0.41 \text{ V}$$

Therefore, the **initial** electrolysis reaction would be the sum of these half-reactions:

$$2\,H_2O_{(\ell)} + 2\,I^- \rightarrow I_{2\,(s)} + H_{2\,(g)} + 2\,OH^-;\ \mathcal{E}^0_{cell} = -0.95 \text{ V}$$

The negative $\mathcal{E}^0_{cell}$ indicates that the reaction is **not** spontaneous and that a **minimum** of 0.95 V must be **applied** to start the reaction. Note also that $Q$ increases as the electrolysis proceeds; hence, >0.95 V would be needed to sustain the electrolysis. In fact, **other** reactions may become predominant at the later stages of an electrolysis, forming other species that can result in a mixture of products. The best answer is (A), since $OH^-$ is formed as a result of the electrolysis.

### 17.D. Chemical Change Produced by a Given Current Flow—Faraday's Law

Assume that the electrolysis reaction from 17-C is carried out, using 1.0 L of solution, until the pH = 13.00. A steady current of 5.0 amp is drawn. How many grams of $I_2$ would be formed?

(A) 10. g (B) 5.0 G (C) 13 g (D) 25 g (E) 26 g

#### *Solution*

If the pH = 13.00, the pOH = 1.00 and $[OH^-] = 0.10\ M$. In a 1.0-L volume this corresponds to production of 0.10 mole of $OH^-$, which required 0.10 mole of electrons (0.10 faraday) at the cathode. This means that 0.10 faraday must have been withdrawn from the anode. Thus the anode reaction had to occur to this extent:

$$(0.10\ \cancel{\text{mole}}\ e^-)\left(\frac{1\ \text{mole}\ I_2}{2\ \cancel{\text{mole}}\ e^-}\right) = 0.050\ \text{mole}\ I_2$$

This corresponds to 12.7 g of $I_2$, or to 2 significant figures, 13 g. [C]

---

## Exercises on General Concepts of Electrochemistry and Electrolysis. See Examples 17-C and 17-D.

1. The product "amperes × seconds" is equal to the number of:

   (A) coulombs transferred (B) electrons transferred
   (C) faradays transferred (D) volts (E) **moles** of electrons transferred

2. 1.00 faraday of electricity is:

   (A) an ampere/second (B) 96,500 coulombs/second
   (C) $6.02 \times 10^{23}$ volts (D) $6.02 \times 10^{23}\ e^-$
   (E) $9.65 \times 10^4$ moles of electrons

3. Which of the following processes could occur at the cathode of an electrochemical cell?

   (A) $Cu^{2+} + 2e^- \rightarrow Cu$ (B) $Zn^{2+} + 2e^- \rightarrow Zn$ (C) $Zn \rightarrow Zn^{2+} + 2e^-$
   (D) $Cu \rightarrow Cu^{2+} + 2e^-$ (E) both A and B

4. A current of 10. amperes is passed through molten magnesium chloride for 3.0 hours. How many moles of magnesium metal could be produced by this electrolysis?

   (A) 1.1 (B) 2.2 (C) 0.56 (D) 0.37 (E) 0.22

## DRILL ON ELECTROLYSIS

***Student's Name*** ______________________

Complete the following table by entering the current, the time the current needs to flow, the mass of the element produced at the anode and the mass of the element produced at the cathode. You may be asked to turn in this sheet.

| System | Current, amps | Time, hours | Anode Element, grams | Cathode Element, grams |
|---|---|---|---|---|
| Excess molten $Al_2O_3$ | 10.0 amp | 1.00 hr | | |
| Excess molten $MgCl_2$ | 100. amp | | | 15.0 g |
| Excess saturated $CuSO_{4(aq)}$ | | 24.0 hr | 14.2 g | |
| Excess saturated $NaNO_{3(aq)}$ | 5.00 amp | 12.0 hr | | |
| Excess saturated $AgF_{(aq)}$ | | 6.00 hr | 0.179 g | |
| | | | | |
| | | | | |
| | | | | |

5. What is the anode reaction during electrolysis of aqueous copper(II) sulfate?

   (A) $2\,H_2O \rightarrow O_2 + 4H^+ + 4\,e^-$ (B) $2\,SO_4^{2-} \rightarrow S_2O_8^{2-} + 2\,e^-$
   (C) $2\,H_2O + 2\,e^- \rightarrow H_2 + 2\,OH^-$ (D) $2\,H^+ + 2\,e^- \rightarrow H_2$
   (E) $Cu^{2+} + 2\,e^- \rightarrow Cu$

6. What mass of Cu could be plated out by electrolyzing aqueous $CuSO_4$ for 12 hours at 2.0 amperes?

   (A) 58 g (B) 28 g (C) 120 g (D) 430 g (E) 860 g

7. The product produced at the anode during electrolysis of aqueous $CuI_2$ is:

   (A) $H_2$ (B) $I_2$ (C) Cu (D) $OH^-$ (E) $H_2O$

8. How many seconds would it take a 10.0-amp current to produce enough aluminum to make a 27.0-g beer can?

   (A) $2.89 \times 10^4$ (B) $9.65 \times 10^3$ (C) $3.22 \times 10^3$ (D) $9.65 \times 10^4$
   (E) $9.65 \times 10^5$

9. How many electrons pass by a point in a wire carrying a current of 1.0 ampere in a second's time?

   (A) $6.0 \times 10^{23}$ mole (B) $1.0 \times 10^{-5}$ mole (C) $9.6 \times 10^4$ mole
   (D) 1.0 mole (E) $3.2 \times 10^{-9}$ mole

10. Two electrolytic cells, one electrolyzing molten sodium chloride, the other electrolyzing a chromium(III) nitrate ($Cr(NO_3)_3$) solution to deposit metallic chromium, were connected in series, *i.e.*, exactly the same current passes through both cells. If 1.0 g of metallic sodium is deposited in the one cell, how many grams of chromium would be deposited in the other cell?

    (A) 0.33 g (B) 0.75 g (C) 1.0 g (D) 1.5 g (E) 3.0 g

11. How many coulombs of electric current would be needed to reduce the aluminum in 1.00 mole of $Al_2(SO_4)_3$ to aluminum metal?

    (A) 3 (B) 6 (C) $5.79 \times 10^5$ (D) $2.90 \times 10^5$ (E) 2

---

## Exercises on Reduction Potentials and Galvanic Cells

12. Which of the following is the best reducing agent at standard-state conditions and 25°C?

    (A) 1 *M* HCl (B) $Mn^{2+}$ (C) $Zn^{2+}$ (D) $Zn_{(s)}$ (E) $Ag_{(s)}$

13. Which of the following is the best oxidizing agent at standard-state conditions and 25°C?

    (A) $H_3O^+$ (B) $H^+$ (C) $H_{2(g)}$ (D) $Ag^+$ (E) $H_2O_{(\ell)}$

14. From the table of standard potentials it can be concluded that:

    (A) $Zn^{2+}$ reacts spontaneously with $H_{2\,(g)}$
    (B) $Ag_{(s)}$ reacts spontaneously with 1 $M$ $H^+$ (as $HClO_4$)
    (C) $Ag_{(s)}$ reacts spontaneously with $Zn^{2+}$
    (D) $Ag_{(s)}$ is spontaneously oxidized by nitrate ion in acidic solution
    (E) $Zn^{2+}$ will liberate $H_2$ from 1 $M$ $H_3O^+$ (as aq. HCl)

15. From the table find a species that will convert $Cu^{2+}$ to Cu but **will not** convert $Fe^{2+}$ to Fe.

    (A) $Ag^+$ (B) $H^+$ (C) Cd (D) Zn (E) Ag

16. Which of the following is the strongest reducing agent?

    (A) Fe (B) Ag (C) $H_2$ (D) $Ag^+$ (E) $Fe^{2+}$

17. Of the species given below which is the best reducing agent?

    (A) $Mn^{2+}$ (B) Cu (C) $Cl^-$ (D) Pb (E) $Ag^+$

18. A galvanic cell employs the following reaction:

$$Sn_{(s)} + 2\,Ag^+ \rightarrow Sn^{2+} + 2\,Ag_{(s)}$$

    The voltage produced by this cell under standard-state conditions and 25°C is:

    (A) 0.66 V (B) 1.46 V (C) 0.52 V (D) 0.94 V (E) 1.74 V

19. The voltage of the cell, $Zn|Zn^{2+}$ (*0.0010 M*)$\|Cu^{2+}$ (*2.0 M*)$|Cu$, at 25°C would be:

    (A) between 0.76 and 1.10 V (B) >1.10 V
    (C) between 0.34 and 0.76 V (D) <0.42 V
    (E) between 0.00 and 0.76 V

20. What is the voltage of the cell: $Cd|Cd^{2+}$ (*1 M*)$\|Ni^{2+}$ (*1 M*)$|Ni$?

    (A) -0.65 (B) 0.15 (C) 0.65 (D) -0.15 (E) -0.10

21. The potential of the cell in Question 20 would be made more positive by increasing the concentration of:

    (A) $Cd^{2+}$ (B) Cd (C) $Ni^{2+}$ (D) Ni (E) none of these

22. The emf of the cell, $In|In^{3+}$ (*1.00 M*)$\|Cu^{2+}$ (*1.00 M*)$|Cu$, is 0.68 V. What is the standard potential for $In^{3+} \rightarrow In$?

    (A) +0.34 V (B) -0.34 V (C) +1.02 V (D) 0.00 V (E) none of these

23. If the above cell is discharged until the $[Cu^{2+}]$ = 0.010 $M$, what voltage will the cell have?

    (A) 0.62 (B) 0.50 (C) 0.74 (D) 0.56 (E) 0.68

## GENERAL EXERCISES ON CHAPTER 17–ELECTROCHEMISTRY

24. A steady current of 5.00 amp flowing for 3860 sec corresponds to:

    (A) a flow of $1.93 \times 10^4$ electrons (B) 0.200 volt (C) 0.200 coulomb (D) 0.200 faraday (E) a transfer of $1.93 \times 10^4$ faradays

25. If the current-time data given in Question 24 applies to the electrolysis of aqueous, $NiSO_4$, in which a 10.0-g necklace is to be nickel-plated, how much does the necklace weigh after the electrolysis?

    (A) 15.9 g (B) 21.7 g (C) 11.2 g (D) 12.9 G (E) 10.2 g

26. How many moles of $e^-$ are required per mole of nitrate ion in the half-cell reaction in which $NO_3^-$ is reduced to $NH_4^+$?

    (A) 2 (B) 3 (C) 4 (D) 6 (E) 8

27. A 1.0 *M* solution of NaOH was electrolyzed for 2.0 hr using a current of 5.0 amp. How many moles of gas were produced at the **anode**?

    (A) 0.37 (B) 0.11 (C) 0.67 (D) 0.025 (E) 0.093

28. How many faradays are required to reduce a mole of $Fe^{3+}$ to $Fe^{2+}$?

    (A) 5 (B) 3 (C) 2 (D) 1 (E) 0

29. What is $\mathcal{E}^0$ for the cell: $Co|Co^{2+}||Fe^{3+}, Fe^{2+}|Pt$?

    (A) 1.82 V (B) 0.49 V (C) 1.05 V (D) 0.72 V (E) 1.26 V

30.*The solubility product of AgCl, as calculated from $\mathcal{E}^0$ values is:

    (A) $1.2 \times 10^{-10}$ (B) $1.6 \times 10^{-10}$ (C) $7.7 \times 10^{-10}$ (D) $4.8 \times 10^{-11}$ (E) $1.2 \times 10^{-9}$

**Questions 31 to 33 refer to the following: A voltaic cell is made by placing a Cd electrode into a solution with $[Cd^{2+}] = 1.0 \times 10^{-4}$ *M* in one compartment, and a Ni electrode in the other compartment dipping into a 0.10 *M* $Ni^{2+}$ solution.**

31. What is the value of $\mathcal{E}^0$ for this cell and which electrode is positive?

    (A) 0.65 V, Ni (+) (B) 0.15 V, Ni (+) (C) 0.00 V (D) 0.15 V, Cd (+) (E) 0.65 V, Cd (+)

32. What is the emf of the cell at the given concentrations?

    (A) 0.24 V (B) 0.07 V (C) 0.45 V (D) 0.30 V (E) 0.73 V

33. What is the equilibrium constant for the spontaneous cell reaction?

    (A) $9.3 \times 10^{21}$ (B) $8.3 \times 10^{-6}$ (C) $1.2 \times 10^5$ (D) $1.0 \times 10^{-5}$ (E) $2.9 \times 10^{-3}$

34. The state of "charge" of a lead storage battery can be determined by measuring the density of the electrolyte solution because:

(A) $PbO_2$ is formed on discharge (B) the emf is 2.00 V
(C) $PbSO_4$ is formed on charging (D) lead is heavy
(E) sulfuric acid is consumed on discharging

35. If a current of 1.00 ampere is drawn from a Daniell cell ($Zn|Zn^{2+}||Cu^{2+}|Cu$) for 1.00 hour, how many electrons have been forced through the external circuit?

(A) $1.3 \times 10^{19}$ (B) $2.2 \times 10^{22}$ (C) $1.3 \times 10^{15}$ (D) $9.0 \times 10^{22}$
(E) $3.6 \times 10^{23}$

36. For the discharge reaction in Q-35, suppose that the initial $[Zn^{2+}]$ had been 0.100 *M* and the initial $[Cu^{2+}]$ had been 2.000 *M*. If the volume of the solution containing the zinc ions were 2.00 L, the $[Zn^{2+}]$ would have been:

(A) decreased by 5% (B) increased by 19% (C) increased by 13%
(D) decreased by 0.1% (E) increased by 9%

37. As a result of the discharge in Q-36, the cathode would:

(A) increase in mass by 1.2 g (B) increase in mass by 0.64 g
(C) increase in mass by 2.4 g (D) decrease in mass by 1.2 g
(E) decrease in mass by 0.64 g

38. If 0.30 L of 1.0 *M* $CuSO_4$ is electrolyzed until the pH = 1.00, how many moles of Cu will be produced?

(A) 0.0075 (B) 0.060 (C) 0.015 (D) 0.083 (E) 0.030

39. A current of 30. amp is passed through molten $Mn_3O_4$ for 10. minutes. How many moles of Mn metal could be produced?

(A) 0.070 (B) 0.19 (C) 0.062 (D) 0.37 (E) 0.093

40. If we apply the Nernst equation to the galvanic cell,

$$Pt|H_{2\,(g)}, H_3O^+||Cl^-, Cl_{2\,(g)}|Pt$$

the reaction quotient, *Q*, would take the form:

(A) $\dfrac{[H_3O^+][Cl^-]}{p_{H_2} \cdot p_{Cl_2}}$ (B) $\dfrac{[Cl^-]p_{Cl_2}}{[H_3O^+]p_{H_2}}$ (C) $\dfrac{[Cl^-]^2[H_3O^+]^2}{[H_2][Cl_2]}$ (D) $\dfrac{[Cl^-]}{[H_3O^+]}$
(E) $\dfrac{[Cl^-]^2[H_3O^+]^2}{p_{H_2} \cdot p_{Cl_2}}$

41. What is the value of $\mathcal{E}$ for the half-cell:

$MnO_4^-(0.010\,M) + 8\,H^+(0.20\,M) + 5e^- \rightarrow Mn^{2+}(0.020\,M) + 4\,H_2O$?

(A) 1.50 V (B) 1.86 V (C) 1.44 V (D) 1.58 V (E) 1.52 V

42. Which of the following would release $H_2$ from 1 *M* HCl?

(A) $Pb^{2+}$ (B) Ag (C) $Zn^{2+}$ (D) Fe (E) $Cl^2$

43. Which of the following would reduce $Fe^{3+}$ to $Fe^{2+}$ **but not** $Fe^{2+}$ to metallic iron?

(A) $Ni^{2+}$ (B) Cu (C) $Zn^{2+}$ (D) Zn (E) $I_2$

44. If powdered Pb and Fe are added to a solution that is 1.0 *M* in both $Fe^{2+}$ and $Pb^{2+}$, a reaction should occur in which:

(A) additional $Fe^{2+}$ and $Pb^{2+}$ are formed
(B) additional $Fe^{2+}$ and Pb are formed
(C) additional Fe and Pb are formed
(D) additional Fe and $Pb^{2+}$ are formed
(E) Fe is deposited on the Pb

**Questions 45 to 48 refer to the cell**

$$\mathrm{Fe}|\mathrm{Fe}^{3+}\ (0.010\ M)\|\mathrm{Cl}^-\ (0.020\ M),\ \mathrm{Cl}_2\ (0.1\ atm)|\mathrm{Pt}$$

**for which the emf was found to be 1.51 V at 25°C**

45. As this cell discharges,

(A) electrons flow from the Pt electrode to the Fe electrode
(B) gaseous hydrogen is formed and mixes with the chlorine
(C) the Pt is oxidized (D) iron(III) is reduced to iron metal
(E) none of the above occurs

46. What is the standard emf for this cell?

(A) 1.37 V (B) 1.40 V (C) 1.16 V (D) 1.31 V (E) 1.28 V

47.*What is $\mathcal{E}^0$ for the half-reaction, $Fe^{3+} + 3e^- \rightarrow Fe$?

(A) 0.33 V (B) -0.04 V (C) -0.33 V (D) 0.08 V (E) 0.20 V

48. The approximate value of the equilibrium constant for the cell discharge reaction is:

(A) $10^{10}$ (B) $10^{47}$ (C) $10^{71}$ (D) $10^{24}$ (E) $10^{142}$

49.*From the potentials involving cobalt in the table, the standard reduction potential for $Co^{3+} + 3e^- \rightarrow Co$ would be calculated as:

(A) +1.56 V (B) +1.28 V (C) +0.43 V (D) -1.56 V (E) none of these

50. How long would it take to plate out 40. g of copper by electrolysis of a $CuSO_4$ solution if a current of 3.0 amp is used?

(A) 24 minutes (B) 11 hours (C) 48 minutes (D) 12 minutes
(E) 5.5 hours

51. What minimum number of 1.5-volt dry cells would need be connected in series to initiate electrolysis of a neutral 1 *M* NaCl solution?

    (A) one (B) two (C) three (D) four (E) five

52. What is the value of $\Delta G^{\circ}_{rxn}$ for the Daniell cell, $Zn|Zn^{2+}||Cu^{2+}|Cu$, at 25°C?

    (A) -106 kJ/mol (B) -31.8 kJ/mol (C) -20.1 kJ/mol
    (D) -53.1 kJ/mol (E) -212 kJ/mol

53. When mixed, which of the following pairs of reactants should give a spontaneous, redox reaction?

    (A) $Ni^{2+}$; $Cu^{2+}$ (B) Co; $Sn^{2+}$ (C) $H_2$ ; $Cd^{2+}$ (D) $H^+$, $NO_3^-$; $Co^{2+}$
    (E) $I_2$ ; $Br^-$

54. When mixed, **all** of the following pairs of reactants should result in a spontaneous oxidation-reduction EXCEPT:

    (A) Ag; $H^+$, $NO_3^-$ (B) $Cl_2$ ; $Br^-$ (C) Fe; $Fe^{3+}$ (D) $Fe^{3+}$; Hg
    (E) Cu; $Fe^{3+}$

55. When a large excess of finely divided lead is stirred with a solution containing $Fe^{2+}$, $Pb^{2+}$, $Ni^{2+}$, $Mn^{2+}$, and $Fe^{3+}$, which of the following should occur?

    (A) the $[Fe^{2+}]$ and $[Pb^{2+}]$ should increase
    (B) iron should be deposited, while the $[Pb^{2+}]$ increases
    (C) the $[Pb^{2+}]$ increases, while metallic iron, nickel, and manganese are formed
    (D) $Pb^{2+}$ and metallic iron are formed
    (E) there should be no net reaction

56. What is $\mathcal{E}^0$ for $Pb_{(s)} \rightleftharpoons Pb^{2+}$ as calculated from the thermodynamic data in the Appendix? (‡)

    (A) +0.126 (B) -0.126 (C) -0.130 (D) -0.0301 (E) +0.301

57. What is $\Delta \overline{G}^{\circ}_f$ for $Ag_2CrO_{4(s)}$ at 25°C if $\mathcal{E}^0$ = 0.47 volt for $Ag_2CrO_{4(s)}$ + $2\,e^- \rightleftharpoons 2\,Ag_{(s)}$ + $CrO_4^{2-}$? (‡)

    (A) -379 kJ/mol (B) -1,590 kJ/mol (C) -824 kJ/mol
    (D) -646 kJ/mol (E) -1,200 kJ/mol

58. Calculate $\Delta \overline{G}^{\circ}_f$ for $Ag_2CrO_{4(s)}$ at 25° from its $K_{sp}$ and other tabulated thermodynamic data. The $K_{sp}$ at 25°C = $9.0 \times 10^{-12}$. (‡)

    (A) -646 kJ/mol (B) -1,590 kJ/mol (C) -1,200 kJ/mol
    (D) +1,590 kJ/mol (E) -379 kJ/mol

59. What maximum mass of silver could be plated out by electrolysis of a saturated solution of silver nitrate for 1.0 hr at 1.0 ampere?

    (A) 0.037 g (B) 1.1 g (C) 1.7 g (D) 4.0 g (E) 6.3 g

60. For the electrolysis in Q-59, it would be predicted that:

(A) silver would be deposited at the anode
(B) water would be oxidized at the cathode
(C) water would be reduced at the cathode
(D) water would be reduced at the anode
(E) oxygen would be formed at the anode

61. For this electrolysis (Q-59), the net cell reaction would involve the production of:

(A) one mole of oxygen per four moles of silver
(B) one mole of silver per four moles of oxygen
(C) one mole of silver per four moles of hydronium ion
(D) one mole of hydronium ion per four moles of silver
(E) one mole of oxygen per two moles of hydroxide ion

62. A *Nicad* battery involves the following cell-discharge reaction:

$$NiO_{2(s)} + Cd_{(s)} + 2\,H_2O_{(\ell)} \rightarrow Ni(OH)_{2(s)} + Cd(OH)_{2(s)}$$

Such a battery rated at "1.00 amp-hr" (1.00 amp × 1.00 hr) would have to contain a **minimum** of ______ g of nickel(IV) oxide. (Fill in the blank.)

(A) 0.91 (B) 1.7 (C) 2.3 (D) 4.6 (E) 9.1

63. If $\mathcal{E}^0$ = 1.30 volts for the *Nicad* battery (above) and if $\mathcal{E}^0$ = 0.49 volt for

$$NiO_{2(s)} + 2\,H_2O_{(\ell)} + 2\,e^- \rightarrow Ni(OH)_{2(s)} + 2\,OH^-$$

what is $\mathcal{E}^0$ for

$$Cd(OH)_{2(s)} + 2\,e^- \rightarrow Cd_{(s)} + 2\,OH^-$$

(A) −1.79 V (B) −0.81 V (C) +0.81 V (D) +1.30 V (E) +1.79 V

64. Commercially, copper is purified by depositing it electrolytically. It is desired to construct electrolytic cells capable of plating 100. g of copper metal per minute from a copper sulfate solution. How many amperes current would the cells draw?

(A) 84.3 amp (B) $5.06 \times 10^3$ amp (C) $2.62 \times 10^{-2}$ amp
(D) $2.53 \times 10^3$ amp (E) 1.67 amp

65. Two electrochemical cells were connected in series. At one of these 1.00 g of Ag metal was deposited from a silver nitrate ($AgNO_3$) solution. Simultaneously, 0.445 g of iridium was deposited in the other cell from an $Ir^{n+}$ solution. What is the charge on the iridium ion in solution?

(A) +1 (B) +2 (C) +3 (D) +4 (E) +6

66. A current passing through a solution of silver nitrate for a given time causes deposition of 0.7500 g of silver at the cathode. The same quantity of electrical charge passed through a solution containing $X^{2+}$ ions caused deposition of 0.2041 g of metal X. What is the identity of metal X?

(A) Cd (B) Zn (C) Mn (D) Cu (E) Ni

**Questions 67 through 70 refer to the following cell at 25°C,**

$$Tl|Tl^+(0.0010\ M)||Cu^{2+}(0.10\ M)|Cu$$

**for which the emf is $\mathcal{E}_{cell} = 0.83$ V.**

67. What is the standard emf, $\mathcal{E}^0_{cell}$, for the cell?

    (A) 0.98 V (B) 0.83 V (C) 0.68 V (D) 0.42 V (E) 0.25 V

68. What is the standard reduction potential of $Tl^+$? Refer to the table on p. 415 for any additional information needed.

    (A) -0.34 V (B) +0.34 V (C) -1.02 V (D) +1.02 V (E) -0.56 V

69. At equilibrium, the ratio $[Tl^+]^2/[Cu^{2+}]$, would be approximately:

    (A) $10^{-5}$ (B) $10^5$ (C) $10^{11}$ (D) $10^{23}$ (E) $10^{14}$

70. The emf of the cell could be **increased** by:

    (A) increasing the mass of the Tl electrode
    (B) increasing the $[Tl^+]$
    (C) saturating the cathode compartment with $H_2S$, precipitating CuS
    (D) increasing the volume of both solutions, but keeping the concentrations of all ions constant
    (E) none of these

---

## ANSWERS TO CHAPTER 17 PROBLEMS

| | | | | | | |
|---|---|---|---|---|---|---|
| 1. A | 11. C | 21. C | 31. B | 41. C | 51. A | 61. A |
| 2. D | 12. D | 22. B | 32. A | 42. D | 52. E | 62. B |
| 3. E | 13. D | 23. A | 33. C | 43. B | 53. B | 63. B |
| 4. C | 14. D | 24. D | 34. E | 44. B | 54. D | 64. B |
| 5. A | 15. C | 25. A | 35. B | 45. E | 55. A | 65. D |
| 6. B | 16. A | 26. E | 36. E | 46. B | 56. A | 66. E |
| 7. B | 17. D | 27. E | 37. A | 47. B | 57. D | 67. C |
| 8. A | 18. D | 28. D | 38. C | 48. E | 58. A | 68. A |
| 9. B | 19. B | 29. C | 39. A | 49. C | 59. D | 69. D |
| 10. B | 20. B | 30. B | 40. E | 50. B | 60. E | 70. E |

# 18
# *Organic Chemistry*

The chemistry of carbon (*organic chemistry*) is particularly complicated because of the enormous number of known carbon compounds; it is particularly important because carbon compounds are the building blocks of all living organisms, *e.g.*, cellulose, starch, protein, and fat. The property of carbon giving rise to this unique behavior is the ease with which carbon forms bonds to other carbon atoms (*catenation*). The carbon-carbon bonds may be single, double, triple, or have a fractional bond order, and may be linked to form chains, rings, or branched structures. Indeed, several compounds can have the same molecular formula, yet be linked together in different ways; these substances (*isomers*) will be different compounds and may have very different physical and chemical properties.

Carbon forms compounds with most other elements with the guiding principle regarding the formulation of compounds being that **each carbon atom is associated with four pairs of electrons** (octet). The great majority of carbon compounds contain the element hydrogen, although the addition of oxygen and/or nitrogen adds markedly to the diversity and number of known compounds. Carbon (2.5) and hydrogen (2.2) have somewhat similar electro-negativities so that compounds containing only these elements (*hydrocarbons*) are essentially nonpolar and, hence, are insoluble in solvents such as water, but soluble in nonpolar solvents like benzene and carbon tetrachloride. Water solubility of organic compounds tends to increase with incorporation of atoms such as oxygen and nitrogen that form polar bonds with carbon. Thus, compounds like $CH_3OH$, $CH_3NH_2$, and $C_{12}H_{22}O_{11}$ (cane sugar) are quite soluble in water. Compounds of large molecular weight (many *polymers,* for example) are frequently not particularly soluble in any solvent. A property of most organic compounds is that they will burn in oxygen, and in the case of compounds containing only C and H, or C, H, and O, the products of combustion in excess air (or oxygen) are carbon dioxide and water.

Two general categories of hydrocarbons are recognized: (a) *aliphatic*–

compounds in which the carbon-carbon bonds are all single, double, or triple bonds, and (b) *aromatic*—compounds where some of the carbon atoms are in rings in which the carbon-carbon bonds have a nonintegral bond order (frequently 1-1/2). Examples of aliphatic hydrocarbons are:

(*a*) $CH_3-CH_2-CH_2-CH_2-CH_2-CH_3$

(*b*) $CH_3-CH_2$ and $CH_3-CH_2$, both joined to $CH-CH_3$, i.e. $(CH_3-CH_2)_2CH-CH_3$

(*c*) a six-membered ring of $CH_2$ groups: $CH_2$, $H_2C$, $CH_2$, $H_2C$, $CH_2$, $CH_2$

(*d*) $CH_3-CH_2-CH=CH-CH_2-CH_3$

(*e*) a five-membered ring: $CH_2-CH_2$ over $CH_2-CH_2$, closed through $CH-CH_3$

Examples of aromatic hydrocarbons include:

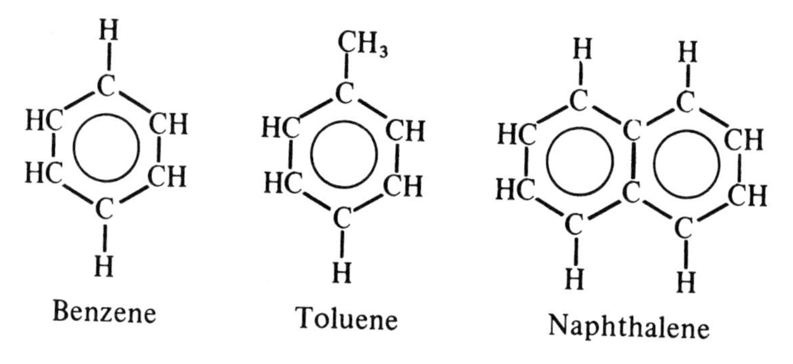


Benzene Toluene Naphthalene

These frequently are written in abbreviated forms, respectively,

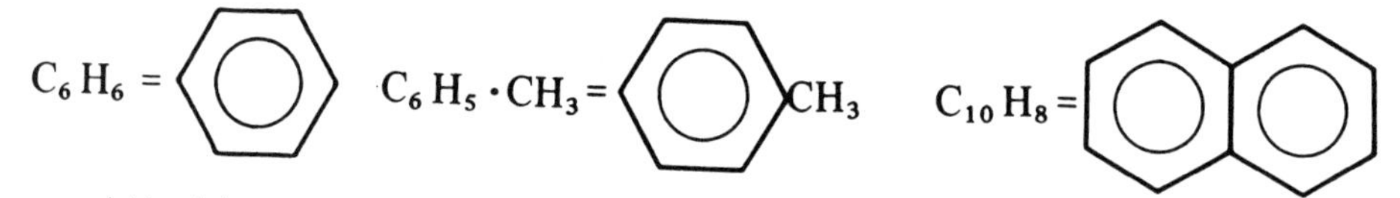


Attaching various groups of atoms (*functional groups*) in place of hydrogen on a hydrocarbon imparts properties to the resulting compound characteristic of the particular group. Thus, attaching $-COOH$ gives a substance that is an acid, *e.g.*, $CH_3-COOH$, acetic acid, and $C_6H_5-COOH$, benzoic acid. Some of the more common functional groups are $-O-$, ether; $-OH$, alcohol (or on an aromatic nucleus, phenol); $>C=O$, ketone; $-CHO$, aldehyde; $-NH_2$, amine; $-COO-$, ester; $-CO\cdot NH_2$, amide.

## DRILL ON ORGANIC FUNCTIONAL GROUPS

***Student's Name*** ____________________

For each of the given formulas indicate which functional groups are present by placing a check mark in the appropriate square. You may be asked to turn in this sheet.

| FORMULA \ Functional Group | Primary Alcohol | Secondary Alcohol | Alkene Group | Alkyne Group | Amide | Amine | Carboxylic Acid | Ester | Ether | Ketone |
|---|---|---|---|---|---|---|---|---|---|---|
| $CH_3OCH_3$ | | | | | | | | | | |
| $(CH_3)_3CCOOH$ | | | | | | | | | | |
| $HOCH_2COOH$ | | | | | | | | | | |
| $(CH_3)_2CHOH$ | | | | | | | | | | |
| $C_6H_5COCH_3$ | | | | | | | | | | |
| $p$-$H_2NC_6H_4COOH$ | | | | | | | | | | |
| $p$-$H_2NC_6H_4CONH_2$ | | | | | | | | | | |
| $CH_2{=}CHCHO$ | | | | | | | | | | |
| $H_2N(CH_2)_4COOCH_3$ | | | | | | | | | | |
| $CH{\equiv}CC(CH_3)_2CH{=}CH_2$ | | | | | | | | | | |
| $C_6H_5OCH_2CH_2COOCH_3$ | | | | | | | | | | |
| | | | | | | | | | | |

## GENERAL EXERCISES ON CHAPTER 18—ORGANIC CHEMISTRY

1. Compound *(b)* (p. 354) is called:

   (A) 3-ethylbutane (B) 3-hexane (C) methylhexane
   (D) 3-methylpentane (E) 1,1-diethylethane

2. Compound *(e)* (p. 354) is called:

   (A) cyclohexane (B) 1,3-hexane (C) 3-ethylbutane (D) 2,4-hexane
   (E) methylcyclopentane

3. Consider only the five compounds on page 354 that are labeled *(a)*, *(b)*, *(c)*, *(d)*, and *(e)*. Which of these are isomers of one or more compounds also listed?

   (A) (*a*) with (*d*) (B) only (*a*) with (*b*) (C) (*b*) with (*e*)
   (D) all five are isomeric with each other
   (E) (*a*) with (*b*), and (*c*) with both (*d*) and (*e*)

4. Complete combustion of a mole of compound (*e*) in excess air would yield ______ moles of water.

   (A) 6 (B) 8 (C) 10 (D) 12 (E) 24

5. Consider only the five compounds on page 354 that are labeled *(a)* through *(e)*. Optical isomers would exist for which of these?

   (A) all (B) only (*e*) (C) (*b*) and (*e*) (D) only (*d*)
   (E) none of these has optical isomers

6. Again considering only compounds (*a*) through (*e*), which of these would react readily with hydrogen bromide?

   (A) (*a*) (B) (*b*) (C) (*c*) (D) (*b*) and (*e*) (E) (*d*)

7. The sum of the coefficients in the balanced equation for the complete combustion of propane in excess air is:

   (A) 4 (B) 8 (C) 12 (D) 13 (E) 22

8. Which of the following formulas can represent more than one structural isomer?

   (A) $C_2H_4$ (B) $C_2H_2$ (C) $C_2F_6$ (D) $C_2H_4F_2$ (E) $C_2H_5F$

9. Which of the following exhibits optical isomerism?

   (A) $CHFCl{-}CF_3$ (B) $F_2C{=}CCl_2$ (C) $CHF{=}CHF$ (D) $CHF_2{-}CHF_2$
   (E) $CH_2F{-}CHF_2$

10. Which of the formulas listed in Question 9 could represent stereoisomers?

    (A) (B) (C) (D) (E)

11. In the refining of petroleum, "cracking" refers to:

(A) separation of compounds by boiling point (distillation)
(B) a careful, selective oxidation
(C) a catalytic reduction of unsaturated compounds
(D) making small molecules out of large ones
(E) the catalytic rearrangement of molecules to increase the amount of isooctane

12. Propanoic acid may be prepared by oxidation of

(A) $CH_3 \cdot CO \cdot CH_3$ (B) $CH_3 \cdot CH(OH) \cdot CH_3$ (C) $CH_3CH_2OH$
(D) $CH_3CH_2CHO$ (E) $CH_2(OH) \cdot CH_2 \cdot CH_2(OH)$

13. Which of the five compounds, *(a)* through *(e)* given on p. 354, would react with bromine in the absence of light?

(A) *(a)* (B) *(b)* (C) *(c)* (D) *(d)* (E) *(e)*

14. Which of the five compounds, *(a)* through *(e)* given on p. 354, would be water soluble?

(A) only *(b)* and *(e)* (B) all are readily soluble in water (C) only *(d)*
(D) none are appreciably water soluble
(E) *(d)* is soluble and *(c)* is somewhat soluble in water

15. Which of the five compounds, *(a)* through *(e)* given on p. 354, are soluble in concentrated sulfuric acid?

(A) only *(b)* and *(e)* (B) only *(c)* (C) only *(d)*
(D) none are soluble in conc. $H_2SO_4$
(E) all are decomposed by conc. sulfuric acid and the decomposition products then dissolve in the acid

16. The formula for propyne is:

(A) $CH_3CH_2CH_3$ (B) $CH_3CH{=}CH_2$ (C) $CH_3(CH_2)_3CH_3$
(D) $CH_3C{\equiv}CH$ (E) $CH_3CH_2CHO$

17. Treatment of propane with bromine in carbon tetrachloride yields:

(A) 1-bromopropane (B) 2-bromopropane (C) cyclopropane
(D) 1,2-dibromopropane (E) 1,1,2,2-tetrabromopropane

18. The major product formed by treatment of $(CH_3)_2CH{-}Br$ with hot alcoholic base is:

(A) $CH_3CH{=}CH_2$ (B) $CH_3C{\equiv}CH$ (C) $(CH_3)_2CH{-}OH$
(D) $CH_3CH_2CH_3$ (E) cyclopropane

19. Which of the following compounds is a primary alcohol?

    (A) $CH_3CH_2CHO$ (B) $CH_3 \cdot CH(OH) \cdot CH_3$ (C) $CH_3 \cdot CO \cdot CH_3$
    (D) $CH_3CH_2CH_2OH$ (E) $CH_3 \cdot O \cdot CH_3$

20. Which of the following is most soluble in water?

    (A) $CH_3CH{=}CH_2$ (B) $CH_3C{\equiv}CH$ (C) $CH_3CH_2CH_2OH$
    (D) $CH_3CH_2CH_3$ (E) cyclopropane

21. Mild oxidation of $CH_3CH_2CH_2OH$ yields:

    (A) $CH_3CH_2CHO$ (B) $CH_3 \cdot CH(OH) \cdot CH_3$ (C) $CH_3 \cdot CO \cdot CH_3$
    (D) $CH_3CH{=}CH_2$ (E) $CH_3C{\equiv}CH$

22. Under somewhat more vigorous oxidative conditions the product of the oxidation of 1-propanol is:

    (A) 2-propanol (B) acetone (C) propanoic acid (D) propene
    (E) propanone

23. Mild oxidation of 2-propanol yields:

    (A) $CH_3CH_2CHO$ (B) $CH_3 \cdot CO \cdot CH_3$ (C) $CH_3CH_2COOH$
    (D) $CH_3C{\equiv}CH$ (E) $CH_3CH_2CH_2OH$

24. Mild oxidation of propanal yields:

    (A) $CH_3CH_2CHO$ (B) $CH_3 \cdot CO \cdot CH_3$ (C) $CH_3CH_2COOH$
    (D) $CH_3C{\equiv}CH$ (E) $CH_3CH_2CH_2OH$

25. The average oxidation number of carbon in $CH_3CH_2CHO$ is:

    (A) -4 (B) -4/3 (C) 0 (D) +3/4 (E) +2

26. In going from $CH_3CH_2CHO$ to $CH_3CH_2CH_2OH$ there is

    (A) a gain of 4/3 of an electron per molecule
    (B) a gain of 2 electrons per molecule
    (C) no loss or gain of electrons since these are covalent substances
    (D) a loss of 2 electrons per molecule
    (E) a loss of 4/3 of an electron per molecule

27. Hydrolysis (saponification) of a fat would yield:

    (A) water and an alkene (B) ethanol and propanoic acid
    (C) glycerol and soap (D) ethanol and a soap
    (E) a triester of glycerol with fatty acids

28. The organic starting materials for the preparation of an ester could be:

    (A) an acid and an alcohol (B) a ketone and an alcohol
    (C) an alkane and a ketone (D) only an acid (E) an amine and an acid

29. Which is the strongest base?

 (A) $CH_3CH_2CHO$ (B) $CH_3CH_2CH_2OH$ (C) $CH_3CH_2CH_2NH_2$
 (D) $CH_3CH_2COOH$ (E) $CH_3C{\equiv}CH$

30. Hydrolysis of a protein would yield:

 (A) alcohols and acids (B) glycerol and a soap (C) $\alpha$-amino acids
 (D) aromatic hydrocarbons (E) none of the above

31. Polymerization of the monomer $CF_2{=}CF_2$ gives a commercial product called:

 (A) Teflon (B) polyvinyl (C) Styrofoam (D) Bakelite
 (E) polyethylene

32. $C_6H_5OH$ is called:

 (A) benzyl alcohol (B) hydroxyhexane (C) phenol (D) hexyl alcohol
 (E) benzene hydroxide

33. Treatment of propene with concentrated sulfuric acid in the presence of $Hg^{2+}$ followed by addition of water yields:

 (A) 1-propanol (B) 2-propanol (C) propanal (D) propanone
 (E) propanoic acid

34. Treatment of propyne with sulfuric acid (conc.) in the presence of $Hg^{2+}$ followed by addition of water yields:

 (A) 1-propanol (B) 2-propanol (C) propanal (D) propanone
 (E) propanoic acid

35. The hybrid orbitals used by carbon in benzene are:

 (A) $sp$ (B) $sp^2$ (C) $sp^3$ (D) $dsp^3$ (E) $d^2sp^3$

36. There are ______ isomeric trimethylbenzenes.

 (A) 2 (B) 3 (C) 4 (D) 5 (E) 6

37. Vigorous oxidation of *p*-dimethylbenzene would yield:

 (A) a monoprotic acid (B) a diprotic acid (C) a di-alcohol
 (D) *p*-dihydroxybenzene (E) phenol

38. Vigorous oxidation of ethylbenzene would yield:

 (A) a monoprotic acid (B) a diprotic acid (C) a di-alcohol
 (D) *p*-dihydroxybenzene (E) phenol

39. Addition of HBr to propene in the presence of a peroxide catalyst yields:

 (A) $CH_3CHBrCH_3$ (B) $CH_3CHBrCH_2Br$ (C) $CH_3CH_2CH_3$
 (D) $CH_3CH_2CH_2Br$ (E) $CH_3BrCH{=}CH_2$

40. Which of the following is a disaccharide?

    (A) glucose (B) starch (C) sucrose (D) cellulose (E) dextrose

41. Breakdown of which of the following would yield only glucose?

    (A) cellulose (B) sucrose (C) protein (D) fat (E) soap

42. The isoelectric point of an amino acid is:

    (A) the neutral pH; 7.00
    (B) the pH where all of the acid is in the anion form
    (C) the pH where all of the acid is in the cation form
    (D) the pH at which the acid is least water soluble
    (E) none of the above

---

## ANSWERS TO CHAPTER 18 PROBLEMS

1. D
2. E
3. E
4. A
5. E
6. E
7. D
8. D
9. A
10. C
11. D
12. D
13. D
14. D
15. C
16. D
17. D
18. C
19. D
20. C
21. A
22. C
23. B
24. C
25. B
26. B
27. C
28. A
29. C
30. C
31. A
32. C
33. B
34. D
35. B
36. B
37. B
38. A
39. D
40. C
41. A
42. D

# *Appendix Tables*

## SELECTED VALUES OF STANDARD THERMODYNAMIC PROPERTIES AT 298.15 K

| Substance | $\Delta\overline{H}^\circ_f$ | | $\Delta\overline{G}^\circ_f$ | | $\overline{S}^\circ$ | |
|---|---|---|---|---|---|---|
| | kJ/mole | *(kcal/mole)* | kJ/mole | *(kcal/mole)* | J/mole-K | *(cal/mole-K)* |
| **Bromine:** | | | | | | |
| $Br_{2(\ell)}$ | 0 | *(0)* | 0 | *(0)* | 152.23 | *(36.384)* |
| $Br_{2(g)}$ | 30.91 | *(7.387)* | 3.14 | *(0.751)* | 245.35 | *(58.641)* |
| $Br_{(g)}$ | 111.88 | *(26.741)* | 82.429 | *(19.701)* | 174.91 | *(41.805)* |
| $HBr_{(g)}$ | -36.4 | *(-8.70)* | -53.43 | *(-12.77)* | 198.59 | *(47.463)* |
| **Calcium:** | | | | | | |
| $CaO_{(s)}$ | -635.5 | *(-151.9)* | -604.2 | *(-144.4)* | 40. | *(9.5)* |
| $Ca(OH)_{2(s)}$ | -987 | *(-236)* | | | | |
| $CaCO_{3(s)}$ | -1,206.9 | *(-288.45)* | -1,128.8 | *(-269.78)* | 92.9 | *(22.2)* |
| **Carbon:** | | | | | | |
| $C_{(s, graphite)}$ | 0 | *(0)* | 0 | *(0)* | 5.740 | *(1.372)* |
| $C_{(s,\ diamond)}$ | 1.897 | *(0.4533)* | 2.900 | *(0.6930)* | 2.38 | *(0.568)* |
| $C_{(g)}$ | 716.681 | *(171.291)* | 671.289 | *(160.442)* | 157.987 | *(37.7597)* |
| $CO_{(g)}$ | -110.52 | *(-26.416)* | -137.15 | *(-32.780)* | 197.56 | *(47.219)* |
| $CO_{2(g)}$ | -393.51 | *(-94.051)* | -394.36 | *(-94.254)* | 213.6 | *(51.06)* |
| $CCl_{4(g)}$ | -103 | *(-24.6)* | | | 309.7 | *(74.03)* |
| $CCl_{4(\ell)}$ | -135.4 | *(-32.37)* | | | 216.4 | *(51.72)* |

| | | | | | | |
|---|---|---|---|---|---|---|
| $HCN_{(g)}$ | 135 | *(32.3)* | 125 | *(29.8)* | 201.7 | *(48.20)* |
| $CH_4{}_{(g)}$ | -74.9 | *(-17.9)* | -50.6 | *(-12.1)* | 186.15 | *(44.492)* |
| $C_2H_2{}_{(g)}$ | 226.7 | *(54.19)* | 209.2 | *(50.00)* | 200.8 | *(48.00)* |
| $C_2H_4{}_{(g)}$ | 52.28 | *(12.496)* | 68.12 | *(16.28)* | 219.5 | *(52.45)* |
| $C_6H_6{}_{(g)}$ | 82.927 | *(19.820)* | 129.66 | *(30.989)* | 269.2 | *(64.34)* |
| $C_6H_6{}_{(\ell)}$ | 49.028 | *(11.718)* | 124.50 | *(29.756)* | 172.8 | *(41.30)* |
| $C_8H_{18}{}_{(\ell)}$ | -226 | *(-54.0)* | | | | |
| **Chlorine:** | | | | | | |
| $HCl_{(g)}$ | -92.30 | *(-22.06)* | -95.31 | *(-22.78)* | 186.80 | *(44.646)* |
| **Chromium:** | | | | | | |
| $Cr_2O_3{}_{(s)}$ | -1,138 | *(-272)* | -1,059 | *(-253)* | 81.2 | *(19.4)* |
| $(NH_4)_2Cr_2O_7{}_{(s)}$ | -1,807 | *(-432)* | | | | |
| **Copper:** | | | | | | |
| $Cu_{(s)}$ | (0) | *(0)* | (0) | *(0)* | 33.15 | *(7.923)* |
| $CuSO_4{}_{(s)}$ | -771.36 | *(-184.36)* | -661.9 | *(-158.2)* | 110 | *(26)* |
| **Fluorine:** | | | | | | |
| $F_2{}_{(g)}$ | 0 | *(0)* | 0 | *(0)* | 202.7 | *(48.44)* |
| $F_{(g)}$ | 78.99 | *(18.88)* | 61.92 | *(14.80)* | 158.64 | *(37.917)* |
| $HF_{(g)}$ | -271 | *(-64.8)* | -273 | *(-65.3)* | 173.67 | *(41.508)* |

(Continued)

## SELECTED VALUES OF STANDARD THERMODYNAMIC PROPERTIES AT 298.15 K (Continued)

| Substance | $\Delta\overline{H}_f^\circ$ | | $\Delta\overline{G}_f^\circ$ | | $\overline{S}^\circ$ | |
|---|---|---|---|---|---|---|
| | kJ/mole | *(kcal/mole)* | kJ/mole | *(kcal/mole)* | J/mole-K | *(cal/mole-K)* |
| **Hydrogen:** | | | | | | |
| $H_{2(g)}$ | 0 | *(0)* | 0 | *(0)* | 130.57 | *(31.208)* |
| $H_{(g)}$ | 217.97 | *(52.095)* | 203.26 | *(48.581)* | 114.60 | *(27.391)* |
| $H_2O_{(\ell)}$ | -285.83 | *(-68.315)* | -237.2 | *(-56.687)* | 69.91 | *(16.71)* |
| $H_2O_{(g)}$ | -241.82 | *(-57.796)* | -228.59 | *(-54.634)* | 188.7 | *(45.104)* |
| **Iodine:** | | | | | | |
| $I_{2(s)}$ | 0 | *(0)* | 0 | *(0)* | 116.14 | *(27.757)* |
| $I_{2(g)}$ | 62.438 | *(14.923)* | 19.36 | *(4.627)* | 260.6 | *(62.28)* |
| $I_{(g)}$ | 106.84 | *(25.535)* | 70.283 | *(16.798)* | 180.68 | *(43.184)* |
| $HI_{(g)}$ | 26.5 | *(6.33)* | 1.71 | *(0.41)* | 206.48 | *(49.351)* |
| **Iron:** | | | | | | |
| $Fe_{(s)}$ | 0 | *(0)* | 0 | *(0)* | 27.3 | *(6.52)* |
| $Fe_2O_{3(s)}$ | -824.2 | *(-197.0)* | -742.2 | *(-177.4)* | 87.40 | *(20.89)* |
| **Nitrogen:** | | | | | | |
| $N_{2(g)}$ | 0 | *(0)* | 0 | *(0)* | 191.5 | *(45.77)* |
| $N_{(g)}$ | 472.704 | *(112.979)* | 455.579 | *(108.886)* | 153.19 | *(36.613)* |
| $NO_{(g)}$ | | | | | 210.7 | *(50.347)* |
| $NH_{3(g)}$ | -46.11 | *(-11.02)* | -16.5 | *(-3.94)* | 192.3 | *(45.97)* |

| | | | | | | |
|---|---|---|---|---|---|---|
| **Oxygen:** | | | | | | |
| $O_{2(g)}$ | 0 | *(0)* | 0 | *(0)* | 205.03 | *(49.003)* |
| $O_{(g)}$ | 249.17 | *(59.553)* | 231.75 | *(55.389)* | 160.95 | *(38.467)* |
| **Silicon:** | | | | | | |
| $SiO_{2(s)}$ | -910.94 | *(-217.72)* | -856.67 | *(-204.75)* | 41.84 | *(10.00)* |
| $SiH_{4(g)}$ | 34 | *(8.2)* | 56.9 | *(13.6)* | 204.5 | *(48.88)* |
| **Silver:** | | | | | | |
| $Ag_{(s)}$ | 0 | *(0)* | 0 | *(0)* | 42.55 | *(10.17)* |
| $Ag_2S_{(s)}$ | -32.6 | *(-7.79)* | -40.7 | *(-9.72)* | 144.0 | *(34.42)* |
| **Sulfur:** | | | | | | |
| $S_{(s,\ rhombic)}$ | 0 | *(0)* | 0 | *(0)* | 31.8 | *(7.60)* |
| $SO_{2(g)}$ | -296.83 | *(-70.944)* | -300.19 | *(-71.748)* | 248.1 | *(59.30)* |
| $SO_{3(g)}$ | -395.72 | *(-94.580)* | -371.1 | *(-88.69)* | 256.6 | *(61.34)* |
| $H_2S_{(g)}$ | -20.6 | *(-4.93)* | -33.6 | *(-8.02)* | 205.7 | *(49.16)* |
| $H_2SO_{4(\ell)}$ | -813.989 | *(-194.548)* | -690.101 | *(-164.938)* | 156.90 | *(37.501)* |

## THERMODYNAMIC DATA FOR SPECIES IN AQUEOUS SOLUTION

| Species | $\Delta\bar{H}_f^o$ kJ/mole | $\Delta\bar{H}_f^o$ *(kcal/mole)* | $\Delta\bar{G}_f^o$ kJ/mole | $\Delta\bar{G}_f^o$ *(kcal/mole)* | $\Delta S^o$ J/mole-K | $\Delta S^o$ *(cal/mole-K)* |
|---|---|---|---|---|---|---|
| $H^+_{(aq)}$ † | 0† | *(0)*† | 0† | *(0)*† | 0† | *(0)*† |
| $OH^-_{(aq)}$ | -229.94 | *(-54.957)* | -157.30 | *(-37.595)* | -10.5 | *(-2.52)* |
| $HC_2H_3O_{2(aq)}$ | -488.44 | *(-116.74)* | -399.6 | *(-95.51)* | | |
| $C_2H_3O_2^-{}_{(aq)}$ | -488.86 | *(-116.84)* | -372.5 | *(-89.02)* | | |
| $NH_{3(aq)}$ | -80.83 | *(-19.32)* | -26.6 | *(-6.36)* | 110. | *(26.3)* |
| $NH_4^+{}_{(aq)}$ | -132.8 | *(-31.74)* | -79.50 | *(-19.00)* | 112.8 | *(26.97)* |
| $Ag^+_{(aq)}$ | 105.9 | *(25.31)* | 77.11 | *(18.43)* | 73.93 | *(17.67)* |
| $Tl^+_{(aq)}$ | 5.77 | *(1.38)* | -32.45 | *(-7.755)* | 127 | *(30.4)* |
| $Pb^{2+}_{(aq)}$ | 1.6 | *(0.39)* | -24.31 | *(-5.81)* | 21 | *(5.1)* |
| $Cl^-_{(aq)}$ | -167.46 | *(-40.023)* | -131.17 | *(-31.350)* | 55.2 | *(13.2)* |
| $CrO_4^{2-}{}_{(aq)}$ | -894.33 | *(-213.75)* | -736.8 | *(-176.1)* | 38 | *(9.2)* |

†By definition

## STANDARD ENTHALPIES OF COMBUSTION AT 298.15 K

| Substance | $\Delta\overline{H}^\circ_{comb}$ kJ/mole | (kcal/mole) |
|---|---|---|
| $H_{2(g)}$ | - 285.83 | (- 68.315) |
| $C_2H_{2(g)}$ | - 1,305 | (- 312.0) |
| $C_2H_{6(g)}$ | - 1,541 | (- 368.4) |
| $C_2H_5OH_{(\ell)}$ | - 1,371 | (- 327.6) |
| $C_2H_4O_{(\ell)}$ | - 1,167 | (- 279.0) |
| $C_3H_{8(g)}$ | - 2,220.0 | (- 530.59) |
| $C_6H_{6(\ell)}$ | - 3,268 | (- 781.0) |
| $C_8H_{18(\ell)}$ | - 5,494 | (- 1,313.2) |

## THREE-PLACE LOGARITHMS

| | 0 | 1 | 2 | 3 | 4 | 5 | 6 | 7 | 8 | 9 |
|---|---|---|---|---|---|---|---|---|---|---|
| 1 | 000 | 041 | 079 | 114 | 146 | 176 | 204 | 230 | 255 | 279 |
| 2 | 301 | 322 | 342 | 362 | 380 | 398 | 415 | 431 | 447 | 462 |
| 3 | 477 | 491 | 505 | 519 | 532 | 544 | 556 | 568 | 580 | 591 |
| 4 | 602 | 613 | 623 | 634 | 644 | 653 | 663 | 672 | 681 | 690 |
| 5 | 699 | 708 | 716 | 724 | 732 | 740 | 748 | 756 | 763 | 771 |
| 6 | 778 | 785 | 792 | 799 | 806 | 813 | 820 | 826 | 833 | 839 |
| 7 | 845 | 851 | 857 | 863 | 869 | 875 | 881 | 887 | 892 | 898 |
| 8 | 903 | 909 | 914 | 919 | 924 | 929 | 935 | 940 | 945 | 949 |
| 9 | 954 | 959 | 964 | 969 | 973 | 978 | 982 | 987 | 991 | 996 |

The above table gives "base-10" logarithms ($\log_{10}$). This is the power (exponent) to which ten must be raised to equal a given number. Before evaluating a logarithm, it is helpful to express the number in scientific notation, *e.g.*,

$$\log 20. = \log (2.0 \times 10^1) = \log 2.0 + \log 10^1$$

When broken down in this way, the log of the first term will always be a decimal fraction (from the table) and that of the second term will be the exponent of 10. Thus,

$$\log 20. = 0.301 + 1 = 1.30$$

Therefore, $10^{1.30} = 20$. Some examples follow with the underline indicating the number of significant figures:

(a) $\log \underline{190}. = \log (1.90 \times 10^2) = \log 1.90 + \log 10^2 = 0.279 + 2 = 2.\underline{279}$

(b) $\log 0.000\underline{59} = \log (5.9 \times 10^{-4}) = \log 5.9 + \log 10^{-4} = 0.771 - 4 =$

$-3.\underline{23}$

Often it is desired to obtain a number when given its logarithm. For example, what is $x$ if $\log x = 4.\underline{74}$? This may be rewritten as

$$x = 10^{4.74} = 10^4 \cdot 10^{0.74}$$

The first term will remain as is for the value expressed in scientific notation, but it is necessary to find the value for $10^{0.74}$ within the body of the log table. It is seen that $10^{0.74} = 5.5$ and, therefore, $x = \underline{5.5} \times 10^4$. A log table only contains positive values, so that if $\log y = -4.\underline{74}$, then

$$y = 10^{-4.74} = 10^{-5} \cdot 10^{0.26} = \underline{1.8} \times 10^{-5}$$

# *Reporting Numerical Results*†

Although an arithmetic result can be calculated to as many decimal places as one might desire (particularly with the advent of the electronic calculator), the number of decimal places that *should be reported* in a calculated result is dependent upon the **accuracy** of the numbers that enter into the calculation. The rigorous fixing of the accuracy of a calculated answer can be a very complicated procedure; however, answers that take into account the least accurate measurement are usually considered realistic and acceptable. The number of *significant figures* included in a number is customarily taken to imply the number's approximate accuracy. The following discussion of significant figures presents a set of rules which, although they are not rigorous, when followed gives an answer reflecting the accuracy of the data. The answers reported in this book are based upon the application of these rules.

Any calculated numerical result stems from mathematical operations involving other numbers. These numbers are of two types:

1. *Measured numbers* are numbers that are limited in accuracy by the measuring device. Unless otherwise stated, these numbers usually are considered as having an inaccuracy of ±1 in the last decimal place. Thus, 1.1034 g should be taken as meaning 1.1034 ± 0.0001 g and the number is said to have five significant figures.

2. *Exact numbers,* or counting numbers, are defined numbers. These have no error and, hence, imply an infinite number of significant figures, *e.g.,* 12 for the number in a dozen and 2 hydrogen atoms per

---

†Excerpted, with the authors' permission, from F. C. Hentz and G. G. Long, *Experiments with Chemical Reactions,* **3rd Edition,** Paladin House Publishers, Geneva, Ill., (1985) pp. 9–11.

water molecule. A calculated result is not limited in the number of ***reportable*** significant figures by the use of an exact number.

The number of significant figures reported in a calculated result depends upon the number of significant figures in the various numbers that enter into the calculation. The following rules will give a calculated result that is in reasonable agreement in accuracy with the data:

1. *Counting the number of significant figures in a measured number.* If the number contains no zeros, the number of significant figures is the same as the number of decimal places occupied, *e.g.*, 423, 41.7, and $1.29 \times 10^{-3}$ all have 3 significant figures; 0.2112, 2112, and $2.112 \times 10^2$ have 4 significant figures. There is ambiguity with regard to numbers that have no decimal point and have one or more zeroes to the right of the last nonzero place. Frequently such zeroes are not significant and are used merely to hold the decimal place. Examples are 620, which may contain 2 or 3 significant figures and 3000, which has as many as 4 and as few as 1 significant figures. Such numbers are best expressed as a multiple of a power of ten so as to clearly indicate the correct number of significant figures, *e.g.*, $6.2 \times 10^2$ and $3.0 \times 10^3$ clearly indicate 2 significant figures, while $6.20 \times 10^2$ and $3.00 \times 10^3$ show 3 significant figures. A number ending in one or more zeroes followed by a decimal point indicates that the zeroes are significant figures, *e.g.*, 200. has 3 significant figures, 200 has as few as 1 and as many as 3 significant figures. In all other cases, zeroes occurring to the right of any nonzero figure **are significant**; those occurring to the left of the first nonzero figure **are not** significant, *e.g.*, 0.212, 0.202, 21.0 and 0.000212 have 3 significant figures; 5000., 0.5005, 0.5000, and $5.000 \times 10^6$ have 4.
2. *Rounding off numbers.* One or two extra figures may be carried through a calculation and then the number of figures is reduced in the final answer by rounding off extraneous figures so that the answer contains the correct number of significant figures. If the number following the last figure to be retained is greater than 5, this last figure is increased by 1, and the number is said to have been rounded up; if the number following the last figure to be retained is less than 5, all numbers to the right of the last figure to be retained are dropped, and the number is said to have been rounded down; if the figure following the last figure to be retained is exactly 5, the last figure is unchanged if it is even, increased by 1 if it is odd. As examples, the numbers 35.21, 45,201, $3.1550 \times 10^4$, 0.007150, and 32.500 when rounded to 2 significant figures would be 35, $4.5 \times 10^4$, $3.2 \times 10^4$, 0.0072 or $7.2 \times 10^{-3}$, and 32; rounded off to 3 significant figures, these would be 35.2, $4.52 \times 10^4$, $3.16 \times 10^4$, 0.00715 or $7.15 \times 10^{-3}$, and 32.5.

3. *Calculations involving multiplication, division, and/or square root.* The answer resulting from any one or from a series of these operations should contain the same number of significant figures as that of the *measured number* entered into the calculation that has the least number of significant figures, *e.g.*, $3.00 \times 2.1 = 6.3$. The answer has 2 significant figures since the two numbers making up the product have 3 and 2 significant figures, respectively and, hence, the accuracy of the result is limited by the number 2.1, the number containing 2 significant figures. In the calculation, $\frac{0.0252 \times 3}{5.1 \times 10^3} = 0.000015$ (or better expressed as $1.5 \times 10^{-5}$), the answer should be given to 2 significant figures if the 3 is an *exact number,* limited by the 2 significant figures of the number in the denominator.

4. *Calculations involving addition and/or subtraction.* Numbers involved in these operations should be rounded off so that all numbers have the same number of significant figures to the right of the decimal point. The result then will have the correct number of significant figures, the same number of figures to the **right** of the decimal point as all of the other numbers after rounding off, *e.g.*, 27.0 - 1.29 need be rounded off to 27.0 - 1.3, which gives the result 25.7; the sum of $4.2 \times 10^2 + 1.8 + 2107$ would first require that these numbers be rounded off to $(4.2 \times 10^2) + (0.0 \times 10^2) + (21.1 \times 10^2)$, which would then yield an answer of $25.3 \times 10^2$. In both cases the answers obtained have the correct number of significant figures. It should be noted that the total number of significant figures can be increased in an addition; decreased in a subtraction.

5. *Chain calculations involving both addition/subtraction and multiplication/division.* In these cases, the addition/subtraction must be performed first following rule 4. The sum or difference then becomes one term containing a specific number of significant figures. This term is then used in the multiplication/division steps as in rule 3. Thus, for the calculation

$$\frac{(25.3 \text{ mL} - 25.0 \text{ mL})\left(0.1186 \dfrac{\text{mg X}}{\text{mL}}\right)}{4.1305 \text{ gY}} = 0.008614 \frac{\text{mg X}}{\text{gY}} .$$

However, this number needs to be rounded to 0.009, one significant figure, due to the subtraction operation.